THE DESERT ENCYCLOPEDIA

An A–Z Compendium of Places, Plants, Animals, People, and Phenomena

---

ROBERT HAUPTMAN

LYONS PRESS

ESSEX, CONNECTICUT

An imprint of Globe Pequot, the trade division of
The Rowman & Littlefield Publishing Group, Inc.
4501 Forbes Blvd., Ste. 200
Lanham, MD 20706
www.rowman.com

Distributed by NATIONAL BOOK NETWORK

British Library Cataloguing in Publication Information available

Library of Congress Cataloging-in-Publication Data
Names: Hauptman, Robert, 1941– author.
Title: The desert encyclopedia : an A-Z compendium of places, plants, animals, people, and phenomena / Robert Hauptman.
Description: Essex, Connecticut : Lyons Press, [2024] | Includes bibliographical references and index. | Summary: "This reference work will be the first English-language A-Z compendium on all topics related to deserts, including geography, geology, meteorology, climatology, hydrology, botany, zoology, anthropology, art, music, film, culture, sports, as well as the specific and diversely different deserts that one finds in all parts of the world"— Provided by publisher.
Identifiers: LCCN 2023053203 (print) | LCCN 2023053204 (ebook) | ISBN 9781493072811 (hardcover) | ISBN 9781493086375 (epub)
Subjects: LCSH: Deserts—Encyclopedias. | Desert ecology—Encyclopedias. | Desert people—Encyclopedias. | LCGFT: Encyclopedias
Classification: LCC QH541.5.D4 H38 2024 (print) | LCC QH541.5.D4 (ebook) | DDC 577.5%3—dc23/eng/20231205
LC record available at https://lccn.loc.gov/2023053203
LC ebook record available at https://lccn.loc.gov/2023053204

♾️™ The paper used in this publication meets the minimum requirements of American National Standard for Information Sciences—Permanence of Paper for Printed Library Materials, ANSI/NISO Z39.48-1992.

*Dedicated to the Australian Aborigine, Amazigh, Bedouin, Berber,
Hopi, Navaho, San, Tohono O'odham, Tuareg, and other nomadic and
sedentary indigenous peoples who live in harmony with the desert
and its arroyos, wadis, rivers, mountains, plants, and animals.*

Es un desierto circular el mundo, el cielo
está cerrado y el infierno vacío.
—Octavio Paz

I don't see the desert as barren at all; I see it as full and ripe. It
doesn't need to be flattered with rain. It certainly needs rain,
but it does with what it has, and creates amazing beauty.
—Joy Harjo

Everything that ever happened to me that was
important happened in the desert.
—Michael Ondaatje

Magic of the shadows can best be seen in the deserts.
—Mehmet Murat Ildan

Between two rivers
White sand oblivion
Blinds the fury dance
We are weakened in the shadow world of dunes
Drift between the strains unturned
Demotic script through sifting fingers
Leaves the web marks Yucca brushed
Before the ebbing.
In this day between two rivers
White sand dissolves to dust
Inquiries destined to reveal
The wind is still with cedar sage
We lose ourselves in the
Shadow world of time.
—"Between Two Rivers," Terry Hauptman

Contents

List of Maps . . . . *vi*

Foreword . . . . *vii*

Introduction . . . . *viii*

The Entries 1

Bibliography 225

Index 238

Acknowledgments and Thanks 245

About the Author 246

List of Maps

Trokenklimate Arid Climates ix

African Desert 7

Antarctic Convergence, 2014 13

Arabian Peninsula 16

Arctic (relief) 19

Atacama 23

Atacama Desert 24

Australia Deserts 25

Great Basin 29

Californian Deserts Region 41

Chihuahuan Desert 47

The Four Major American Deserts 72

The Empty Quarter in the Arabian Desert 79

Gobi Desert 94

Great Basin 97

Kalahari Basin, Distribution of Three Major Language Families 117

Kalahari Basin, Distribution of Three Major Language Families 123

Mojave and Sonoran Deserts 135

Namibia Rivers 143

Israel Outline, North Negev 146

Exodus Route 146

Nile River from Alexandria to the First Cataract, 1911 149

Deserts of Peru, Based on Ecological and Geological Characteristics 158

Sechura Desert Ecoregion 186

Near East (topographic), Syrian Desert 188

Sonora 191

Mojave and Sonoran Deserts 192

Taklamakan Desert (location) 199

Thar Desert (location) 203

Tuareg 209

Foreword

The sheer variety of our fragile blue planet's ecosystems is simply amazing: the immense oceans, impressive mountain ranges, remarkable polar icecaps, vibrant archipelagos, vast grasslands, and extraordinary volcanic landscapes are part of this incomparable palette of Earth landscapes. This diversity would not be whole without deserts, which are abundant, extreme, mysterious, and foreboding. *The Desert Encyclopedia* is a wonderful elegy to these unique landscapes, their flora and fauna, history and geography, orography, and the enormous role they have played in religion, from Ur to Luxor and Canaan. We discover therein that these special, sacred places of the planet, that initially may seem barren and devoid of life, are in fact keenly interesting, and, literally, thought-provoking. This has been noticed for millennia, for literature, both profane and sacred, has told many tales, poems, fables, and parables that take place in the arid emptiness; movies have also used such wilderness to great effect. In fact, alien landscapes are often imagined as cousins of Earth's desolate tracts, and NASA-trained astronauts in similar environments.

Revelation occurs in the barren forever; humility strikes those confronted with the immense emptiness, whether under the katabatic winds of Antarctica, the Sirocco of the Sahara, or the Simoom. The night brings a sharp thermal schism that offers opportunities for the nocturnal creatures that subsist and thrive under the enormously challenging conditions of desertic regions. Similarly, plants, algae, and microbes have evolved awesome abilities to cope with the extreme conditions presented from Atacama to the Gobi, and Antarctica, across the Sonoran and Mojave deserts, in the Negev, the Danakil region of the Afar depression. or the Wadi Rum. Deadly winds, searing heat, apocalyptic cold, salt lakes, acidic soils, and near absolute dryness reign in the desert; life therein is sublimated, hidden, fantastical, and mystical.

In the desert, ruins are hauntingly beautiful or strangely misplaced, human traces vanish, and time returns, indomitable.

These aspects of the desert landscape, and a great many more, are beautifully and meticulously described in this lovely book. Dear Reader, start a revealing voyage within the barren regions of Earth, guided by the author, to profound places and surprising facts; witness the overwhelming power of Nature and the patient resilience of life; behold the mystical powers that inspired man's religions. "All men dream, but not equally" said T. E. Lawrence; *The Desert Encyclopedia* will let you dream with open eyes and a keen spirit.

Fred V. Hartemann
Irvine, California
April 4, 2024

Introduction

The desert is anomalous: In various incarnations, it covers a third of the planet's surface and appears at times to be barren, devoid of water, and lifeless but in reality often teems with animal, insect, and plant life; wind in precipitating storms; and even water, which is evident in those places where its inhabitants know to search. Even outsiders are aware of wells, oases, and wadis. When a desert blooms with thousands of flowers of different species after a spring rainfall, as occurs in Death Valley, for example, one might think they were in the Fertile Crescent or an arboretum.

Still, deserts are very different from California's fecund Central Valley or the Kansas plains, both replete with lettuce and cantaloupe or wheat and corn. Contrastingly, mile after mile of Australia's Gibson and Simpson Deserts are devoid of meaningful plant life but rather covered by what appears to be inedible scrub. The only remarkable things here are the enormous termite mounds, 4 to 20 feet high. It is probable that on a pleasant day, were the temperature to drop from 110 to 80 degrees Fahrenheit, one might encounter lizards, skinks, dingoes, camels, and other fauna. All I saw in the torrid heat were a single live and a dead kangaroo and, incredibly, an Aborigine stretched out asleep in the roadway. The bus driver nonchalantly drove around him.

Snakes are a special case because one must never forget that hiking in these deserts can be quite dangerous. One hundred of this land's 170 snake species are venomous and 9 of the world's 10 most virulent snakes can be found slithering along here. Most people do not carry antivenin, and a snakebite kit will have little effect on a neurotoxic snake's strike. Thirty minutes later the victim will be dead.

The desert plays an inordinate role in the life of its indigenous inhabitants—the Aborigine, the Berber, the San—but it also insinuates itself into the culture of those folks living in Sydney or Cairo or Pretoria. Novels, films, and even architecture can be influenced by the scrub and sand and dunes and life forms that one finds in the deserts in many parts of the world. A desert is likely to surprisingly appear where one least expects it: Little Sahara in Oklahoma or the barren eastern halves of Oregon and Washington, for example.

Indigenous inhabitants but also those who have chosen to make the desert their temporary or permanent home, either in nature or in a community built up slowly into a megalopolis such as Las Vegas, do not consider the metaphysical aspects that so intrude into the lives of scholars, thinkers, memoirists, diarists, novelists, and poets. These folks often appreciate their desert surroundings, but some few seem to associate the desert with death. This is most unfair. There is no more death in the desert than there is in the oceans or the mountains or the frozen northlands. In order for death to obtain, there must also be life, and the desert, as noted, teems with both botanical and zoological life forms that flourish and prosper.

There exist many hundreds of entities, events, and peoples who are influenced by and who, in turn, influence the desert in which they exist or live. Eight of these are of crucial importance and, therefore, call for brief comment.

Without *water*, there can be no life. And so it is either carried along by caravans or nomads, found at wells or oases, or extracted from cacti. Where there is little or no water, in parts of the Atacama or Sahara Desert, for example, few people, plants, or animals are found.

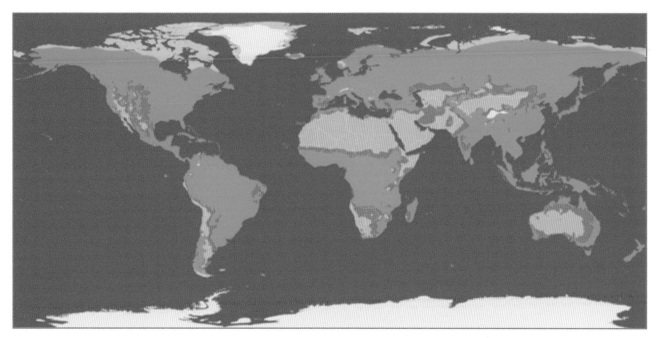

Trokenklimate Arid climates (Deserts of the World). LORD TORAN, 2007. CREATIVE COMMONS ATTRIBUTION-SHARE ALIKE 3.0

The *weather* is crucial, since it brings intense heat or cold, rain or snow, deadly flash floods, and wind that whips the sand into a frenzy, which may even negatively affect distant locations such as Beijing or Phoenix.

Desertification is a major problem, particularly in North Africa where the Sahara is encroaching on Sahel countries, gobbling up fertile land and making life more difficult for inhabitants in Niger, Mali, or Chad, for example, which is one of the world's poorest countries. The Gobi is expanding, and historically the US Dust Bowl resulted from poor farming techniques. Oasification can rescue these parched lands. Note that some scholars contend that desertification is a myth. (See *agriculture* below.)

Countless *plants, animals, and indigenous and other peoples* have adapted to the desert's harsh conditions, have learned to travel across or through the desert with impunity, or made it their home. When mistakes are made by individual San or Aborigine, traditional camel caravans, or modern mechanized vehicles, tragedy ensues. Plants' seeds patiently wait for rain, germinate when it arrives, bloom and flower magnificently, and then die. Succulents store water and can last for long periods, but they too must have moisture or succumb to the forces of nature. Animals of every sort, from beetles and flying insects to cold-blooded lizards, skinks,

and snakes to warm-blooded birds and mammals, have adapted to the diverse conditions that prevail in the desert, survive within their natural habitat, and thrive, at least until man comes along. Most indigenous peoples live harmoniously in their limited environments, although exceptions such as the expansive Inca do exist. Desert invaders are quite different and have dramatically altered the environment, in Los Angeles, Las Vegas, Dubai, or Abu Dhabi, for example. And when such large urban settlements replace desert sands, enormous quantities of water become a crucial necessity.

In order to survive, both herbivores and carnivores *hunt* for plants or other creatures. This is naturally also true for humans, and traditional desert-dwelling nomads seek edible plants as well as prey such as antelope, kangaroo, or lizard. Some groups domesticate goats, yaks, and/or camels and these herds travel with them.

Agriculture is either a concomitant to or a substitute for hunting. For people who settle down—at oases, for example—cultivated plants such as dates, olives, figs, wheat, and corn provide sustenance. Israel and Saudi Arabia have reclaimed desert lands and produce great quantities of agricultural products, some of which are actually exported.

War always has a devastating effect on inhabitants and their environment. The Khanian Mongols laid waste wherever they ravaged, Rommel (the Desert Fox) fought for the Nazis in North Africa, and T. E. Lawrence did the same for Arabian tribes. Saddam Hussein wreaked real havoc in Kuwait, where the gas well fires that his troops lit as they fled burned incessantly and the oil released polluted and killed wildlife, thus wasting a nonrenewable resource.

One might not think that *corporate exploration and development* would be of such moment—that is, would deserve a dishonorable place in this list—but consider that petroleum companies drill and pollute; conglomerates dig enormous open-pit mines, pollute, and mistreat their indigenous workers; transportation companies overburden the desert and its arteries with road trains; and travel agencies encourage hordes of tourists to invade pristine areas (just how horrific this has become generally can be seen in Venice's unprecedented decision to insist that tourists register before visiting and *pay a fee!*).

The Entries

Notes:

Cross-references within the entries are indicated by **boldface**.

Entry selection: There exist many, many thousands of desert minerals, ores, rocks, plants, flowers, animals, peoples, locations, books, films, and other entities; for example, some 2,500 species of birds live in arid lands; the Israeli Negev, which appears empty, is replete with multiple physical locations; and the diminutive (119-square-mile) Jordanian Dana Nature (Biosphere) Reserve contains at least sixteen named wadis. It would be impossible to include every desert location in a single-volume encyclopedia and so only a limited number of all of the above are listed individually. Germane Youtube videos are appended to a limited number of entries.

Desert size: The size of deserts, in square miles, varies from source to source, sometimes dramatically. I have chosen the one (or two or three) that appears to be most consistent and reasonable.

Nomenclature: Just as different languages alter the names of countries and cities—the United States = die Vereinigte Staaten; Deutschland = Germany; Praha = Prague; Firenze = Florence—so may deserts have translated or modified names: the Rub' al Khali = the Empty Quarter; the Thar = Tar (in Russian, because the Cyrillic alphabet does not have the *th* sound); and Gobi = Gobiöknen (in Swedish). Additionally, terminology can sometimes

be confusing: It is generally considered incorrect to pluralize Himalaya but many scholars (including Michael Martin, 96, 98, 125 and Jake Page, 49, 50, 53) do so anyway. And the major deserts often subsume minor ones: The Sahara contains the Libyan; the Arabian, the Syrian; the Australian, the Gibson; the Gobi, the Taklamakan (but not so for everyone); and so on.

Abbreviations
Following are some common abbreviations used in this enyclopedia:
- HIGW: human-induced global warming
- UAE: United Arab Emirates
- US: United States of America
- UVM: University of Vermont

Language: In historical overviews of the world's deserts, it is probable that one will encounter terminology that is no longer considered acceptable because it is prejudicial, discriminatory, pejorative, or derogatory. Such terms include Navaho (Dine' is preferred), Apache (N'de), squaw (referred to as the s-word), Bushman (now San), Hottentot (now Khoikhoi), and Eskimo (now Inuit). If an author of a 2010 article in the *New York Times* (Barry Bearak, "For Some Bushmen, a Homeland Worth the Fight") chooses to use such a term, there is nothing I can do to sanitize matters. (See the Racial Slur Database at http://www.rsdb .org and Survival International, "Terminology," at https://www.survivalinternational.org/info/terminology for a list of some very surprising problematic terms.) Incredibly, difficulties also occur in animal and plant terminology; e.g., *Erythrina caffra* (African coral tree) of South Africa uses a racial slur (from Kaffir). See "Racism Lurks in Many Plant and Animal Names" at https://www.snexplores.org/article/racism-lurks-plant-animal-names.

Language abbreviations: Ten common, primarily geographic terms (cairn, camel, cave, desert, dune, nomad, petroglyph, salt, sand, and river) include equivalents in the following major European and additional relevant languages:

- German: G
- French: F
- Spanish: S
- Icelandic: I
- Afrikaans: Af
- Hebrew: H (first transliteration is traditional, second is modern—when necessary)
- Arabic: Ar
- Mongolian: M
- Chinese: C (traditional characters and pinyin)

Islamic terminology: Because many Muslims inhabit deserts, some Islamic terminology is included here.

Sidebars: The sidebars contain unadorned lists of connected materials. Since, in some cases, there are many hundreds or thousands of, for example, books or films or wildflowers or birds, only a limited number of these are included; they sometimes, though not always, lead back to individual entries.

Sources: Substantive material for this encyclopedia derives from the author's experience and knowledge as well as the following. Note that academic institution and public library catalogs lead not only to hard-copy books in collections but also to countless online full-text books and (at times abstracted) articles, some, though not all, of which are immediately accessible to the public.

- Books: The most important monographs were located in my personal collection; library catalogs including the Library of Congress (10,000), the New York Public (17,607), and UVM (6,272); online booksellers; Google Books; and so on, and can be found listed in the bibliography. See also Books and the books sidebars.
- Newspapers and magazines: the *New York Times*, *Wall Street Journal*, *National Geographic*, and others.
- Journals: Scholarly articles were located through UVM's combined catalog (ca. 35,000 open access) and the databases noted below. Extremely esoteric technical articles on, for example, bacterial diversity, microbial ecology, isotopes in precipitation, radiative effects of mineral dust, or ancient climates are elided here. The most popular topics, among thousands of articles, seem to be on locust genetics, newly discovered bacteria, and aspects of various creatures, e.g., the ant or the biology of the slender Kalahari mongoose.
- Online sources: Google Scholar ("desert" brings up millions of records); Google Books; Project Gutenberg (contains many germane, full-text books, https://www.gutenberg.org), and many databases (Proquest, Gale, Science Direct, Springer, Wiley, etc.) whose listings are agglomerated in the UVM library's online catalog. Some unascribed material is derived from basic internet information.

Photographs: Many indigenous peoples do not wish to be photographed. In my tour of the four major American deserts, the Negev, and the Australian deserts, I do not believe that I photographed a single indigenous person. The images I use were taken by others long ago, and I include them with respect and to honor the people who inhabit desert environments. Many of the diverse photos included here are from Wikimedia Commons. Those ascribed to Robert Hauptman were all shot in 2004.

Abal is a 4-foot bush that grows in the **Arabian Desert** (https://arabiandesert1.weebly.com/plants.html).

Abbey, **Edward.** See *Desert Solitaire*.

Ablation is the loss of **snow** and ice from a **glacier** through melting or evaporation and, by analogical extension, loss of arable land to **desertification** or **sand** through wind erosion.

Aborigine. Although the word means "original inhabitant" generally, it commonly refers to the indigenous Australian peoples who have lived there for tens of thousands of years; they thrive in the desert environment. Their **population** was decimated by white settlers, the earliest of whom were British convicts. Even today, these ca. 500,000 people do not participate fully in Australian life, but rather live on closed reservations, although Michael Welland claims that 70 percent reside in towns and cities (Welland, *The Desert*, 129) and are often quite poor. Thirty years ago, in Cairns, there was a contingent of derelicts; those in charge had requested that no alcohol be served to them in restaurants and so there was a big sign above a bar indicating this, something that perhaps should not exist in a democracy. Yan Zhuang points out that a general ban in certain communities was lifted but in 2023 it was reinstated, because of crime, so that Aborigines, even those who served in the armed forces in Vietnam, cannot buy a beer. (The

Aborigine kangaroo hunters in Queensland, Australia, in 1916.

Australian Aboriginal culture.

neighboring Maori in New Zealand have fared much better.) *The Australian Aborigines*, A. P. Elkin's 1938 study (reissued in 1964), presents a comprehensive anthropological overview of these peoples, but impossibly the chapter on the land fails to take it into account, and the index does not contain the **desert**. (See **Aborigine: art**, **languages**, and **tales**; **Archeology**; **Dreamtime**; **Mardu**; **Songlines**; and **Stolen generation**.)

AUSTRALIAN ABORIGINAL TRIBES
(partial listing)

There are some 500 Australian Aboriginal tribes. A complete list can be found at http://www.fact -index.com/l/li/list_of_australian_aboriginal _tribes.html.

- Koori (or Koorie)
- Murri
- Nunga
- Noongar
- Wiradjuri

Aborigine: art. The famous paintings that are sold to collectors are a new development. Stretching back 60,000 years, Australian **Aborigines** have painted on rock faces, bark, or their own bodies. They also carve, sculpt, and weave. (See **Dot paintings** and **Sand paintings**.)

Aborigine: languages. Only 120 of the original 250 languages that the **Aborigines** speak are still in use and some are heading for extinction. They have many dialectical variations and are difficult for outsiders to learn as are the whistle, click, and Navaho languages.

> *Gheerlayi ghilayer,*
> *Wahi munnooomerhdayer.*
> Kind be,
> Do not steal. (Langloh Parker, 25)

Aborigine: tales. *Wise Women of the Dreamtime* presents K. Langloh Parker's Aborigine tales, which she collected from her indigenous Australian friends in the late 19th century. These are stories of ancestral, animal, and magical powers as well as healing—dreamtime **myths** and legends in which,

for example, "Moodoobahngul, the Widow" harms a spirit (90ff.).

Abrahamic religions. See **Religion**.

Abrasion is the wearing away of any surface. (See **Ablation**.)

Abstract Expressionism (AE). David Jasper, in *The Sacred Desert*, declares that "**desert art** then bears with the burden of a deep truth that requires that it moves always towards abstraction" (118), a claim that does not necessarily follow from his previous remarks. He continues, "Some of the greatest desert art of the twentieth century is to be found in the pure abstractions of the artists of the New York School of the American Abstract Expressionists, painters and sculptors such as Jackson Pollock, Mark Rothko, and Barnett Newman, who have entered the desert of the mind" (118). Nothing whatsoever in the illustrations he provides (120, 123, 125) nor anything, I insist, in any of the abstract works of these and other AE artists signifies, symbolizes, implies, manifests, represents, indicates, suggests, delineates, exemplifies, nor illustrates a desert environment (even if the artists happen to note that this is what they intend, which I do doubt they ever have). This is reminiscent of Kazimer Malevich's 1918 *White on White* and Robert Rauscheberg's 1951 *White Painting*, repainted by Brice Marden (created with "latex house paint applied with a roller and brush on canvas"!). Agnes Martin would have been a better choice here: Her desert canvases are painted a solid tan color and look like a desert floor.

Abu Simbel is the site of two monumental temples carved into a rock face in southeastern **Egypt**. It was lost in shifting **sands** until the 19th century when it was discovered by **Johann Ludwig Burckhardt**. The damming of the **Nile** threatened to submerge the temples and so in 1964 they were deconstructed and moved inland and higher on the cliff ("Abu"). This was a truly astonishing archeological and engineering accomplishment. (See videos at https://www .youtube.com/watch?v=Nk3gwFsb9eo, https://www .youtube.com/watch?v=CgoGYYa71Us, and https:// www.youtube.com/watch?v=l4O4pCRm2xY.)

Abydos is an ancient Egyptian city located south of **Cairo**, discovered under desert sands, and excavated. A 5,000-year-old group of royal boats was discovered here.

Acacia (timbe). This **plant** comes in many species; is between 8 and 15 feet tall; has pods and usually thorns; and is useful against conjunctivitis, dysentery, and sore throat and as a sedative (Moore, 11–12). It can also be found in the **Arabian Desert**, where leaves and seedpods are eaten by **animals** and people (https://arabiandesert1.weebly.com/plants.html).

Academic programs that lead to masters and/or doctorates in appropriate areas include the University of Arizona's Arid Lands Resource Sciences (PhD) and Japan's Tottori University's Arid Land Research Center (PhD).

Access to publications: sources for this encyclopedia. See comment in the preliminary notes to the entries.

Acoma (Sky City), west of **Albuquerque**, is a **pueblo** built on a mesa almost 400 feet above the valley floor at about 7,000 feet above sea level. It is

Acacia tree in Makhtesh Gadol, Negev Desert. WILSON44691. PUBLIC DOMAIN

the oldest continuously inhabited city in the **US** and known for its exquisite **pottery**. Visitors are adjured to follow very strict regulations (brochure). (In order to take **photographs** for this encyclopedia, I had to meet with the elders; I declined to do so.) (See https://www.youtube.com/watch?v=Ro__ocGdjg0.)

Acoma Pueblo. LIBRARY OF CONGRESS

Adaption. "Adapt or perish," H. G. Wells tells us, "now as ever, is nature's inexorable imperative." All terrestrial **flora** and **fauna** adapt to their specific environments. A mild Mediterranean or southern US **climate** requires far fewer adaptions than does the harsh environs one finds in any **desert**. For a compendium of specific plant and animal but not human adaptions, see *Desert*, Roslynn D. Haynes's extremely informative and wonderfully illustrated volume. Consider the **camel** as an exemplar of how this works. The dromedary protects its ingested **water** through minimal excretion; it can survive 113-degree **body temperature**; it does not sweat; it can diminish body weight and then rehydrate quickly even with salty water; its eyes, ears, and nostrils are specially equipped to protect in sandy conditions; its fur insulates against the cold; and so on (Haynes, *Desert*, 56–57). Joana L. Rocha and her colleagues discuss the genomics of adaption in **mammals** over extended periods of time. Mammals present "adaptive traits that have evolved repeatedly and independently in different species across the globe and in response to similar selective pressures of extreme **temperatures**, **aridity**, and water and food deprivation." So, different creatures in diversely different periods have made

the same adaptions to the desert environment in, for example, fat metabolism and insulin signaling (Rocha, "Life"). John Sowell notes some amazing analogous adaptions; these paired **animals** resemble each other: the fennec fox in the **Sahara** and the kit fox in US deserts as well as the jerboa in **China** and **Mongolia** and the **kangaroo rat** in the **US** (138). (See also *A Desert Calling* and **Survival**.)

Addax (white or screwhorn **antelope**) inhabits the **Sahara**. It has large, extremely curved horns.

Adder (Peringuey's adder) is a poisonous **snake** (a sidewinder) found in the **Namib Desert**, among other locations.

Adiabatic cooling occurs when "warm, moist air masses forced up and over **mountains** experience decreasing atmospheric pressure and, as a result, expand. This process causes the expanding air mass to release a portion of its latent **heat**, resulting in a decrease in air **temperature**" (West Virginia Geological and Economic Survey).

Adobe (S: mud-brick) is a **soil** with a high clay content; it is used to make sun-dried bricks that the

Adobe wall. LIBRARY OF CONGRESS

Typical contemporary adobe structure, Santa Fe. ROBERT HAUPTMAN

African desert map. NASA BLUE MARBLE, NASA IMAGE BY RETO STÖCKLI AND ROBERT SIMMON, DESERTS BY PALOSIRKKA (2013). PUBLIC DOMAIN

southwestern US Indians employed to build **dwellings**. It is also the name of the bricks. (When I lived in Oklahoma, I was told that if one dug a hole mixed in a bit of the removed red earth and added **water**, a brick would emerge.) (See **Mogollon**.)

The Adventures of Priscilla, Queen of the Desert is a camp **film** concerning a drag queen, a cross-dresser, and a transsexual. Rotten Tomatoes gave it a five-star rating (https://www.rottentomatoes.com/ m/adventures_of_priscilla_queen_of_the_desert). See the trailer at https://www.youtube.com/watch ?v=QgFDIinCeYI. The entire film is available on Youtube if one has a Google account. (Not to be confused with ***Queen of the Desert***.)

Aerial observation. Viewing the **desert** from above offers a very different perspective than the one that is evident when a person is on the ground, a camel, or a horse. Photographic, X-ray, infrared, and thermal images and mapping record these points of view for others. Satellites present an even broader range and depth. (See **Nazca Lines**.)

Aerial tramway. The Sandia Peak Tramway in **Albuquerque**, at 2.7 miles, claims to be the world's longest (although there seem to be other contenders for this honor, including one in Armenia). At any rate, it can hold 50 people and goes up to 10,378 feet (brochure) at a very steep angle, and the cliff below is often vertical. A brochure claims **mountain lion**, **bear**, skunk, ca. 200 species of **birds**, and 33 species of **reptiles** for the Sandia Peak area.

Afar Depression (Afar Triangle) is a geological depression of desert scrubland in Ethiopia.

Afforestation is the planting of **trees** (in desert plantations) where none had existed in the past.

Africa is one of the world's seven continents, each of which contains at least one **desert**. In Africa one finds the **Sahara**, **Kalahari**, and **Namib**. A most unusual desert journey, "By Bicycle from Capetown to Cairo," is available at https://www.youtube.com/ watch?v=KDOhQ-hyI5w. (See https://www.travel awaits.com/2663066/best-experiences-in-africas -ten-unique-deserts for very brief remarks on 10 major and minor African deserts and "The World's Most Important Deserts" sidebar.)

African peyote cactus is a Sahara Desert **plant** that stores **water** in its stem. The single flower resembles a daisy. Tribal peoples use the peyote in their rituals (https://saharadesertproject1.weebly.com/ plants.html).

African sycamore tree is a large **plant** found in the **Negev Desert**; its bark is covered in a furry coating that defends against water loss (https://deadsea .com/explore/outdoors-recreation/nature-hiking/ein -gedi-botanical-garden).

Afton Canyon (Grand Canyon of the **Mojave**) is located about 35 miles east of Barstow, **California**, and is popular for bird-watching, **hiking**, **hunting**, vehicle touring, rockhounding, horseback riding, stargazing, photography, natural history study, and camping (https://www.blm.gov/visit/afton-canyon).

Agave (century plant) is similar to **yucca** in appearance; its leaves and roots are effective medicinally against various ailments (Moore, 12–13). *Agave palmeri* of the southwestern **US** has yellow flowers that attract **insects** and **birds** such as male hooded orioles and hummingbirds (*Desert Bird Gardening*, 8).

Agave. LIBRARY OF CONGRESS

Agriculture followed in the wake of the life cycles of hunter-gatherers and **nomads** who eventually began local farming in and around **deserts** in **Jordan**, **Morocco**, the southwestern **US**, and **Saudi Arabia**, among other places. (See **Israel**, **Kibbutz**, **Negev**, and **Oasification**.)

Aïr Mountains are located in Niger's **Sahara**.

Ait Atta are a large Berber tribal confederation of southeastern **Morocco**.

Ajo lily (desert lily) is a spectacular species with large white flowers. It is only found in limited areas of **Arizona**, **California**, and Sonora, **Mexico** (*Desert Wildflowers*, 35).

Alamogordo is a city in New Mexico's **Chihuahuan Desert**.

Albuquerque is a major city in **New Mexico**.

Alcohol. See **Aborigine**.

Al Dhafra Festival is a Bedouin festival that takes place two hours from Abu Dhabi, near Madinat Zayed, in close proximity to the **Empty Quarter**. In addition to other things, it presents "camel beauty contests" for which camel tailors bedeck the creatures with blankets and necklaces. Winners' owners share $16 million in prize money (Streitberger). The beauty of a camel is in the eye of the beholder!

Algeria, the largest African country, lies along the Mediterranean Sea, is home to a substantial portion of the **Sahara Desert**, and is rich in **petroleum**.

Alice Springs is a small city in central **Australia** from which trips to **Uluru** begin. It can be stiflingly hot but also quite cold, both depending on the season. (See **Larapinta Trail**.)

All-American Canal. Located in the **Sonoran Desert**, it is the largest irrigation canal in the world. This 82-mile-long aqueduct conveys **water** from the **Colorado River** into the Imperial Valley and to nine cities.

Allen, Benedict (1960–), is an **explorer** who traveled along Namibia's **Skeleton Coast** and crossed the entire **Namib** on foot.

All-American Canal. LIBRARY OF CONGRESS

Allenby Bridge (King Hussein Bridge) crosses from **Israel**/Palestine to **Jordan**.

Alluvial fan is the area where a **river** opens up into a delta and debris (silt, sand, gravel) collects, at the Nile delta, for example.

Al Marmoom Desert Conservation Reserve is located in Dubai, **UAE**.

Almásy, Count László (1865–1951) (**Sahara**), was a desert **explorer** from Hungary who worked for the Nazis and is featured in *The English Patient*.

Alone: The Classic Polar Adventure is Richard Byrd's extraordinary account of his 1934 multi-month isolation in a small hut near the South Pole. He suffered especially from smoke inhalation as well as psychologically and refused to contact his men because he did not want to put them at risk. (See also **Books** and the books sidebars.)

Sir Douglas Mawson. BAIN NEWS SERVICE, 1916

Alone on the Ice is David Roberts's enthralling account of Douglas Mawson's epic adventure, what Sir Edmund Hillary called "the greatest survival story in the history of **exploration**." (See *Life and Death in Antarctica*, a brief 2008 documentary, https://www.youtube.com/watch?v=nE9erdT0f00&list=PLCZIrbk4cofvmcM8M0AwUPPlJpRzCihxF, and a trailer for a **film** of the same title that records Tim Jarvis's attempt to repeat Mawson's long trek, https://www.youtube.com/watch?v=tP3-rX9Ap8A&list=PLRwOQaoL-1ds-rqRB5C7UyeLoi9aEjKZV; also see *Mawson's Will* and the "Desert Explorers" sidebar.)

Altai is a mountain range in Russia, **China**, **Mongolia**, and Kazakhstan.

Altiplano (Puna) is a high plateau in Argentina, Bolivia, **Chile**, and **Peru**.

Amazigh. See **Berber**.

Amphibians are a class of cold-blooded **animals** such as **frogs**, toads, salamanders, and newts that often can be found in **deserts**.

DESERT AMPHIBIANS, INSECTS, AND ARACHNIDS

(partial listing)

• Ant	• Locust	• Tarantula
• Bee	• Newt	hawk wasp
• Beetle	• Salamander	• Termite
• Cicada	• Scorpion	• Toad
• Frog	• Spider	• Wasp

Amundsen, **Roald** (1872–1928), was the first explorer to arrive at both the South and North Poles. His books include *The South Pole*. (See *The Last Place on Earth: Scott and Amundsen's Race to the South Pole*, **Explorers**, and the "Desert Explorers" sidebar.)

Anasazi is the ancient name of the Rio Grande Pueblo, Hopi, and Zuni peoples who lived in the southwestern **US**. Neighboring groups include the Hakataya, **Hohokam**, and **Mogollon**. (See **Ancient desert peoples**, **Crafts**, and **Montezuma Castle National Monument**.)

Ancient desert civilizations include Sumer and Akkad, **Egypt**, Baghdad, **Mongolia**, **Mali**, and

Roald Amundsen. BILINMIYOR, 1911. CREATIVE COMMONS ATTRIBUTION-SHARE ALIKE 4.0

Mesopotamia (Allaby, *Deserts*, 118–21) as well as those in the southwestern **US**. (See **Ancient desert peoples** and **Desert peoples**.)

Ancient desert peoples. Stephen Plog's *Ancient Peoples of the American Southwest* covers these many groups over an extended period stretching back to Paleo-Indians (hunter-gatherers) (9500–7000 BC and subsequently), then early Puebloans (1800 BC), through the Basketmakers (600 BC onwards) (9, 37ff.). Next come the named groups—**Hohokam**, **Mogollon**, **Anasazi**, Mimbres—who began to inhabit villages (56ff.) followed by the cliff dwellers. Eventually, many settlements were abandoned (116). (See **Architecture**, **Kachina**, **Kiva**, **Pottery**, as well as **Chaco Canyon**, **Mesa Verde**, **Montezuma's Castle**, and many others through the "Southwest US Indian Ruins" sidebar.)

Andes is a major mountain range that runs along the western coast of **South America**.

Andrews, **Roy Chapman** (1884–1960), explored the **Gobi**, where he found dinosaur eggs; he used **automobiles** in his expeditions. He was also the director of the American Museum of Natural History. His many books include *All About Dinosaurs* and *On the Trail of Ancient Man*.

Animal rights. Some philosophically minded people object that **animals** cannot have rights because although they are sentient, they are not invested with those special qualities inherent in being human (although some folks also denied rights to certain ethnic groups of people). Nevertheless, these same

Reconstruction of a room, Montezuma's Castle. ROBERT HAUPTMAN

Montezuma's Castle National Monument, an extraordinary structure built into cliffs. ROBERT HAUPTMAN

Vicuñas, Atacama Desert. JESS WOOD, 2017. CREATIVE COMMONS ATTRIBUTION 2.0

humans have come a long ethical way from earlier periods when Colosseum games, bear baiting, destroying America's enormous herd of bison (for pleasure), or killing thousands of **birds** and other animals (in order to paint them) obtained. Desert creatures ranging from diminutive **kangaroo rats** to enormous **elephants** and all creatures in between have the right to thrive in their appointed environments. It is unimaginable that humans in the 12th century would have formed organizations to protect **donkeys** or parrots or spend their Saturday mornings rescuing **desert tortoises**. Today we take this type of activity for granted, whether animals have "rights" or not.

Animals. Though **deserts** may appear to be lifeless, they are in fact home to many thousands of bird, mammal, reptile, amphibian, arachnid, arthropod, and insect species. Animals, like **plants**, have adapted to the desert environment by, for example, moving in unusual ways, storing **water**, and **hunting** at night when the sun has gone down. Edmund C. Jaeger, in *Desert Wildlife*, lovingly describes his scrupulous observations and interactions with more than 40 diverse creatures that inhabit southwestern US deserts; these include the kit and gray fox, **coyote**, desert lynx, caracal (also found in **Africa**, the Middle East, and **Asia**), **desert bighorn sheep**, **pronghorn**, **bat**, skunk, prairie falcon, cactus wren, **raven**, **insects**, **desert tortoise**, and **lizard** (passim). And surprisingly, Anne Woodin recounts a life in the **Sonoran Desert** in which her family opened its home to a broad assortment of wild desert animals including bobcat, wolf, coatimundi, **owl**, raven, kingsnake, lizard, and tarantula; they often came in, returned to the wild, and then revisited (passim). There exist countless **books** on various animals in specific countries' deserts. (See **Climate change** and individual species.)

Anna Creek Station in **Australia** is the world's largest (cattle) farm.

Ant is an **insect** that is ubiquitous in all environments. The scholarly journal literature on desert ants is enormous.

Antarctic

| 5.4 (5.5) million square miles | Antarctica |
| Cold desert | No precipitation |
| No indigenous people |

Although the continent of Antarctica consists of land covered with many millions of gallons of frozen fresh **water** (and therefore it is not arid), it is considered a **desert** because there has been little or no **precipitation** for millions of years and very little grows here. Much of the world's fresh water is contained in the continent's ice. Antarctica is getting warmer, which may melt the ice and result in an ecological disaster. **Sir Ernest Shackleton** died on his third Antarctic expedition. Nathaniel Harris points out that ice shelves or sheets act as bridges across bays and seas, some active **volcanoes** exist, and dinosaur **fossils** have been found near the South Pole. Living entities include **lichen**, mold, **moss**, fungi, seal, porpoise, dolphin, walrus, and whale (166–67) as well as albatross and **penguin**. Kim Heacox, in his beautifully illustrated overview, notes that 90 percent of the world's ice and about 70 percent of its fresh water as well as petrel and 15,000 scientists, tourists, and other human beings can be found here (12, 110, 131, 160).

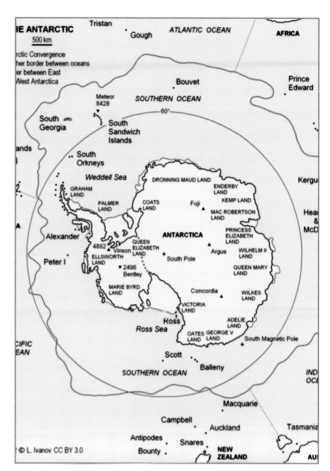

Antarctic convergence map, 2014. CREATIVE COMMONS ATTRIBUTION-SHARE 3.0

Double Mountain, Paradise Bay Coral Princess. AMANDSEN2, 2019. CREATIVE COMMONS ATTRIBUTION-SHARE 2.0

Ross Ice Shelf at the Bay of Whales, Antarctica. MICHAEL VAN WOERT, 1998.
PUBLIC DOMAIN

Crabeater seals. LIAM QUINN, 2011.
CREATIVE COMMONS ATTRIBUTION-SHARE 4.0

Penguins. CREATIVE COMMONS
ATTRIBUTION-SHARE 2.0

Extent: Antarctica covers an enormous area, more than 5 million square miles, and contains much of the world's fresh water in frozen form.

Environment: It is extremely cold and only hardy creatures can survive; the most incredible is the emperor penguin.

Surfaces: Ice, snow, and melted water

People: No indigenous people live here, but there are national outposts for scientists and tourists who flock to Antarctica despite the high cost.

Animals: Seal, porpoise, dolphin, walrus, penguin, and whale

Plants: Lichen, mold, moss, and fungi

Resources: Fresh water, fish, and some minerals (antimony, chromium, copper, gold, lead, molybdenum, tin, uranium, and zinc) but in quite small deposits

(See **Climate change**, **Explorers**, **Guidebooks**, **Maps**, and "The World's Most Important Deserts" sidebar.)

Antelope is a deerlike **animal** that comes in at least 70 species (including dik-dik, **gazelle**, impala, **pronghorn**, and reedbuck) and inhabits desert environments.

Antelope horns (spider milkweed) is a **plant** topped by a purple-white umbel with many flowers (similar in appearance to wild carrot). It sprouts pods that eventually exude silklike seeding strings and is found in the **Sonoran Desert** (Gerald A. Rosenthal, 66).

Anthropogenic influence. A study determined that human activity in Egypt's **Eastern Desert** is more influential than soil properties in the "destruction of plant cover, soil erosion, and degradation of natural habitat," which are due to "over-grazing, road construction, over-collection of **plants**, salinization, over-cutting, military activities, urbanization, and industrialization" (Hussein abstract).

Anthropology is the discipline that studies what used to be called "primitive" societies (some few **desert peoples**), but this term is no longer politically correct. So anthropology is now generalized to human beings, but this discipline does not consider America's wealthy or poverty-stricken. This is the concern of sociology or, despite the foolish misnomer, urban anthropology. "The idea of a primitive society is a delusion." How then could Claude Lévi-Strauss, were he so wonted, differentiate between the Stone Age Tasaday and the Palo Alto Googlers? Thomas Widlok claims that major changes in hunter-gatherer studies have occurred: The Bushmen (now **San**) in the **Kalahari** neither stayed put nor retreated to uninhabited desert; in reality they moved out of the desert (17, 18). Kathryn Przywolnik concludes her paper with another controversy, based on fieldwork in northwestern Australia, viz, hunter-gatherer societies become more complex, sedentary, and agriculturally oriented, but the author demurs and sees "social change as non-directional and reversible" (202). Ethical quandaries in anthropology are legion and include deception in ethnographic observation and false accusations (Derek Freeman's aimed at Margaret Mead or Patrick Tierney's at Napoleon Chagnon). Carolyn Fluehr-Lobban, who did fieldwork in the **Sudan**, discusses deception,

harm-avoidance, informed consent, vile cultural practices that a visiting anthropologist may have to appear to condone while doing fieldwork, and institutional review boards, all of which may obtain in desert settings. (See also **Archeology**.)

Antiquities. Alterations in **climate** and human activities are affecting Egyptian temples, as **water**, **salt**, **pollution**, and **tourism** cause harm. Karnak and Medinet Habu are now protected by pumps that help keep the encroaching water at bay (Yee, "Climate"). (See **Abu Simbel**.)

Antivenin (antivenom) is the medicine administered to a snake bite (or strike) victim that in many cases will save their life. For hemotoxic snakes such as the copperhead, one has a few hours to get to a hospital. With neurotoxic creatures such as the coral or cobra, the medicine must be administered within a 30-minute window. Sadly, there is no antivenin for certain snakes.

Anza-Borrego Desert (930 square miles) lies in southwestern **California** and occasionally erupts with countless **wildflowers**. The author knew a person who spent a month wandering here; all he carried was flour, **water**, and other basic foodstuffs from which he cooked his meals.

Anza-Borrego Desert dandelion, desert sand verbena, and desert sunflower. JOANNA GILKESON, 2017. PUBLIC DOMAIN

Anza-Borrego Desert State Park, the largest **state park** in **California**, contains **sand**, gravel, **cactuses**, **bighorn sheep**, and fields of large **boulders** that present a truly bizarre and inhospitable landscape, but also a palm tree–studded **oasis** complete with a waterfall.

Apache (N'de) is an Indian tribe native to **Arizona** and **New Mexico**.

Aquifer is an underground containment area of porous limestone, **sand**, or gravel where **water** has accumulated, often for many thousands of years. In **deserts**, the water comes to the surface on its own or in dug wells. In other areas, humans now drain aquifers through deep wells, and surface rainwater cannot resupply them fast enough so they move toward depletion. This is the case, for example, with the enormous Oglala aquifer in the midwestern **US**. Farmers use the water to irrigate their crops. This does not occur in California's fertile valleys where it rains infrequently. Instead, here **rivers** are diverted into channels, diminutive canals, that crisscross the enormous fields where lettuce or cantaloupe are grown. Eventually, this will have to be done in empty aquifer areas. An aquifer some 3,000 feet below the surface of the **Negev Desert** supplies water for fish farming and is then used for **irrigation** (Kraft).

Arabah (Araba, Arava), an area south of the **Dead Sea**, is located in both **Israel** and **Jordan**.

Arabian Desert

900,000 (714,800, 1,600,000) square miles | Saudi Arabia and other lands | Hot desert | Minimal precipitation | Bedouin

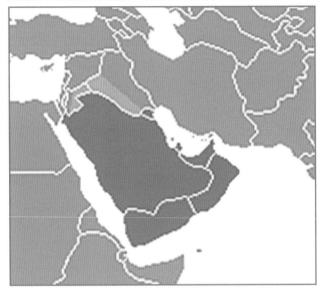

Arabian Peninsula. CIA WORLD ALMANAC

This enormous **desert** in **Saudi Arabia** and surrounding countries, of which the **Empty Quarter** is a part, is subdivided into minor deserts, including the **Syrian** and Al Nafud's red **sands**, and is extremely rich in oil. **Animals** found here include cape hare, sand cat, striped hyena, red fox, caracal, Arabian wolf, and **camel**, plus **reptiles**, multiple bird species, **insects**, and **arthropods** (https://arabiandesert1.weebly.com/animals.html). (See individually listed **plants**.)

Extent: Between 714,800 and 1,600,000 square miles depending on the source

Environment: The Arabian Desert is an enormous tract of empty and barren land, second only to the Sahara; its Empty Quarter is considered the most desolate area in the world.

Surfaces: Sand, rock, and gravel

People: Many indigenous Bedouin of sometimes warring tribes, now united by King Saud and Saudi Arabia

Animals: Striped hyena, red fox, caracal, Arabian wolf, and camel, plus reptiles, multiple bird species, and others (see above)

Plants: Acacia, alfalfa, date palms, desert rose, juniper, salt-bush, and tamarisk, plus vegetables in oases

Resources: Oil, natural gas, phosphates, and sulfur, plus gold, silver, and copper

(See **Great Nafud**, **Slavery**, and **Wadi Hajr**.)

Arabian Sands is one of Wilfred Thesiger's many books, but most interesting because he begins by informing readers that "I never thought that I would write a book about my **travels**. Had I done so, I should have kept fuller notes" (xiii). Had he "kept fuller notes," this tome would have been a lot longer than its 326 detailed pages, and his output more than the 15 books that he eventually produced. Most of his books deal with the **Arabian Desert**, and this one is no exception. Here he visits Abyssinia and the **Sudan**, but the **Empty Quarter** and Abu Dhabi predominate. He also notes that the desert that he encountered during the 1940s no longer exists because of the influx of modern conveniences and the physical junk foreigners left behind (xiii). Thesiger was a prolific photographer and so the 69 trenchant **photographs** are probably his own.

Arachnids such as **spider**s and **scorpions** can be found in many **deserts**.

Aral Sea was an enormous body of **water** lying between Kazakhstan and Uzbekistan. It has been drained by irrigation projects and is now a **desert**, an ecological disaster.

ARAMCO, the Arabian American Oil Company, produced **petroleum** in **Saudi Arabia** between 1933 and 1990. In 1988 it altered its name to the Saudi Arabian Oil Company, or Saudi Aramco, and is now entirely owned by the Saudi Arabian government.

The Arboretum at Flagstaff's stated goal is to increase the understanding, appreciation, and **conservation** of **plants** and plant communities native to the **Colorado Plateau**. (See also **Desert Botanical Garden** and **Tucson Botanical Gardens**.)

Archeological finds. Deserts are replete with animal skeletons, shards and other human **artifacts**, **rock art**, **caves** filled with treasure (such as the **Dead Sea Scrolls**), and lost cities. Jeremy Swift discovered at least 22 colorful, variously shaped Neolithic arrowheads in the **Sahara** (22). Upon returning by ship to Istanbul from Haifa, I was approached by a fellow hippie who inquired, "Did you get any?" I was mystified. He explained by taking out a handful of ancient Roman coins and said that all you had to do was kick the Negev sand and they would appear. That was not my experience kicking and luckily so because it is against the law to remove artifacts from **Israel**.

Archeology. Archeologists who work in **deserts** dig at **ruins**, examine and analyze **artifacts**, and search for lost cities buried in the **sands**. Archeological endeavors in certain deserts make sense since they harbored ancient civilizations and their architectural and artifactual remnants are available for investigation in, for example, Egypt's **Sahara** (see David Wengrow's *Archeology of Early Egypt*), the southwestern US's **Mojave**, and perhaps to a slightly lesser extent, Mongolia's and China's **Gobi**.

But what can one make of such an enterprise in Australia's deserts, inhabited by nomadic **Aborigines** often with no real permanent **dwellings**? Mike Smith's *Archaeology of Australia's Deserts* attempts to answer this enigmatic question in 406 oversize pages supported by a 42-page bibliography. Here there were prehistoric settlements, Late Pleistocene settlers, and eventually hunter-gatherers (1). Before the white man arrived, there were no more than 100,000 people living in the deserts (10), and from this early period, the Late Pleistocene, there is little if any wood, bone, fiber, or plant material left. Archeologists must depend on past climates, settlement, and stone tools, although "the archeology of these hunter-gather systems relies as much on context as on material remains." Most sites have few artifacts except for rock shelters and paintings (13). This does not augur well for creating a meaningful archeological/historical record. There are some 50 early sites (79) but the yielded shells, silcrete flakes, burnt emu eggshells, bones, charcoal, and an ochered corpse, among other things (82, 83, 93) are slim pickings. Holocene **middens** yielded gastropods (172) and other sources flaked stone (184), points (191), seed-grinding implements (198), and **dingo** remains (206). Stone axes and stone knife blades were manufactured for trade and exchange (287, 294), and Smith claims that **rock art** plays a critical role here (212ff.).

Consider a typical research paper, which tells us that beginning 46,000 years ago, Riwi, a **cave** on the edge of Australia's **Great Sandy Desert**, was occupied by Aborigines; artifacts indicate that it was used especially during wet periods (Balme abstract). Not to denigrate the successes of Australian archeology, but excavations in other deserts are far more productive, presenting countless meaningful artifacts and ruins, e.g., ships along with **gold** in the **Namib** and **Casa Grande** in the **Sonoran Desert** (https://www.youtube.com/watch?v=WFqQstZ4hg8). A collection that, to some extent, affirms this is *Desert Peoples: Archeological Perspectives* (Veth). The 16 papers

by various **scholars** cover Australia, to be sure, but also global deserts, hunter-gatherers in the **Patagonian** and **Kalahari**, and people of the coastal **Atacama**.

It may come as a surprise, even to the initiated, that like **anthropology**, archeology is riddled with controversy. Lynn Meskell's edited *Archeology Under Fire* includes 12 papers that focus, in part, on the Middle East and thus the desert. Political ideologies and nationalism as well as colonialism continue to influence archeological practices (8), some of which have resulted in the removal of **antiquities**, although this probably occurs far less frequently today than in the distant past. If the legal transportation of statuary, **pottery**, **obelisks**, and other artifacts has diminished, there still exists an illegal trade in (looted) antiquities. (See **Archeological finds**, **Dryland archeology**, **Ethics**, **Julian D. Hayden**, **Theft**, and the "Desert Explorers" sidebar.)

Arches National Park. LIBRARY OF CONGRESS

Arches National Park is located in **Utah** and contains some 2,000 extremely beautiful, red-hued sandstone arches.

Architecture. Humans have built desert dwellings, buildings, **pyramids**, and **ziggurats** in desert environments for millennia. These structures, in turn, have influenced building techniques and shapes in non-desert environments. Kazimierz Butelski discusses the relationship between the **Atacama Desert** and nearby buildings such as **museums**, observatories, and hotels constructed in "new architectural forms built in the desert and using desert materials" (Butelski abstract). And Mohamed El Amrousi and his coauthors remark on the Oasis City of Al Ain in Abu Dhabi, which presents an "oasis/villa experience, a form of revival to Andrea Palladio's mansions" with "about 11,000 large Mediterranean-styled villas with extensive green landscaping and an artificial canal in the desert." This brings up sustainability issues, since **water** is lacking and grass lawns are not really acceptable (Amrousi abstract). (See *The Invention of the American Desert*.)

Arcosanti, founded in 1970 by Paolo Soleri (PhD), it is a futuristic, ecologically oriented city, an "urban laboratory" in Arizona's **Sonoran Desert**. Employees and volunteers reside here (only 100 people, more or less, out of a potential 5,000 are in residence) and the organization holds workshops. At Arcosanti, one will find wetlands, a rainforest, grassland, desert, coral reef, and **ocean**. The **architecture** is amazing, which is understandable since Soleri worked for **Frank Lloyd Wright**. Cosanti, located 65 miles south of Arcosanti, combines desert landscape and unusual architecture. (See https://www.arcosanti.org and **Earthships**.)

Arctic

5.4 million square miles | Arctic | Cold desert | Some precipitation (rain is increasing as snow diminishes) | Inuit

Arctic relief map. UWE DEDERING, 2020. WIKIMEDIA COMMONS

The Arctic stretches as far south as Greenland, and the Arctic Ocean (5.4 million square miles) lies at its center. It is watery and icy, but it is a **desert** because **precipitation** and vegetation are scarce, especially on Greenland's enormous **glacier**. Animal life includes **polar bear**, caribou, musk ox, wolf, **fox**, saiga, wolverine, ptarmigan, and crane. The Arctic (which is shrinking) holds large oil and gas reserves (Nathaniel Harris, 164–65).

Extent: 5.4 million square miles

Environment: Cold

Surfaces: Snow, ice, and water

People: Inuit

Animals: Polar bear, caribou, musk ox, wolf, fox, wolverine, and others (see above)

Plants: Arctic daisy, arctic poppy, arctic willow, purple saxifrage moss, lichen, white arctic heather, and many others

Resources: Bauxite, copper, diamond, iron ore, nickel, petroleum, and phosphate

(See **Climate change**, Barry Lopez and Markus Rex in the bibliography, and "The World's Most Important Deserts" sidebar.)

North Pole, Arctic Ocean. MATT & KELLIE, 1990. CREATIVE COMMONS ATTRIBUTION-SHARE 4.0

Arctic fox. EMMA, 2011. CREATIVE COMMONS ATTRIBUTION-SHARE 2.0

Inuit women. ANSGAR WALK, 1999. CREATIVE COMMONS ATTRIBUTION-SHARE 2.5

Two rare ivory gulls in the drift ice of the Arctic. ANDREAS WEITH, 2015. CREATIVE COMMONS ATTRIBUTION-SHARE 4.0

Musk oxen on Arctic National Wildlife Refuge. ROBIN WEST, 2013 TRANSFERRED. PUBLIC DOMAIN

Arctic National Wildlife Refuge is a protected area (an intact **ecosystem**) in the Arctic **desert**. It is the largest such refuge in the **US**. Drilling for **oil** has been limited, but in 2023 President Joe Biden announced that he would allow it.

Area 51 is the controlled government area near Roswell, **New Mexico**, where aliens and their spaceships are secreted. This, at least, is what the conspiracists think. The bizarre **films** *Men in Black* (and its sequels) and especially *Independence Day* confirm this, although they are presumably fiction.

Arid Land Research and Management (1987–) is one of a group of similar journals that offer extremely esoteric papers on, for example, "Efficiency of MGDA and GLDA ligands in extracting plant-available Zn from calcareous soils: Kinetics and optimization of extraction conditions." Some open-access material is available. (See also *Journal of Arid Land*, *Journal of Arid Environments*, and **Journals**.)

Aridity is the quality of being dry: There is little (or no) **precipitation** in arid areas, which cover a third of the earth's surface. Semiarid land is less dry, whereas **desert** is more extreme.

Arizona is a state in the southwestern **US** that is home to the **Sonoran Desert**. Amazingly, the US's three other major deserts (**Great Basin**, **Chihuahuan**, and **Mojave**) all touch Arizona. Some people consider Arizona to be the most beautiful US state, although others would give this honor to Hawaii, Alaska, **Utah**, or even Vermont. (See https://www.youtube.com/watch?v=VcoTgPB9ON0.)

Arizona Highways (1925) is a magazine that celebrates the beauty of **Arizona** in text and **photographs** (https://www.arizonahighways.com).

Arizona-Sonora Desert Museum. Although called a museum, this wonderful Tucson institution resembles a zoo with many desert creatures (ca. 300

Owl in docent's arm, Arizona-Sonora Desert Museum.
ROBERT HAUPTMAN

animals and 1,200 **plants**) including raptors, a saw-whet **owl** that rides on the wrist of a docent, and a **mountain lion** in attendance. A botanical garden and an aquarium add to the treasures of the **Sonoran Desert**, which the museum aims to protect (https://www.desertmuseum.org). (See also **Pueblo Grande Museum and Archeological Park** and the "Museums" sidebar.)

Arizona State Museum in **Tucson** collects archeological and anthropological **artifacts** from the southwestern **US**. (See the "Museums" sidebar.)

Arroyo is a declivity like a **gully** that remains empty during the dry season but which fills with a torrent of rushing **water** when it rains. It is a very dangerous place to be when water, often from distant **storms**, comes cascading down. (See **Water** for synonyms in other languages.)

Art. The **desert** inspires artists, film directors, and other creators. Nineteenth-century French artists Jean-Léon Gerome, Eugène Delacroix, and Théodore Géricault are well known for their desert paintings. **Georgia O'Keeffe** is an excellent example of

a painter who lived in and painted desert images. Artists have taken up residence in the desert near **Joshua Tree National Park**, and works are displayed out of doors (Chaplin). *the desert*, a catalog of a photographic exhibition held in Paris in 2000, is a coffee-table volume whose texts and eccentric images are surprisingly unappealing. Participants include William Eggleston, Lee Friedlander, Paul Virilio, and an interview with **Wilfred Thesiger**.

The African Origin of Civilization, an ongoing exhibit at New York's Metropolitan Museum of Art, includes works from **Egypt** and sub-Saharan areas. Roslynn D. Haynes, in *Desert: Nature and Culture*, has excellent discussions of art (88ff., 175ff.). (See **Abstract Expressionism**, **Art: indigenous**, **Crafts**, **Egyptian art**, **Egyptian sculpture**, *The Invention of the American Desert*, **Petroglyphs**, and **Pictographs**.)

Art: indigenous. Indigenous peoples have been creating various forms of art for tens of thousands of years. (The parietal works at Lescaux, France, and Altamira, Spain, attest to this.) (See also **Aborigine: art** and **Pueblo**.)

Mimbres bowl   USA37   Kutenai parfleche USA37   Tlingit sculptures USA37   Ho-Chunk bag   USA37   Seminole doll   USA37

Mississippian effigy USA37   Acoma pot   USA37   Navajo weaving   USA37   Seneca carving   USA37   Luiseño basket   USA37

Indigenous art.

Arthropod. There are four types of invertebrate arthropods: **insects**, myriapods, **arachnids**, and crustaceans, all of which can be found in the **desert**.

Artifacts. See **Archeological finds**.

Asia. One of the world's seven continents, each of which contains a **desert**, here the **Gobi** and the **Thar**. (See "The World's Most Important Deserts" sidebar.)

Ass. See **Donkey**.

Aswan High Dam was built in 1960 on the **Nile**. It caused some archeological sites to be flooded. (See **Abu Simbel**.)

Rare snow in Atacama. NASA

Atacama Desert

54,000 (41,000, 77,000, 140,000) square miles | Chile, Peru | Hot desert | Basically no precipitation | Atacameño, Aymara

Atacama map. COBALTCIGS, 2010. CREATIVE COMMONS ATTRIBUTION-SHARE ALIKE 3.0

This is an extremely arid **desert** between the Pacific Ocean and the Andes in **Chile** and contains salt flats and **dunes**. Like the Namib's **Skeleton Coast**, it is a coastal desert. Sadly, its **indigenous peoples** were mistreated by the Inca and the Spanish. The 22,000-foot **volcano** Llullaillaco lies to its east. Although it has not rained here in eons, surprisingly, **water** flows down from distant **mountains** and feeds lagoons. It is a source for **guano**, saltpeter, **copper**, and much of the world's lithium. Telescopic observatories are built here in the desert. The Atacama is also the location where solar surface irradiance is highest. (See **Architecture**.)

　　Extent: 54,000 (140,000) square miles
　　Environment: Coastal yet extremely arid and hot
　　Surfaces: Lava, salt lakes (salares), sand, and rock
　　People: Atacameño and Aymara
　　Animals: Condor, flamingo, guanaco, iguana, lava lizards, and many others
　　Plants: Black sage, buckwheat bush, ferns, little-leaf horsebrush, rice grass, salt bush, tufted grass, and many others

Flowers, Atacama Desert in Chile and Peru. DICK CULBERT, 2013.

Atacama Desert map. WIKIMEDIA COMMONS

Resources: Boron, copper, gold, iron, lithium, silver, sodium nitrate, and potassium

(See **Aymara**, **Boulders**, **Fog**, **Geoglyphs**, **Pan de Azúcar National Park**, and **Temple of Tulán-54** as well as the 2010 **documentary** *Nostalgia for the Light*, which was shot in the Atacama and deals with astronomy, **archeology**, and human rights. See the **film** at https://www.you tube.com/watch?v=ztoJl7USiNQ.)

Atlas Mountains is a mountain chain that runs east–west through **Morocco**, **Tunisia**, and **Algeria**. It separates the **Sahara** from the Mediterranean.

The Atlas of a Changing Climate is a handsome volume that emphasizes visualizations with more than 100 colorful and often oversize illustrations and **maps**. A small section covers the **desert** (Buma).

Atlases. Various types of atlases present cartographic and textual data and information on **deserts** and desert environments. A superb example is Benchmark Maps' *New Mexico Road and Recreation Atlas*, which divides the state into many sections and thereby presents a very detailed overview. Just as useful are its many descriptive lists of, for example, parks and monuments, wildlife areas, attractions and **trails**, and **pueblos** and tribal lands. A useful index gathers all of this material together and offers handy access. This publisher also produces similar atlases for **Arizona**, **Nevada**, **Utah**, and other states. Pointed atlases are rare but see Fretwell, Nathaniel Harris, Hayes, Kumar, Mendelsohn, Middleton, Zell, and *An Atlas of the Sahara-Sahel* in the bibliography. (See also ***The Atlas of a Changing Climate***, **Conservation**, and **Maps**.)

Austin, **Mary Hunter** (1868–1934), was a feminist desert naturalist. Her *Land of Little Rain*, which discusses nature and indigenous cultural and social issues in the **Mojave Desert**, is a classic. It is available online at https://www.gutenberg.org/files/365/365-h/365-h.htm.

Australia is one of the world's seven continents and the only one that is also a country. It is the driest inhabited continent and is replete with **deserts** (80 percent of Australia is desert) that often contain enormous **stations** (cattle farms) that are measured in square miles rather than acres. The desert land, however, is so barren that it is hard to believe that it can support **livestock**. Other areas of the country are more fertile. Its 10 deserts include the **Great Victoria**, **Gibson**, and **Simpson**. It is well known for its past gold and current diamond production. Plant (15,000 species) and animal life include **acacia** and eucalyptus and mole, **frog**,

snake, magpie, galah, and **emu** (Nathaniel Harris, 163, 162). (See **Australian Desert**, **Outback**, and "The 10 Australian Deserts" and "The World's Most Important Deserts" sidebars.)

Australian Arid Lands Botanic Garden is a botanical collection in Port Augusta, **Australia**. (See https://www.aalbg.org.)

Australian Desert

890,000 (ca. 980,000) square miles | Central Australia | Hot desert | Little precipitation followed by drenching, flooding downpours | Aborigines

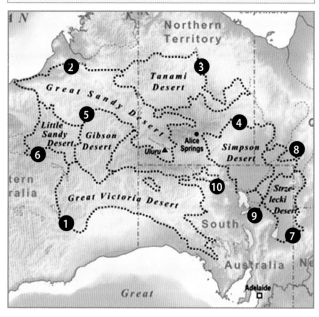

Deserts in Australia map. LENCER, 2013. CREATIVE COMMONS ATTRIBUTION-SHARE ALIKE 3.0

Kangaroo Flat. A musk lorikeet *(Glossopsitta concinna)* feeding on desert ash fruit. PATRICK_5K9, 2013. CREATIVE COMMONS ATTRIBUTION 2.0

This is the general term for the many deserts, contiguous with each other, including the **Great Victoria**, **Great Sandy**, **Gibson**, and **Simpson**, in central **Australia**.

THE 10 AUSTRALIAN DESERTS

From larger to smaller:

1. Great Victoria Desert
2. Great Sandy Desert
3. Tanami Desert
4. Simpson Desert
5. Gibson Desert
6. Little Sandy Desert
7. Strzelecki Desert
8. Sturt Stony Desert
9. Tirari Desert
10. Pedirka Desert

Automobiles/trucks. Because people drive much too fast and really badly, the carcasses of old, abandoned, and wrecked vehicles litter highways, **roads**, and routes in many **deserts**. Multiple photographic images are available to confirm this.

Ayers Rock. See **Uluru**.

Ayers Rock–Mount Olga National Park. See **Uluṟu-Kata Tjuṯa National Park**.

Aymara are a people who live especially in Bolivia as well as in the **Atacama Desert**. (Note that my brother-in law was a full-blooded Aymara Indian.)

Azalai is the salt caravan route in the **Sahara**.

Aztec Ruins National Monument lies in the northwestern corner of **New Mexico**. It contains "great houses," **pueblos**, **kivas**, and **artifact**s including 800-year-old wooden beams (brochure).

Bab'Aziz: The Prince Who Contemplated His Soul is a **film** about a man and his granddaughter who travel through the harsh but beautiful **desert**. He entertains her by telling her a story about a prince. (See the trailer at https://www.youtube.com/watch ?v=IPYjenA3VFg.)

Baboon. The chacma baboon can be found in the **Kalahari Desert**.

Bacteria are single-cell, microscopic creatures (microbes) that do many things, e.g., help cows digest **grass** and grain, **termites** digest cellulose, and humans maintain a healthy gut. They also cause ailments and diseases such as E. coli and salmonella. The **Atacama Desert** contains many new strains, and the literature is replete with descriptive papers on their discovery. (See **Microorganisms**.)

Badain Jaran Desert (Mysterious Lakes) is located in **China** and Inner Mongolia; it has the world's tallest dunes at 1,600 feet.

Badiya (Badu). In Arabic, like **Sahara**, it means **desert** (people of the desert).

Badlands are arid, eroded areas with little plant life.

Badwater Basin, in **Death Valley**, at 282 feet below sea level, is the lowest elevation in the **US**. (Coincidentally, the highest point in the lower 48 states, Mount Whitney at 14,505 feet, is just 135 miles away near Lone Pine, **California**.)

Badwater Basin, Death Valley. ROBERT HAUPTMAN

Flowering barrel cacti with other wildflowers in desert.
GENTRY GEORGE, 2013 TRANSFER. PUBLIC DOMAIN

Baby barrel cactus clump, Mojave Desert. JESSE EASTLAND, 2016.
CREATIVE COMMONS ATTRIBUTION-SHARE ALIKE 4.0

Bagnold, Ralph Alger (1896–1990) (**Sahara-Libyan**). A founder of **geomorphology**, Bagnold practiced scientific experimentation, joined the military, and was posted to various locations. He explored that part of the **Sahara** known as the **Libyan Desert** using specially equipped, early-model Ford vehicles. He was able to cross great dune fields easily, although at times as many as 13 men had to pull and push the car. He did excellent archeological work during his many expeditions. His books include *Libyan Sands* and *The Physics of Blown Sand and Desert Dunes* (Goudie, *Great*, 232–37). (See **Explorers** and the "Desert Explorers" sidebar.)

Baja California is a Mexican peninsula south of **California** that contains two **deserts**: the Vizcaino and the San Felipe.

Bajada is a rocky slope below a **mountain**, like a scree or talus field.

Bandelier National Monument is a large park near **Los Alamos, New Mexico**, that preserves the multiple rock/cave homes of ancient pueblo dwellers. The higher **caves** can only be reached by climbing rickety ladders.

Baobab ("tree of life") is a large **succulent** (**tree**) found in African **deserts** as well as the **Negev**. (See also **Boab**.)

Barchan. Sickle **dune**. (See **Dune types**.)

Barrel cactus is a large, round, multi-ribbed **plant** found in the **Chihuahuan Desert**.

Barth, Heinrich (PhD) (1821–1865). An excellent **explorer**, Barth traveled the **Sahara** to **Lake Chad** and **Timbuktu**. He wrote a ca. 3,600-page "indispensable" report, *Travels and Discoveries in North and Central Africa*, on the Sahara (Michael Martin, 357–58).

Great Basin. KMUSSER, 2020. CREATIVE COMMONS ATTRIBUTION SHARE-ALIKE 3.0

Basin and range are the alternating valleys and **mountains** in the western **US**.

Bat. There are some 2,000 species of bats, some of which live in the **desert**; they hunt at night and hibernate when conditions deteriorate (Jaeger, 124, 128). In Cairns, **Australia**, located to the distant east of the desert, enormous fruit bats (with wingspans of almost 5 feet) flit around the city the way pigeons do in New York or Venice. (See also **Carlsbad Caverns National Park**, **Cave**, **Guano**, and https://www.youtube.com/watch?v=yLufIO5fZ6o.)

Battuta, Muhammad Ibn (1304–1368?), was an intrepid medieval traveler who wrote the *Riḥlah*, which describes his 75,000 miles of adventures from **Morocco** to **China** and **Mali**. (See **Explorers**, **Maps: history**, and the "Desert Explorers" sidebar.)

Bauxite is a mineral found in desert environments. It yields aluminum.

Beach grass (*Halopyrum mucronatum*) can be found on the **dunes** in Egypt's **Eastern Desert** along the Red Sea coast (Zahran, 134). It occurs in tussocks that can be 3 feet in height.

Bear do inhabit desert environments. The black bear is ubiquitous and the least dangerous to humans; grizzlies are more aggressive but playing dead may deter them. The brown bear is to be feared and playing dead will lead to a calamity. **Polar** and Kodiak bears live in the Arctic **desert**. (See **Gobi bear**.)

Bear River National Wildlife Refuge (also, Bear River Migratory Bird Refuge) in **Utah** encompasses marshland and mudflats on the edge of the **desert** (land once home to the **Shoshone**, **Paiute**, Bannock, and Ute) where hundreds of thousands of **birds** of 222 species live, breed, or pass through on their migrations. Sago pondweed may provide **food** for loon, grebe, egret, heron, tundra swan, eagle, duck, avocet, ibis, and meadowlark; weasel, skunk, marmot, and garter snake can also be observed here (Wall, 65–71). This is one of the US's most important breeding grounds and halting places for migratory birds. Surprisingly, **hunting** is allowed. (See https://www.fws.gov/refuge/bear-river-migratory-bird, https://www.youtube.com/watch?v=0zg8IEyw17I, and the "National Wildlife Refuges (NWR): USA" sidebar.)

Beau Geste is a 1939 **film** based on a **novel** by P. C. Wren starring a host of celebrities including Gary Cooper and Ray Milland. Three brothers find themselves in the **French Foreign Legion** in North Africa. It received five stars from Rotten Tomatoes. There are, additionally, 1926, 1966, and 1977 versions.

Beavertail cactus is 6 to 12 inches in height, has reddish flowers, is partially edible, and is found in the **Mojave**, **Anza-Borrego**, and **Colorado Deserts**. (See **Prickly pear**.)

Bedouins (Bedu) (Ar: *badawi*) are a nomadic people, 6 to 7 million of whom inhabit the **deserts** of North Africa, the Arabian Peninsula, **Israel**, **Iraq**, Syria, and **Jordan**. They represent under 10 percent of the Arab world and only 1 percent are real **nomads** (Haynes, *Desert*, 69). James P. Mandaville's discussion of the ethnobotanical aspects of the Arabian Bedouin make it surprisingly clear that botanical concerns play an inordinately crucial role here; indeed, they are just as essential

Syrian Desert Bedouins. YEOWATZUP, 2010.
CREATIVE COMMONS ATTRIBUTION 2.0

as the **animals** the Bedouin nurture. Jibrail S. Jabbur's *Bedouins and the Desert* (based in the **Syrian Desert**) is summative and comprehensive, another masterful work of desert scholarship. Here the author epitomizes, and later explicates in great detail, the four pillars of Bedouin life: a desert area, camels, a tent, and the Bedouin himself. He further insists that unless all four pillars exist simultaneously, genuine Bedouin life is impossible (39–40). There follow many chapters on **trees** and **plants** (64ff.), desert animals (82ff., 96ff.), desert

1912 photo of Bedouin women weaving clothes. PUBLIC DOMAIN

birds (including training falcons) (120ff.), desert **reptiles** and **insects** (146ff.), and domesticated animals (154ff.) (all in detail, noting individual species), as well as the innumerable tribes (259ff., 286ff.), character, appearance, raiding, education, and so on. Nearby, in the **Negev**, the 2007 Bedouin **population** numbered about 172,000, some of whom reside in illegal villages (Zerubavel, 147). (See **Ethnobotany**; **Sinai, Peninsula**; and https://www.youtube.com/watch?v=HALjUHkzdVw.)

Beersheba is a (biblical) city in **Israel** located at the northern terminus of the **Negev Desert**. There is little between it and **Eilat**, Israel's southernmost city. Over the past half century, it has grown from a small desert town into a large urban area.

Beetle. An **insect** of many species that can be found in the **desert**.

Bell, Gertrude Margaret Lowthian (1868–1926) (Syrian, Arabian). Other woman traveled with female companions or their husbands; Bell (like **Freya Stark**) was one of the few independent female desert **explorers** as well as an archeologist and spy. She took scientific measurements and exemplary **photographs**. Some who met her found her disagreeable, but others were quite fond of her and her accomplishments. Her books include *The Desert and the Sown*, which concerns her 1905 trip from Jericho to Antioch (Goudie, *Great*, 67–70). She also visited **Petra**. (See ***Queen of the Desert*** and the "Desert Explorers" sidebar.)

Gertrude Bell in Iraq, 1909. PUBLIC DOMAIN

Benguela current. See Current.

Berber (Amazigh, pl: Imazighen) (from the Latin for "barbarian") (free people). The 30 to 40 million **indigenous peoples** of North Africa (**Algeria, Libya, Mali, Morocco,** and **Niger**) who are primarily Muslim, though there are also some few Jews among them. They speak a host of different languages including Kabyle, Riffi, **Tamazight**, Tarifit, and Tashelhiyt. In *The Berbers*, Michael Brett and Elizabeth Fentress present a complete history of these people from antiquity to the Romans, Islamic unification, Arabization, Ottoman influence, as **nomads**, and the future, at least as it appeared in 1996 (passim). They observe that in the 14th century, the great **scholar** Ibn Khaldun wished to celebrate the achievements of the Berbers (117). Years ago, a transition occurred from tribal politics to nationalism in the **Maghreb** (Hoffman, 1). However, Berber rights and culture were quashed in **Algeria** and so the people have risen up to demand them (Aïtel, 3). To this end, Fazia Aïtel's *We Are Imazighen* is an overview of fairly recent Berber literature and culture, which is surprisingly rich. (See **Tunisia**.)

Besh-Ba-Gowah, southeast of Phoenix, is an archeological park with **ruins, artifacts,** a **museum,** and an ethnobotanical garden. The **Hohokam,** then the Salado, and finally the **Apache** lived here. Unlike at many other southwestern Indian sites, here one may enter the rooms and climb the ladders (brochure). (See https://www.archaeologysouthwest.org/explore/besh-ba-gowah-archaeological-park.)

Bi! bulb plant of the **Kalahari** is a tuber that contains **water**, which is removed by scraping. The process is shown at https://africafreak.com/how-to-find-water-in-the-kalahari-desert.

Bible. The holy book of **Judaism** and **Christianity**. The Old Testament (OT), sacred to Jews, consists of many books including the first five, the Pentateuch (the Torah given to Moses on **Mount Sinai**); the New Testament (along with its predecessor) is sacred to Christians. The patriarchs in both are honored by **Islam**, whose holy book is the Koran.

The first two books of the OT, Genesis and Exodus, describe life and **travel** in the **Negev Desert**, first for Abraham and then when the Israelites leave **Egypt**. The fourth book, Numbers, is called רבדמב (*bamidbor*, but now *bamidbar*), which means in the wilderness, but also in the **desert**.

Bibliography. Desert bibliography is enormous; a comprehensive listing probably does not exist and would be counterproductive unless compiled as a subject listing. One of the most useful is contained in *The Desert: Lands of Lost Borders*, Michael Welland's rather eccentric and generously illustrated 2015 study. The annotated listing contains approximately 300 works and is divided by chapter and then by subject matter thus:

[Chapter] 2 Big and Small, Fast and Slow
• Desert landscapes and processes
• Desert pavements and varnish
• Soils and nitrates
• Floods
• Lakes
• Microbes
• Scale

Material is listed under each heading for a total of 41 items. This series of topics also indicates the diverse nature of Welland's study.

E. I. Edwards's 1958 *Desert Voices: A Descriptive Bibliography* (UVM online) is now quite dated but does contain almost 1,000 entries of the many germane books published in the distant past, some of which are available online. See also Patricia Paylore's three listings: *Desert Research: Selected References 1965–1968* (1969) (410 annotated pages), *Sonoran Desert: A Retrospective Bibliography* (1976), and *Desertification: World Bibliography Update 1976–1980* (1980) (annotated).

Additional pointed bibliographies exist; for example, a 14-page section in the introduction to Jibrail S. Jabbur's *Bedouins and the Desert* that contains a wealth of bibliographic data on the subject in both Arabic and English (14–28); Laurette Isabella Botha's *Namib Desert: A Bibliography*; and the Australian Deserts website (https://bbldeserts .weebly.com/bibliography.html). The individually

authored chapters in the *Reference Handbook on the Deserts of North America* also contain their own often extensive bibliographic listings; for example, Frank S. Crosswhite et al.'s "Sonoran Desert," a 104-page essay (163–266), concludes with a reference list of 750 entries (266–95).

Big Bend National Park (800,000-plus acres) lies along the Rio Grande in western Texas, and here the **desert** blooms with many flowers including bluebonnet, mustard, and **prickly pear**. **Pictographs** and archeological remains attest to the **indigenous peoples** who lived here. (See https://nps.gov/bibe.)

Bighorn sheep. See **Desert bighorn sheep**.

Billy button daisy (drumstick, woolyhead) is an unusual-looking Australian flower with a spherical head containing many flowers.

Bindweed (*Convolvulus ammonia*) is a Gobi plant with pink or white flowers (https://www.discover mongolia.mn/blogs/plants-of-the-gobi-desert).

Bingham Canyon Mine in **Utah**, the largest excavation in North America, yields **copper** as well as **gold** and **silver** (Findley, 84). (See **Mining**.)

Biodiversity is the sum total of the world's diverse animal and plant species. There exist many reasons why species are lost, not the least of which is human agriculture, overfishing, **hunting**, and **mining**, among other things, and this reduces diversity in the **biosphere**. In order to attempt to protect the earth, its living entities, and the food supply, 190 countries approved a United Nations agreement to protect 30 percent of the earth's surface and **oceans** (Einhorn, "Nations").

Biology is the discipline that studies the earth's life forms. Zoology specializes in **animals**, microbiology in very small organisms, and botany in **plants**.

Biome is an area with its own **climate** and plant and animal life. There are five types: aquatic, grassland, forest, **desert**, and **tundra**. (See **Biosphere**.)

Biospeliology studies life forms in **caves**. Aldemaro Romero's *Cave Biology: Life in Darkness* explores the amazing assortment of creatures that inhabit caves, some of which, naturally, can be found in **deserts**. The enormous breadth of this topic can be judged by Romero's extensive 51-page bibliography.

Biosphere is the sum total of all of the **biomes**.

Biosphere reserves. There are 28 UNESCO-protected areas in **hot deserts** and 7 in their cold counterparts including **Big Bend National Park**, Jornada Experimental Range, Mpini Reserve, Desert Experimental Range, Beaver Creek Experimental Watershed, **Organ Pipe Cactus National Monument**, Noatak National Arctic Range, Northeast Greenland National Park, and Northeast Svalbard Nature Reserve (Folch, 257, 458). There is a total of 738 biosphere reserves in 134 countries.

Biosphere 2 is an Earth system science research facility located in Oracle, **Arizona**, north of **Tucson**. Its mission is to serve as a center for research, outreach, teaching, and lifelong learning about Earth, its living systems, and its place in the universe. Its desert habitat is designed to simulate an arid desert scrub **ecosystem**. (See https://biosphere2.org.)

Birds. Many species of birds including the cactus wren, greater roadrunner, western screech owl, **ostrich**, **emu**, **flamingo**, and innumerable others can be found in the **desert**. (See **Guidebooks** and also individual species as well as the "Desert Birds and Adaptions" sidebar.)

DESERT BIRDS AND ADAPTIONS
(partial listing)
- Burrowing owl (*Athene cunicularia*)
- Cactus wren (*Campylorhynchus brunneicapillus*)
- Costa's hummingbird (*Calypte costae*)
- Elf owl (*Micrathene whitneyi*)
- Ferruginous pygmy owl (*Glaucidium brasilianum*)
- Gambel's quail (*Callipepla gambelii*)

(continued)

- Gila woodpecker (*Melanerpes uropygialis*)
- Greater roadrunner (*Geococcyx californianus*)
- Hwamei (*Garrulax canorus*)
- Indigo bunting (*Passerina cyanea*)
- Lucifer hummingbird (*Calothorax lucifer*)
- Phainopepla (*Phainopepla nitens*)
- Pin-tailed sandgrouse (*Pterocles alchata*)
- Rosy-faced lovebird (*Agapornis roseicollis*)
- Verdin (*Auriparus flaviceps*)

(https://www.bioexplorer.net/desert-birds-and-adaptations.html)

Bizarre manifestations found in **deserts** include a fake Prada store in Texas, a sculpture that eventually will decay; a ghost town in the **Namib**, left to its sand-encroaching destiny after the diamonds ran out; **Coober Pedy**, **Australia**, an underground city; **fairy circles**; and many others (https://www.youtube.com/watch?v=-lSxrYB2YRk).

Black blizzards are extreme **dust storms** on southwestern US desert surfaces and elsewhere. (See https://www.youtube.com/watch?v=Ep7-7x2sp8Y.)

Black Gobi. Near **Dunhuang**, **China**, there exists some black Gobi Desert hills.

Black Rock Desert, **High Rock Canyon** (not to be confused with the Black Rock Desert Wilderness). Here in **Nevada** one finds lava beds, **playa** (alkali flats), the **Burning Man Festival**, and the home of the 1997 land-speed record of 766 miles per hour (https://www.blm.gov/programs/national-conservation-lands/nevada/black-rock-desert-high-rock-canyon-emigrant-trails-nca).

The Black Stallion (1979), *The Black Stallion Returns* (1983), and *The Young Black Stallion* (2003). The latter **film**, a prequel, precedes the 1979 version, which deals with a boy and a horse washed up on a desert island. Eventually, he trains the horse to race.

Boab is a **succulent** (**tree**) found in Australia's **Tanami Desert**. (See **Baobab**.)

Body temperature. See **Temperature**.

Bonneville Salt Flats. Located west of **Salt Lake City**, this large, extremely flat area has been used to break land-speed records, which now stands at 766 miles per hour.

Boojum tree (cirio) is a very peculiar-looking **plant** that droops downward; it can be found in Baja California **deserts**. It reaches 50 feet and is related to **ocotillo**.

Books. There exist many thousands of monographs and coffee-table books (in English as well as in innumerable other languages) that devote their pages in whole or in part to the **desert** and its inhabitants (and some extraordinary examples are noted in the text), but the following stand out so blatantly that they deserve special mention and scrupulous perusal, especially for their often magnificent, sometimes enormous **photographs**. First, we have David Miller's *Deserts: A Panoramic Vision* at 12 by 17 inches so that an occasional double-page spread, among its hundreds of exquisite images, measures 12 by 31 inches; Miller emphasizes natural scenery. Next comes Michael Martin's *Deserts of the Earth* (12 by 12 inches); it too contains hundreds of fine photos, which include many people. And then there is Carol Beckwith and Angela Fisher's magnificent two-volume *African Ceremonies*. This same Angela Fisher is responsible for *Africa Adorned* with many awe-inspiring photos. Tiziana and Gianni Baldizzone's *Wedding Ceremonies: Ethnic Symbols, Costumes and Rituals* is replete with Berber, Tuareg, and other desert peoples' ornate and bejeweled wedding dress. Finally, see the stunning ***Pastoral Tuareg: Ecology, Culture, and Society*** (Johannes Nicolaisen and Ida Nicolaisen). These are all large-format, profusely illustrated books; reading their texts and scrutinizing the colorful illustrations is quite different and superior to glancing at diminutive images on the internet, although this too is profitable.

Marco C. Stoppato and Alfredo Bini's *Deserts* is a small **guidebook**, but so diverse, so replete with data and information, and so colorfully illustrated that it also deserves special mention, as does Dennis Wall's *Western National Wildlife Refuges*, with 80 pages of color plates; Nathaniel Harris's *Atlas of the World's Deserts*, whose title is a slight misnomer, since although it does contain a few excellent **maps**, it also provides textual overviews and lists of facts; and Jake Page's *Arid Lands*, which is filled with excellent images and maps but also an incisive and informative text. Finally, Rorke Bryan's overwhelming *Ordeal by Ice: Ships of the Antarctic* is so informative and so replete with innumerable black-and-white and color photos, paintings, schematics, **maps**, appendices, glossary, and **bibliography** that it deserves careful reading by anyone interested in the polar deserts. (See **Desert**; **Jewelry**; **Manuscripts, rescued**; **Reference books**; and various listed titles throughout the text.)

Books: meaningful. E. I. Edwards compiled a list of his most meaningful desert books, those that he would wish to retain if the others were let go. It includes works by **Mary Austin**, Willa Cather, and 23 others (Edwards, *Desert Harvest*).

Borax is a white mineral (sodium tetraborate) mined in **Death Valley**. It is used as an all-purpose cleaner but poses some health risks. (See **Twenty-mule team**.)

Border fencing. See **Borders**.

Borders between countries, especially in large **deserts** such as the **Sahara** or **Gobi**, can be ambiguous. But the one between **Mexico** and the **US** is often clearly marked—by the Rio Grande and fencing, neither of which deters smuggling or illegal **immigration**. In 2006 new fencing had to be okayed by the Tohono O'odham nation (which opposes it for practical and cultural reasons), through whose lands it must travel (Archibold, "Border"). In 2007 a 28-mile "virtual fence" was begun in **Arizona**; it consists of nine ca. 100-foot towers equipped with radar and cameras (Archibold, "28-Mile"). And later, under the Trump administration, new walls were promised and even built. In 2020 problems continued with the Tohono O'odham nation's historic sites under duress as building continued (Carranza). These walls harm animal and plant life. (See https://www.youtube.com/watch?v=Cx-71C4iguuk and **Immigration**.)

Baobab, Botswana. DIEGO DELSO, 2018. CREATIVE COMMONS ATTRIBUTION-SHARE ALIKE 4.0

Boswellia sacra is the **tree** from which we get frankincense. It can be found in the Sahel countries as well as in **Oman** and **Yemen**.

Botany is the discipline that deals with **plants**.

Botswana is a country north of **South Africa**. The **Kalahari Desert** covers 70 percent of its land, which is rich in animal life and diamonds. Overgrazing of vast heads of cattle have harmed its environment (Michael Martin, 277).

Boulders. In the **Atacama Desert** there are thousands of sometimes enormous boulders that are smoothed across their middles, the result of rubbing against each other during earthquakes ("Smooth").

Boyce Thompson Arboretum, east of Tempe, is part of the University of Arizona. It educates the public on the **plants**, **animals**, and **ecosystems** of the **desert**.

Box Canyon. This canyon in a desertlike area of Idaho leads to the turquoise **waters** of the 11th-largest spring in the **US** (https://www.onlyin yourstate.com/idaho/box-canyon-id/).

Bradshaw paintings. Upper Paleolithic ice age rock art paintings in the Kimberley area of **Australia**. There may be as many as 100,000 galleries of paintings (Haynes, *Desert*, 93ff.).

Breakdown is a 1997 thriller/quasi-horror **film** starring Kurt Russell and Kathleen Quinlan, whose car breaks down in the **desert**. Vile people kidnap Quinlan and demand a ransom. There is lots of violence and very bad things happen. Nevertheless, Rotten Tomatoes saw fit to assign it four stars.

Bristlecone pine. The oldest living entity on earth, these 3,000-, 4,000-, or 5,000-year-old **trees** can be found high above the **Great Basin Desert**. In a stand in California's White Mountains, the trees are small, scraggly, and sometimes appear to be dead. One is adjured not to remove even a tiny branch lying on the ground.

Brittlebush is an abundant **shrub** with yellow flowers; the stem can produce a resin that is used as incense. It is found in the **Sonoran Desert** (Gerald A. Rosenthal, 122). It is also used in basketry, against rheumatism, and for cuts and bruises on horses (Rhode, 84–85.).

Brochures. See **National parks and monuments: brochures**.

Brothels. **Nevada** is notorious for being the only US state in which prostitution is legal. What many people do not realize is that it is not allowed in cities such as **Las Vegas** or **Reno** where gambling is permitted. Therefore, brothels such as the Chicken Ranch are located in the **desert**.

A Brush with Georgia O'Keeffe is Natalie Moscow's 2008 play concerning the **desert** landscape painter.

Buddhism is one of the world's seven major religions. In its name innumerable artistic treasures have been created in desert environments. (See **Mogao Caves**.)

Buddhist structures. Buddhists built structures all across **Asia**. Some are enormous Buddhas, while others are **caves** carved into cliffsides. (See **Mogao Caves**.)

Buenos Aires National Wildlife Refuge lies west of **Tucson** and is Sonoran Desert grassland and marsh 1,000 feet above the desert floor. Its 290 bird species include the masked bobwhite (which requires protection), grebe, sandpiper, hawk, **owl**, sparrow, and warbler; other **animals** include **pronghorn**, **mountain lion**, **coyote**, **bat**, **rattlesnake**, **lizard**, skink, and **tortoise**. **Plants** such as **mesquite**, sycamore, **willow**, oak, desert broom, **saguaro**, and **cholla** are all in evidence (Wall, 94–100). The bobwhite is in decline for many reasons, and anyone who has lived or traveled in Iowa, for example, has seen countless bobwhite (or quail) dead along roadsides. Apparently, they (along with other species) have not learned to avoid traffic accidents.

Buildings. See **Dwellings**.

Bukhara, a city now in Uzbekistan, was an important location on the **Silk Road** as well as a center of Islamic culture.

Burckhardt, Johann Ludwig (1784–1817) (**Arabian**). Andrew Goudie's generous overview of Burckhardt's life indicates that he was the first explorer to see **Petra** and **Abu Simbel**; he also visited **Mecca** and Medina. In order to succeed in his endeavors, he studied Arabic and surgery and renamed himself Shaikh Ibrahim ibn Abdullah. His books include *Travels in Nubia* and *Travels in Arabia* (Goudie, *Great*, 16–22). (See **Explorers** and the "Desert Explorers" sidebar.)

Bureau of Land Management (BLM) is a US government agency that manages public lands.

Burning Man Festival takes place annually in **Black Rock Desert** (**Nevada**) and is a gathering of people who manifest creativity, self-expression, cultural differences, and knowledge sharing, thereby releasing social stigma. Only radically bizarre "art cars" are allowed on the streets.

Burro. See **Donkey**.

Burton, Sir Richard F. (1821–1890) (**Arabian, Mecca**). Burton's famous, illustratively footnoted *Personal Narrative of a Pilgrimage to Al-Medinah and Meccah* is a (perhaps *the*) pinnacle of desert travelogues. He crossed the **desert** and visited the holy cities. It should be noted that had it been discovered that this disguised man was an infidel, he would have been killed. He knew 25 languages and published more than 70 books. (See **Explorers** and the "Desert Explorers" sidebar.)

Richard Burton in native dress, before 1904. PUBLIC DOMAIN

Admiral Richard E. Byrd. LIBRARY OF CONGRESS

Bushmen (now derogatory). See **San**.

Byrd, **Richard E., Jr.** (1888–1957) (**Antarctica, Arctic**), was one of the most famous **explorers** of the early 20th century (comparable to Lindbergh). He lived for months in a small cabin in frigid Antarctica, where he almost succumbed to noxious fumes. He ostensibly reached the North Pole by plane. His many books include ***Alone: The Classic Polar Adventure*** and *Little America: Aerial Exploration in the Antarctic, the Flight to the South Pole.*

Cabeza Prieta National Wildlife Refuge (Arizona). Located near **Organ Pipe Cactus National Monument**, in the **Sonoran Desert**, this refuge is harsh and hot (its **road** is called **El Camino del Diablo**), and a permit is required because it is also ensconced within military air space. **Birds** include hawk, dove, quail, **roadrunner**, and kinglet; other **animals** found here are **desert bighorn sheep, pronghorn, mountain lion, bat, rattlesnake, lizard, desert tortoise,** and **scorpion**. **Plants** one may see are **mesquite, willow, ironwood, creosote,** bursage, and **cactuses** such as the **night-blooming cereus**. Even though bighorn are endangered, **hunting** is allowed (Wall, 101–8).

Cactus (plural: cactuses, cacti) is a **succulent** of 1,200 diverse species that thrives in desert environments because it is able to store **water**. Timber Press in Portland, Oregon, publishes books on cacti and succulents including one on the search for Bolivian, Peruvian, and Argentinean cacti; another on agaves and **yuccas**; and a third on *The Cactus Family* with ca. 1,000 color images. (See **Cactus: endangered, Cactus Rescue Crew, Cactus trade, *The Great Cacti*, Plants,** and the "Cactuses" sidebar, which will lead to individually listed species.)

Multiple cactuses in the Tularosa Valley (Basin), Chihuahuan Desert. ROBERT HAUPTMAN

CACTUSES
(partial listing)

• African peyote	• Hoodia
• Barrel	• Jojoba
• Beavertail	• Night-blooming
• Blind	cereus
• Bunny ears	• Organ pipe
• Cardón	• Pincushion
• Cholla	• Prickly pear
• Chuparosas	• Saguaro
• Compass barrel	• Smooth tree pear
• Eve's needle	• Teddy bear
• Hedgehog	

Cactus: endangered. The earth's altering **climate** results in hotter **temperatures** that, in turn, adversely affect the delicate balance necessary for cactuses to thrive. In just 30 years, global warming will threaten 60 percent of cactuses with extinction (Zhong). (See also https://www.lifegate.com/cacti -are-threatened-with-extinction and **Cactus trade**.)

Cactus longhorn beetle is a genus of large beetles, 1,000 species of which live in North American **deserts**. They eat cholla, prickly pear, and saguaro seedlings.

Cactus Rescue Crew is an offshoot of the Tucson **Cactus and Succulent Society**. In the **Sonoran Desert**, where development occurs with great frequency and helps to uproot and destroy cactuses, these volunteers race out in the early morning and dig up young **saguaro**, **barrel**, and **hedgehog** cacti and transplant or sell them. Founded in 2000, by 2006 they had rescued more than 27,000 **plants** (Patricia Leigh Brown).

Cactus and Succulent Society of America (CSSA). Founded in 1929, this organization advances all aspects of these **plants** (https://cactus andsucculentsociety.drcamhosters.com). (See videos at https://www.youtube.com/c/cactusandsuccu lentsocietyofamerica/videos.)

Cactus trade. In most countries, either a permit is required or it is simply illegal to collect and sell cactuses (even when they grow on private land). They are worth a great deal on the illegal black market. In **Saguaro National Park**, some **plants** are protected with microchip tagging devices. In **Arizona** it is a felony to steal or sell these plants without a permit; the penalty is a fine and jail time. Many cactuses are rare, and after the devastating 2020 Dome Fire in the **Mojave**, a million-plus **Joshua trees** were lost. (See also **Gebel Kamil Crater**, **Joshua Tree National Park**, **Ocucaje**, and **Wildlife trade**.)

Caillié, René-Auguste (1799–1838) (**Sahara**). Suffering greatly (from scurvy and other miseries), as many of the desert explorers did, and dying early, Caillié turned himself into a Moor, left the coast and headed north to Kurussa and **Djenné**, and upon reaching **Timbuktu** became the first European to

survive. He then went on to Fez and finally Tangier, an extraordinary journey. His *Travels through Central Africa to Timbuctoo; and across the Great Desert to Morocco Performed in the Years 1824–1828* describes his accomplishments (Goudie, *Great*, 297–302). (See **Explorers** and the "Desert Explorers" sidebar.)

Cairn (G: Steinhaufen; F: cairn; S: mojón; I: vörður; Af: steenhoop; H: גַּלְעֵד [galeed]; Ar: ركام من حجارة [rukam min hijara]; M: кээрн; C: 凱恩 [kǎi ēn]). A group of stones piled (high) on top of each other to mark a route (or location). They are more usually constructed on mountain **trails** but can also be found in **deserts**. Lawrence Hogue, in his personal account of wandering in the **Anza-Borrego Desert**, informs readers that he and his hiking buddies enjoy destroying "ducks," the little cairns that mark trails, claiming they "are poorly placed and almost useless" and ruin the delusion that hikers are the first explorers in the area (79). No matter how many adherents believe in this nonsense, it is harmful and potentially fatal to some folks who follow trails and/or cairns in desolate places, whether in the depths of the **Grand Canyon** or on Mount Katahdin's summit plateau where the enormous cairns undoubtedly save lives on an ongoing basis. I know they saved mine.

Cairo is the capital city of **Egypt**.

Caldera is a large hollow space that forms at the top of a **volcano** after an eruption. This does not occur too often.

Calderan, Max (1967–), is an **explorer** of **deserts** who walked across 700 miles of the **Empty Quarter** in 2020.

Caliche is a sedimentary rock, or crushed stone aggregate.

California is the third-largest state in the **US** and home to three of the four major US **deserts**. (See *The California Deserts* and "The Four Major American Deserts" sidebar.)

California Desert Protection Act of 1994 is a federal law that added 69 new wilderness areas and enlarged and redesignated some parks. It established **Death Valley** and **Joshua Tree** as **national parks**.

California Desert Studies Consortium (CDSC) is a "multicampus consortium committed to advancing education and research opportunities in arid-land **ecosystems**." It operates California State University's Desert Studies Center field station in the Mojave National Preserve, "a premier location and resource for studying the **geology, anthropology**, and **biology**—among many other topic areas—of California's **deserts** and the American West" (https://www.calstate.edu/impact-of-the-csu/research/highlights/pages/desert-studies.aspx)

California Deserts is the term used to designate the deserts that one finds in California: the **Mojave**, **Colorado** (an extension of the **Sonoran Desert**), and **Great Basin**. (See "The Four Major American Deserts" sidebar.)

The California Deserts: An Ecological Rediscovery is Bruce M. Pavlik's superb volume devoted to the **history and ecology** of the **Mojave, Sonoran**, and **Great Basin**, the three semi-contiguous southern **California Deserts**. The author emphasizes the ecological diversity of desert communities, how **plants** and **animals** cope with **heat** and **drought** as well as depletion, and restoration. He notes that here there are 1,836 species of vascular **plants**, 43 of **fish**, 16 of **amphibians**, 56 of **reptiles**, 425 of **birds**, and 97 of **mammals** (152), an extraordinary ecological community. Threats include human incursion, invasive species, and depletion of **resources** (**water**, clean air) (254ff.). Better resource management can help (294ff.). Most pages contain one to four attractive color images, and the 23-page bibliography followed by a 33-page index are invaluable scholarly assets. (See **Desertification: restoration**.)

California poppy is the well-known large, orange flower of the **Mojave**. The Mexican gold poppy is similar. This **plant** has a narcotic effect and can lessen pain (*Desert Wildflowers*, 13).

Map of the Californian Deserts region. CRISTIANO TOMÁS, 2022. CREATIVE COMMONS ATTRIBUTION-SHARE ALIKE 4.0

Poppies, high desert wildflowers. TERRY LUCAS. CREATIVE COMMONS ATTRIBUTION 3.0

Desert Masileh Qom camels. MOSTAFAMERAJI, 2009. CREATIVE COMMONS ATTRIBUTION-SHARE ALIKE 4.0

Camel (G: Kamel; F: chameau; S: camello; I: úlfalda; Af: kameel; H: גָּמָל [gumul/gamal]; Ar: جمل [jamal]; M: тэмээ [temee?]; C: 駱駝 [Luòtuó]). The camel is the semitrailer of the **desert**, capable of carrying heavy loads and traveling for days without access to **food** or **water**. Even today, despite the introduction of cars and trucks, camels are used for **travel** and transport. Two species are the Bactrian (two humps) and the dromedary (one hump). Domesticated and wild camels live in the **Sahara**, **Gobi**, and **Australian Deserts**. New-world camels (the llama, alpaca, vicuña, and guanaco) can be found in South America's Andes. (See **Adaption**.)

Camel thorn tree is a **plant** of the **Kalahari Desert** that can reach more than 50 feet in height; its branches disperse higher up so that it resembles an umbrella. It is used for fuel and construction and as **food** for **elephants** and giraffes (https://kalahari desert12.weebly.com/plants.html).

Canadian desert. The Canadian desert is a small desert located in British Columbia.

Canyon de Chelly National Monument (pronunciation: də-shay') is an enormous tract on Navaho land in **Arizona**. Antelope House, **White House Ruins**, Mummy Cave, and Spider Rock—a 700-foot-high spire—are located here. The spire is an awesome sight when viewed from the canyon rims, where one may hike unimpeded. **Travel** within the canyon, however, is limited to participants in official tours. (See https://www.nps.gov/ cach for information and a brief video, as well as the "Southwest US Indian Ruins" sidebar.)

Canyonlands National Park, on the **Colorado Plateau** in **Utah**, is a high **desert** replete with diverse canyons, mesas, buttes, arches, and **rock art**. Many **indigenous peoples** including **Hopi**, **Navaho**, **Paiute**, Ute, and various **Pueblos** have lived or traveled here (https://www.nps.gov/cany).

Capulin Volcano National Monument is located in northeastern **New Mexico** and celebrates an extinct cinder come. There is a **road** that leads to the summit (https://www.nps.gov/cavo).

Caravan is a group of pilgrims, adventurers, or traders traveling across a **desert** on or with **camels**. At one time, it was a crucial means of conveying trade goods such as **salt**; now it is being replaced by mechanized vehicles. Its expanded meaning includes more-modern conveyances such as trucks or RVs. Interestingly, there exist many musical "caravan" compositions, including those by Duke Ellington and Van Morrison.

Cardón is a large, columnar, multi-stalked **cactus** found in **Baja California**.

Carlsbad Caverns National Park. The many wonders at this New Mexico **national park**—physical (underground corridors, domed caverns) and geological (stalactites, stalagmites, crystals)—are almost incomprehensible in the 30 miles of corridors, some of which are more than 1,000 feet below the surface. Equally amazing is the flight of the Mexican free-tailed **bats**: At about five each evening, many hundreds of thousands of bats fly out of the large cave entrance in search of **food**. The bats overwinter in **Mexico** (brochure).

Cartography. See **Maps** and **Maps: history**.

Casa Grande Ruins National Monument (**Arizona**). The ancient **Hohokam**, who farmed, hunted, made **pottery**, and built irrigation canals in the **Sonoran Desert**, had many settlements along the Gila and Salt Rivers and Casa Grande was one of them (brochure). (See the "Southwest US Indian Ruins" sidebar.)

Casablanca is a major Moroccan city, a few hundred miles north of the **Sahara**. It is also the title of

Casa Grande National Monument sign. ROBERT HAUPTMAN

a famous 1942 **film** starring Humphrey Bogart and Ingrid Bergman.

Casas Grandes (Paquimé) Ruins is a prehistoric archaeological site in the northern Mexican state of Chihuahua that is attributed to the Mogollon culture. It has been designated a UNESCO World Heritage Site.

Casbah (kasbah, qasba, qasaba) is a medina (walled city) or fortress citadel in Moroccan cities and the **Sahara**.

Cat. The sand cat lives in **deserts** in North Africa, the Middle East, and Central Asia; it resembles a domestic house cat. Bobcat and **mountain lion** can also be found in southwestern US deserts.

Cathedral in the Desert (Utah). In 2005, this incredible slot canyon amphitheater returned after more than 50 years. It is an enormous rock **cave** that has been under **Lake Powell** ever since its **water** rose due to the Glen Canyon Dam, but since a **drought** caused the lake to lose 140 feet,

the cathedral was once again visible. Additionally, a natural arch and Anasazi dwellings were also revealed. But the waters returned and it is now again submerged (Price). (The US Army Corps of Engineers should build a coffer dam around it!)

Cave (G: Höhle; F: grotte; S: cueva; I: helli; Af: grot; H: מְעָרָה [m'awraw/m'ara]; Ar: كهف [kahif]; M: агуй [agui]; C: 洞穴 [dòngxué]). Caves abound in the **desert** and provide protection from the elements and wildlife as well as locations for **pictographs** and **petroglyphs**. Broader cave openings have allowed Native Americans to erect buildings or carve them out of the rock walls. The Navaho **White House Ruins** in **Canyon de Chelly** and **Montezuma Castle** in Camp Verde, **Arizona**, are excellent examples. As with many natural wonders, human beings may abuse, harm, or destroy what others wish to conserve and protect. (This is why **Uluru** is now off-limits to climbers.) And so, *Caves: Conservation and Ethics*, an attractive brochure produced by the National Speleological Society and others, admonishes cavers (spelunkers) to act caringly, **leave no trace**, not harm formations, not produce graffiti (no one cares that

Cave interior, southwestern US. ROBERT HAUPTMAN

you and your girlfriend were there), and not remove anything. On one of this author's visits to Mammoth Cave in Kentucky, the National Park Service guide told us that we were not allowed to take even a tiny rock as a souvenir, despite the fact that the cave was littered with many millions of stones. The *Encyclopedia of Caves* is a useful compilation, although inexplicably "desert" does not appear in the index, but entries on **bats** and **beetles** may prove helpful (Culver, passim). (See **Carlsbad Caverns**, **Mogao Caves**, and **Tellem Burial Caves** as well as **Petra** and **Mada'in Saleh**.)

Cenote is a pit or sinkhole (in **Mexico**) that sometimes contains **water**. Four kinds are open, semi-open, cave, and ancient. They were a source of water for people in the past as well as today.

Centipede is a wormlike, poisonous creature with many legs. Different types including the giant desert centipede (Texas redheaded centipede) live in the **Sonoran Desert**.

Central Australian Desert is large area in central **Australia** in which one finds the Tanami, the **Great Victoria**, and especially the **Simpson Deserts** (Nathaniel Harris, 160–61). (See **Western Australian Desert**.)

Central Kalahari Game Reserve. Located in **Botswana**, this reserve was created in 1961 to protect the **San** (formerly Bushman) people. But by 2010, the government was interested in driving the San out of the reserve because, some say, it desires to mine for diamonds and stimulate **tourism**. The government used dastardly tactics such as sealing waterholes, forcing the people to walk for days in order to locate **water** (Barry Bearak). (See the "International Wildlife Refuges" sidebar.)

Ceremonies, which include seasonal rites, initiations, festivals, courtships, weddings, coronations, and other ritualistic celebrations based on beliefs, take place in all cultures and environments including the **desert**. Surprisingly, ceremonies are limited in the **Sahara** but increase in number in the **Sahel**. Beckwith and Fisher's truly extraordinary,

profusely illustrated, two-volume *African Ceremonies* covers the entire continent and offers commentary and vivid **photographs** of ceremonial activity and human decoration in various African desert communities.

In the Sahara we have Berber brides (**Morocco**). A wedding (in a village in the **Atlas Mountains** rather than the deep desert) is a complex business with dancing and lovemaking lasting for days. What stands out here is the **clothing** (even the bride's face is *completely* covered) and an abundance of silver and amber **jewelry** (Beckwith and Fisher, vol. 1, 210, 211). This is the very antipode of other areas where tribal members are naked and at times completely covered either with body paint (usually red ocher) or intricate geometric decorations (usually white). (It is amazing that they can survive this because, normally, overpainting the entire body can halt sweating and cooling and the paint may be toxic; overpainting can lead to death.) The second Sahara ceremony is the Hassania veil **dance** (**Morocco** and other lands). The dancers are at times completely ensconced in flowing cloths, adorned with jewelry (so much on the hanging hair that it must weigh down the head), and have henna-painted hands (vol. 1, 225, 227). They whirl around perhaps like **Whirling Dervishes**. A quarter century ago, only about 60 women could perform the dance to perfection (vol. 1, 224).

There are many tribes with a broad diversity of ceremonies in the Sahel countries. For example, at a Tuareg wedding in **Niger**, groups of people all clad exactly the same, with dark clothing but hanging white head coverings, sit in silence while a second large group wear white clothing and black headgear. Still others, ornately clad and astride **camels** and **donkeys**, cavort joyously (vol. 1, 246–65).

Then there is the seasonal rite of Bobo (or Bwa) Bush Masks in Birkino Faso. These ritual masquerades purify the community and chase away evil. The masks are large, intricate affairs but what is really amazing is the multicolored strands of thick "hair" that cover the entire body and whirl with the spinning dancers, who resemble human haystacks (vol. 2, 98–109).

Finally, among a host of other possibilities, are the **Himba**, cattle herders who live in **Namibia** in close proximity to the **Namib Desert**. For a wedding, there is eating and dancing and then the bride is covered in red ocher. The couple spend their early life in a ceremonial hut. Other than the red paint, there is little concern for body coverings, though the Himba do not seem to condone complete nakedness. The exquisite clothing, cloaks, and facial coverings of the **Sahel** are lacking here. The jewelry is refined and limited (vol. 1, 286–301).

It is, of course, impossible to give a full picture of the incredible diversity of African desert ceremonies. For this one must turn to Beckwith and Fisher. What is truly astonishing is that two foreign, white women (Beckwith and Fisher) were allowed to attend, let alone photograph, many of these very private, personal events, some of which are extremely erotic and perhaps for outsiders, obscene (and one would think embarrassing for the participants who are broadly smiling).

The Baldizzones, in *Wedding Ceremonies*, present similar symbols, costumes, and rituals in the deserts of **Morocco**—henna-bedecked hands, chests heavy with amber, and fully covered faces (33, 30), overburdened with a head and body covering on a litter; the bride may be on the periphery of the desert (165)—and in **Niger**: painted faces and widely opened eyes (121, 154–55). The wedding processes follow very specific sequences.

Outside **Africa** we have Arabian desert weddings (see below) and the Bedouin Laylat Al Henna ceremony, which resembles a bridal shower. In North America there are more than 500 distinct though sometimes related tribes. Those that dwell in desert environments have highly specialized ceremonies such as powwows and vision quests. For example, the **Apache** have the hoop dance and lightning ceremonies. See Frank Waters's *Masked Gods: Navaho and Pueblo Ceremonialism* in the bibliography.

For nomadic and other traditional desert dwellers, ceremonies may play an outsize role in their lives (along with working—acquiring **salt** and transporting it on **camels** that they raise, **hunting**, and gathering **plants**), but those people now living in Abu Dhabi or Medina, for example, obviously have other concerns. This is unequivocally not the case for a limited number of non-desert peoples such as ultra-Orthodox Jews, for whom weddings

and funerals but especially the 10 religious ceremonial holidays—the most well known of which are Hanukkah and Passover, although these are not the most important—are such an integral part of their lives that little else truly matters. And the males often do not work but rather spend their days studying the Torah and Talmud. The same, although to a lesser extent, may be said of Buddhist ascetics, monks, and lamas who spend their time chanting and praying, but who also celebrate 12 holidays. (See **Books** and the video *Saudi Wedding in Desert*, with hundreds of dancing men, horses, **camels**, and cars but few if any women, at https://www.youtube.com/watch?v=rC14BQpNN3c).

Chaco Canyon. See **Chaco Culture National Historical Park**.

Chaco Culture National Historical Park (New Mexico). Just to the south of **Aztec Ruins National Monument** lies **Chaco Canyon**, which was the hub

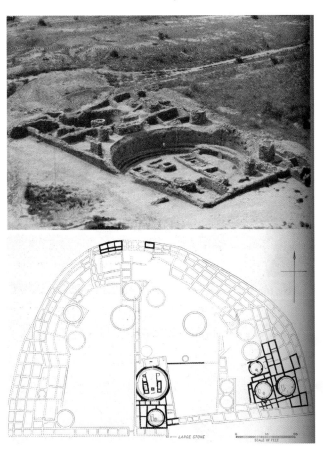

The Great Kiva of Pueblo Bonito and diagram from a 1922 National Geographic article. PUBLIC DOMAIN

of the Chaco culture. **Pueblo Bonito**, with ca. 600 rooms and 40 **kivas**, is at its center; other **pueblos** include Chetro Ketl and Una Vida. The Puebloans built 400 miles of sophisticated roadway that connected their 75 villages to each other, and Aztec Ruins was part of this trading community. Their turquoise **jewelry**, **pottery**, and stonework are outstanding (brochure). (See **Masonry**, https://www.youtube.com/watch?v=Y6XYuWd1foo, and the "Southwest US Indian Ruins" sidebar.)

Chad is a very poor country south of **Libya**, the north of which is in the **Sahara** while the south is in the **Sahel**.

Chalbi Desert. See Danakil Desert.

Chamanchaca. See Garua.

Chang Tang (Lonely Place). A high, arid **desert** with **dunes** in **Tibet**.

Chaparral (**creosote bush**, greasewood). Moore describes these 5- to 12-foot-tall creosote-smelling **plants** as reptilian. They have antimicrobial and analgesic qualities (27–29).

Charco (puddle) is a waterhole.

Charles, Jack (1943–2022), was an Aboriginal actor and activist, known as the "grandfather of Aboriginal theater."

Cheetah is the fastest **animal** on earth, able to run at 70 miles per hour. It can weigh as much as 150 pounds (a **mountain lion** can reach 220). The Asiatic variety lives in the **Iranian Desert**.

Chihuahuan Desert (Desierto de Chihuahua)

200,000 (174,000, 250,000) square miles | SW US, Mexico | Hot desert | Minimal precipitation | Native Americans

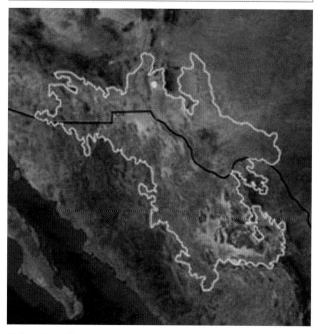

Chihuahuan Desert map. PFLY, 2007. PUBLIC DOMAIN

One of the **four major American deserts**, the Chihuahuan lies in parts of Texas, **New Mexico**, and **Arizona** as well as in **Mexico**, where it is bounded by the Sierra Occidental and Oriental mountain ranges. Living entities include **agave**, **prickly pear**, **cholla**, and **yucca** as well as **lizard**, **snake**, **coyote**, and elk (brochure).

Extent: 250,000 square miles, most in Mexico
Environment: Hot and arid, with cooler winters
Surfaces: Sand and loam
People: Tarahumara (Raramuri), Apache, Comanche, and Guarojío
Animals: Mule deer, quail, pronghorn, rabbit, wolf, kingsnake, flycatcher, and others
Plants: 3,000 species including 500 cactuses: agave, beargrass, tarbush, tree cholla, claret cup cactus, sotol, yucca
Resources: Barite, copper, lead, manganese, oil, sulfur, and zinc
(See "The Four Major American Deserts" sidebar and also a video at https://www.youtube.com/watch?v=r6H-446XMzo&list=PLm1u6z50FrCAhcLOTBdp7tNpriV_KhM3M.)

Chihuahuan Desert Nature Center and Botanical Gardens is located at the Chihuahuan Desert Research Institute in Fort Davis, Texas (http://www.cdri.org). It contains many gardens with beautiful flowering **plants**.

Children's literature. Desert literature, whether of a scholarly or personal nature or fiction, is normally created for adults, but some **books** (and **films**) in this category are aimed at young children, adolescents, and teenagers. (To give readers an idea of the extent of this material, consider that the public library serving South Burlington, Vermont, a small city with a population of little more than 20,000, has 444 books dealing with **deserts** that are aimed at children up to the age of 13.) Examples include M. M. Eboch's beautifully illustrated, 32-page *Desert Biomes around the World*, which can profitably be read by youngsters, as can Peter Murray's 32-page *Deserts*, Peter Benoit's attractive and informative *Deserts*, *Cactus Hotel*, *The Desert Biome*, *The Many Assassinations of Samir, The Seller of Dreams*, and others, naturally. **Films** include *Desert Kids* (2016, **Negev**) and *Rango* (2011, **Mojave**). (See https://books.growingwith science.com/2015/02/13/desert-books-for-kids, https://movieslist.best/list/family-movies-about-the -desert, and the "Children's Literature: Fiction and Nonfiction" sidebar.)

CHILDREN'S LITERATURE: FICTION AND NONFICTION
(extremely limited listing)

- *Alice: A Story of Friendship* (Bill Vermooten)
- *Baby Animals in Desert Habitats* (Bobbie Kalman)
- *Cactus Desert* (Donald Silver et al.)
- *Cactus Hotel* (Brenda C. Guiberson)
- *D Is for Desert: A World Deserts Alphabet* (Barbara Gowan)
- *A Day and Night in the Desert* (Caroline Arnold's Habitats series) (Caroline Arnold)
- *Desert Baths* (Darcy Pattison)
- *Desert Discoveries* (Ginger Wadsworth)
- *Desert Giant: The World of the Saguaro Cactus* (Barbara Bash)
- *Deserts* (Gail Gibbons)
- *Get to Know Gila Monsters* (Get to Know Reptiles series) (Flora Brett)
- *Lets Explore the Desert: Family Go Guide* (Doris Evans)
- *Lizards for Lunch: A Roadrunners Tale* (Conrad J. Storad)
- *Looking Closely Across the Desert* (Frank Serafini)
- *The Night Flower* (Lara Hawthorne)
- *Nobody Hugs a Cactus* (Carter Goodrich)
- *The Secret Lives of Hummingbirds* (David Wentworth Lazaroff)
- *Victor, the Reluctant Vulture* (Jonathan Hanson)
- *Who Lives Here? Desert Animals* (Deborah Hodge)
- *Why Oh Why Are Deserts Dry?* (Tish Rabe)

(From https://books.growingwithscience.com/2015/02/13/desert -books-for-kids, https://www.getepic.com/collection/1024737/ desert, and https://www.desertmuseum.org/books/cat_detail.php ?prcat_id=11.)

Chile is a country in **South America** that hosts the **Atacama Desert**.

Chimayo, New Mexico, is where one will find the famous Ortega family **rugs**. The town is also a pilgrimage site. (See **Crafts**.)

Pencil cholla. ROBERT HAUPTMAN

China is the world's largest country by population (1.4 billion) and third largest in area (only Russia and Canada are larger). Its **deserts** include the **Badain Jaran**, **Gobi**, Gurbantonggut, Hunshandake, Muus, **Taklamakan**, and Tungeli. (See **Silk Road**.)

China–Tibet railway. In 2006 China completed a railway between Beijing and Lhassa, a much touted accomplishment since it reaches an altitude of almost 17,000 feet (oxygen is pumped into the cars). Tibetans are not as elated as the Chinese (MacLeod).

Chinguetti, Mauritania. See **Libraries**.

Cholla (pronunciation: choiya) is a southwestern US **cactus** that has barbed spines that cling to skin and **clothing**. Some types include **silver** and **teddy bear**. **Birds** such as the curved-bill thrasher (as well as cactus wren and mourning dove) nest among its protective spines (*Desert Bird Gardening*, 20).

Chott (shott) is a salt lake that dries up, in North Africa. (See **Sebkha**.)

Christianity is one of the three **Abrahamic religions**, all of which trace their founding to desert environments. (See **Judaism** for a fuller explanation.)

Chuckwalla is a large, nontoxic **lizard** that lives in southwestern US **deserts**. When frightened, it retreats into a rock crevice and inflates its lungs, thus not allowing itself to be removed.

Chuparosa (hummingbird) is a 3-foot **shrub** found in the **Sonoran Desert**.

Cicada. This noisy **insect** comes in 3,000 species, but most interesting are two varieties, the 13- and 17-year types, which emerge in large broods every 13 or 17 years (stragglers may arrive at slightly different times). Some species are found in the **Sonoran Desert**.

Cima Dome (and volcanic field) is a smooth, symmetrical, alluvium-fringed rock dome located in the Mojave National Preserve. This is where the 2020 Dome Fire occurred. (See **Joshua Tree National Park**.)

Cinema. See **Film**.

Circles, **mysterious.** Across **Africa**, in a limited belt from Angola and **Namibia** to **South Africa**, thousands of **fairy circles** ("footprints of the gods"), ranging in size from 6 to 40 feet, appear in the **desert**, apparently the work of **termites** that live underground and turn the desert into grassland (Wilford, "Mysterious"). Some **scholars** discount the termite etiology; indeed, in late 2022, Rachel Nuwer reported that a water-related hypothesis has replaced the termite etiology. Here, **plants** create the circles in order to redistribute **water**. (See also **Termite mounds**.)

Cirio. See **Boojum**.

Cities in the **deserts** of **Egypt**, **Iraq**, Iran, **Saudi Arabia**, **Israel**, **Mali**, **India**, **China**, **Australia**, and the **US**, among other countries, have been constructed, populated, and abandoned for millennia. (See the "Major Cities Associated with Deserts" sidebar for a listing.)

MAJOR CITIES ASSOCIATED WITH DESERTS

- Albuquerque (US)
- Alice Springs (Australia)
- Amman (Jordan)
- Baghdad (Iraq)
- Bend (US)
- Boise (US)
- Cairo (Egypt)
- Casablanca (Morocco)
- Chihuahua (Mexico)
- Ciudad Juárez (Mexico)
- Coober Pedy (Australia)
- Douz (Tunisia)
- Dubai (UAE)
- Dunhuang (China)
- Grand Junction (US)
- Indian Wells (US)
- Jaisalmer (India)
- Kashgar (China)
- Khartoum (Sudan)
- Las Vegas (US)
- Lima (Peru)
- Luxor (Egypt)
- Mecca (Saudi Arabia)
- Memphis (Ancient Egypt)
- N'Djamena (Chad)
- Nouakchott (Mauritania)
- Palm Springs (US)
- Perth (Australia)
- Phoenix (US)
- Reno (US)
- Riyadh (Saudi Arabia)
- San Pedro de Atacama (Chile)
- Santa Fe (US)
- Thebes (Ancient Egypt)
- Timbuktu (Mali)
- Tucson (US)
- Ulaanbaatar (Mongolia)
- Windhoek (Namibia)
- Yakima (US)
- Yuma (US)

City is an enormous (1 by 1.5 miles) megasculpture that took some 50 years and $40 million to create and is still not complete according to the artist. It consists of enormous dirt mounds, buttes, concrete slabs, walls, bars, triangles, and other forms. It lies in a remote Nevada desert valley and may now finally allow visitors. The *New York Times* allocated eight pages to this bizarre enterprise, but it is really impossible to conceptualize even from the extremely large **photographs** (Kimmelman, passim).

Clifford, **Sir Bede Edmund Hugh** (1890–1969), was one of the first white men to traverse the **Kalahari Desert**. He traveled with two large trucks, a group of men, a lot of **water** and gasoline, and other supplies. Though they had problems, they succeeded (Goudie, *Great*, 254–57). (See **Explorers** and the "Desert Explorers" sidebar.)

Climate. See **Desert climate**.

Climate change. The earth's changing climate due to **global warming** is having a devastating effect in both the Antarctic and Arctic **deserts**. In the former location, some groups of **penguins** are having a difficult time and ice shelves are collapsing; in the latter, the ice is melting at an alarming rate and **polar bears** are suffering. And as the earth warms, deserts increase their boundaries. Climate change is implicated in the enormous and ongoing expansion of the **Sahara** due in part to a decrease in rainfall (Thomas and Nigam abstract). Climate change also increases **temperatures** and exacerbates **sandstorms**. According to Dominique Bachelet and her colleagues, alterations in climate have negatively affected the southern California desert with increased temperatures: "Some species may be approaching their physiological threshold" and there is "the probable expansion of barren lands reducing current species survivorship" (Bachelet abstract). Cloud seeding, afforestation, and agroforestry plantations can help to increase rainfall and mitigate climate change, and it is possible that desert air may be able to produce **water** technologically. Note that modeling indicates that the future will be drier and hotter. Ignacio A. Lazagabaster and his colleagues analyzed mammalian remains in **caves** in the **Judean Desert** through radiocarbon dating and concluded that various **animals** decreased (Arabian leopard, **gazelle**) while others increased (**fox**, rock hyrex, etc.) because of an altered Holocene climate as well as an **anthropogenic influence** on food supplies (Lazagabaster abstract). (See *The Atlas of a Changing Climate*, **Desertification**, **Desert weather**, **Human-induced global warming**, **Paleoclimatology**, and Markus Rex in the bibliography.)

Climate engineering. See **Human-induced global warming**.

Clothing. Unless one lives in a consistently hot **climate**, and at times despite this, clothing plays an inordinate role in the lives of most human beings. There are still people who wear little (a loin covering) or nothing at all, and gender does not play a role here, but most traditional desert dwellers do cover themselves, often from head to foot. There are two major reasons for this: Although it can be extremely hot, the sun's rays can parch and harm the skin and the **wind** can whip sand particles into the body, face, and eyes; additionally, those desert dwellers who have converted to **Islam**, in whole or in part, insist on covering their bodies, especially those of women. In some cases the entire face must be ensconced. Clothing takes many forms including togas, swaddlings, robes, burnooses, head cloths (*kufiyyas*), hijabs, and burkas, as well as some unusual materials in strange shapes, especially for **ceremonies**. The colors cover a spectrum of brown, black, blue, purple, and multi-colors. Some peoples prefer white, which does not absorb the **heat**; others claim that black, which does, is superior. Various materials are preferred by different peoples: Wool is excellent, but can be irritating as well as stifling in the heat; cotton is pleasantly practical but in a cold **rain** it can be deadly. "Cotton kills," chanted by hikers and mountaineers, is worth keeping in mind. Modern adventurers (or normal folks) who visit (or live) in the desert wear their cultures' clothes plus specialized items such as Gore-Tex, trekking boots, fleece, safari or cargo pants, and so on. (See **Jewelry**.)

Clouds. The taxonomy of clouds is quite complex, although the various formations occur more frequently in non-desert environments. Nevertheless, it does **rain** (and **snow**) in the **desert** and so clouds such as puffy cumulus, darkening nimbus, wispy cirrus, and combinations such as cumulo-nimbus or alto-stratus may appear from time to time. (See **Fog** and **Precipitation**.)

Coachella Valley Music and Arts Festival is an enormous popular music festival, held since 1999 in Indio, **California**, in the **Colorado Desert**. It returned in 2022 after a two-year hiatus due to the **COVID-19** pandemic. An approximately two-hour video covering two decades of Coachella can be found at https://www.youtube.com/watch?v=jjwilAja7Lc.

Cochise was an Apache chief in the late 19th century.

Cold-blooded. The animals with which people are most familiar are **mammals** and **birds**, which are warm-blooded; that is, their **body temperature** does not vary with the external temperature. Cold-blooded (ectothermic) **reptiles** and **amphibians** cannot keep warm when the external temperature plummets nor do they remain cool in the **heat**. This is why **snakes**, **lizards**, **frogs**, and **desert tortoises** bury themselves in the desert **sands** or burrows, or hibernate or estivate.

Cold desert. See **Hot desert.**

College of the Desert is a community college in **Palm Desert**, **California**.

Colorado Desert (165,000 square miles) is part of the northern **Sonoran Desert**.

Colorado Plateau lies to the east of the **Great Basin** and **Mojave Deserts**, and within it one finds **Monument Valley** and **Canyonlands**.

Colorado River supplies **water** to Los Angeles and Phoenix via an aqueduct, an enormous pipe that runs aboveground, as well as to other cities and especially farmers (who use most of it) in seven states. In 2022 it was suffering from a 20-year **drought** and so Daniel Rothberg notes that it is predicted that a crisis of dire proportions will soon transpire.

Colossal Cave Mountain Park. About a thousand years ago, the **Hohokam** used this crystal-filled cave southeast of **Tucson** for shelter and storage. Many plant, bat, and bird species as well as **tortoises** can be seen here (brochure).

Committee on Desert and Arid Zone Research is a subdivision of the American Association for the Advancement of Science.

Common sowthistle (and other sowthistle species) is a **plant** with a light yellow flower on a thick stem; it is found in the **Sonoran Desert** (Gerald A. Rosenthal, 137).

Common sunflower. Topped by a large yellow flower, this is a tall **plant** that is abundant in desert washes; it is found in the **Sonoran Desert** (Gerald A. Rosenthal, 126).

Compass barrel cactus is a 5-foot **shrub** that looks like a chubby barrel and can be covered with red or yellow flowers. It is found in the **Sonoran Desert**.

Condor is an enormous **bird** with a ca. 10-foot wingspan; it inhabits southwestern US **deserts**.

Conflict. See **War.**

Conservation. Attempts to conserve, restore, or enhance desert lands have been going on since biblical times. These processes are implemented, more or less, depending on many factors including government policy, economics, fertility, greed, and wisdom. *Deserts: The Encroaching Wilderness* emphasizes conservation and contains an exceptional series of **maps**. (See **Desertification**.)

Contamination. See **Pollution.**

Controversies. It is extremely difficult to believe that the neutral and innocuous **desert** creates ongoing controversial debate on, for example, **desertification**, **ethics**, and colonialism, not among normal folks or desert dwellers but within scholarly communities. For hundreds of years virtually all people, regardless of status, position, ethnicity, or any other characteristic, believed that deserts were enlarging, encroaching on forests, meadows, and agricultural land, and in 2024 many still do. But an ongoing countering position has emerged among a group of iconoclastic **scholars** who insist that desertification is a much exaggerated, even meaningless and harmful, concept.

Ethical problems continue to plague **anthropology** and ethnography. Failure to fully protect **petroglyphs** and **pictographs** are an example of a complex issue (for which see **Rock art**).

The colonial mindset is still with us, since it influenced so much of the victims' cultures in the **Sahara**, **Kalahari**, southwestern **US**, and **Australia** (where the **Aborigines** have stood strong against the cultural invaders). Because of this, many scholars (and others) wish to decolonize; that is, rid their respective cultures of all external influence. A few iconoclasts disagree and they raise the hackles of the woke generation. In 2017 Bruce Gilley published "The Case for Colonialism," noting that the colonizers did some good, although he also desired a return to colonialism. The fuss raised was so extreme (with death threats) that he retracted his paper. Those who wish to silence anyone with whom they disagree will have a much harder time harassing Olúfémi Táíwò, whose *Against Decolinsation* makes a similar argument. As an African who teaches at Cornell, he is presumably more familiar with the colonizers and their evil cultural influence than some of Gilley's insulated academic critics.

This naturally is not a defense of desertification, unethical activity, or colonialism and its horrific record of harming both environments and people. Rather, in the latter case, it indicates that some good came out of the invasions and that excising everything would be very harmful to 21st-century **desert peoples**. And that is why this (and the other matters) are controversial. (See **Anthropology** and **Desertification: myth**.)

Coober Pedy is a town in Australia's **Great Victoria Desert** famous for its opals. It was so hot there that I refused to leave the bus. (See **Bizarre manifestations**.)

Copper is an extremely important ore. (See **Atacama Desert**, **King Solomon's Mines**, **Mongolia**, and **Morenci Copper Mine**.)

Copper Mountain College is a community college in **Joshua Tree**, **California**, that provides desert-themed activities for students with interests in

anthropology, **archaeology**, **art**, astronomy, **biology**, **geography**, and **geology**.

Coriolos effect makes things like planes or currents of air traveling long distances around Earth appear to move at a curve as opposed to a straight line. It effects storm systems (https://scijinks.gov/coriolis).

Coronado State Monument, north of **Albuquerque**, celebrates Kuaua, its Puebloans, **pottery**, and kiva's murals (brochure).

Corroboree is a sometimes-sacred ritual dance **ceremony** performed by Australian **Aborigines**.

Cosmos is a Mexican **plant** that grows to 6 feet and has large orange flowers (*Desert Wildflowers*, 11). The domesticated version of this flower is much smaller and is very easy to cultivate.

Cottonwood is a very fast-growing **tree** that can be found in the **Chihuahuan Desert** as well as in the **Grand Canyon**, which touches the **Sonoran** and **Mojave Deserts**.

Countries (see individual entries). Many countries on six continents contain **deserts** or arid desert environments.

COVID-19. Carol Schumacher's family has been devastated by the 2019–2023 pandemic; she lost 42 relatives. Schumacher is a **Navaho**, and **indigenous peoples** have been especially harmed by this ailment. There are many social reasons for this (chronic disease, poverty, toxic environment, alcohol) but the bottom line is that Native Americans have a life span diametrically different from other groups. Asians have a life expectancy of 83.5 years, whereas Native Americans can expect to live only 65.2 years (2021 statistic) (Romero, Rabin, and Walker).

Coyote is a smaller, omnivorous, doglike **animal** that hunts in packs. It often inhabits desert environments. (One once chased me into my house, though not in a **desert**.)

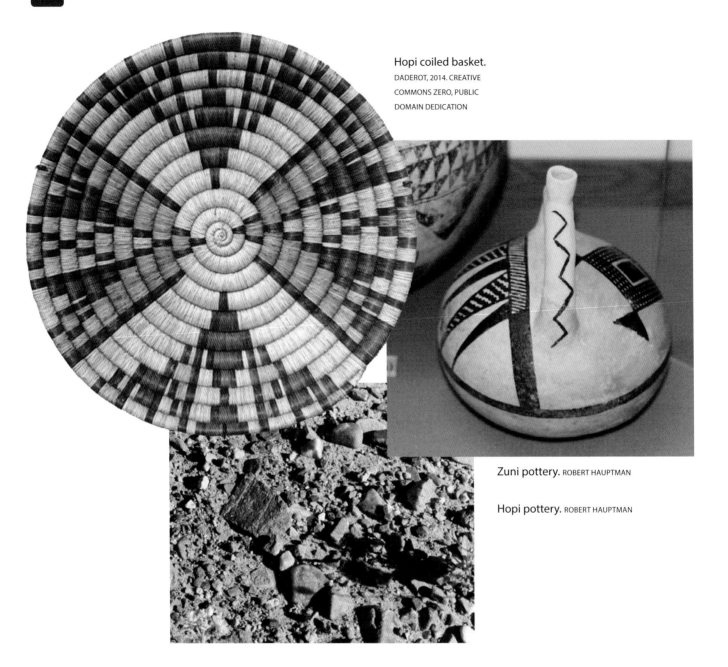

Hopi coiled basket.
DADEROT, 2014. CREATIVE
COMMONS ZERO, PUBLIC
DOMAIN DEDICATION

Zuni pottery. ROBERT HAUPTMAN

Hopi pottery. ROBERT HAUPTMAN

Crafts. Ellen Bradbury points out that the prehistoric peoples (the Anasazi, Hohokam, Mogollon, Basketmaker, and Mimbres cultures) who lived in the southwestern **US** (i.e., the **desert**) created **pottery**, basketry, weaving, and carving, and contemporary peoples continue with these traditions. Pottery is especially important and vibrant. The famous potter Maria Martinez, of the San Ildefonso **pueblo**, produced extraordinarily beautiful black pots; others are colorful and decorated with animal figures or geometric patterns. The **Navaho**, **Zuni**, and **Hopi** produce silver **jewelry**, often embellished with obsidian and especially turquoise. Some of this is created for the tourist trade, and individuals sit around the plaza in **Santa Fe** and offer their rings and bracelets to visitors. Weavers make ponchos, sashes, and especially **rugs**. In **Chimayo**, **New Mexico**, the Ortega and Trujillo families have been weaving for as many as nine generations and the looms can be seen by the public (see https://ortegas weaving.com). Chimayo is also a pilgrimage site. Basketry, sculpture, and leather round out these US desert crafts. The inhabitants of Subarctic and Arctic regions create similar crafts. (See **Art**.)

Craters of the Moon is a large **national monument** in Idaho's high **desert**. It is covered with lava flows, cinder cones, and **sagebrush**. Many videos are available. Consider these: https://www.youtube .com/watch?v=OLUNzZ8bPbI, https://www.you tube.com/watch?v=2UhZK5DMq8o, and https:// www.youtube.com/watch?v=lI99yfzgXf8.

Creosote bush (greasewood, **chaparral**, gobemadura) is an important, abundant, tall, yellow-flowered **shrub** of the southwestern US **deserts** that is useful medicinally for respiratory problems and as an anti-inflammatory, among other ailments (https://www.texasbeyondhistory.net/ethno bot/images/creosote.html). It has a musky odor and some people react allergically to its **toxins**. Michael Allaby points out that some creosote bush clones are almost 12,000 years old (73). David Rhode indicates that in the **Mojave**, it was used for fuel and glue, to build shades and tools, and to cure measles, rheumatism, cramps, and other ailments (46ff.).

Consider that there are almost 400 species of **plants** in the **Sahara**, more than 2,000 in the **Sonoran Desert**, 24,000 across all of **Australia**, and 320,000 in the world. Despite this profligacy—especially, in this case, within deserts—substantial (esoteric) volumes may be devoted to a single species and this is the case with *Creosote Bush: Biology and Chemistry of* Larrea *in New World Deserts* (Mabry), a 284-page edited collection of 10 papers that discuss adaptability, reproductive systems, habitat for **arthropods**, and practical uses (NDGA as an antioxidant, livestock feed, and medicinal applications) inter alia.

Crescent Lake. A falling water table has resulted in the shrinking (falling 25-plus feet) of this important lake in the **Gobi** near **Dunhuang** (Yardley). (See nearby **Mogao Caves**.)

Crow is a corvid, black and smaller than a **raven**, that mingles with sometimes hundreds of fellow creatures. They inhabit many habitats including **deserts**.

Cryptogamic crust. Alga, fungus, **moss**, and **lichen** are beneficial because they allow **water** to penetrate the **soil** rather than running off or evaporating (Jake Page, 144).

Cuesta (slope) is a hill with a gradual slope on one side and a precipitous slope on the other.

Cure Salee Festival (Salt Cure) is a yearly get-together of the nomadic Tuareg and Wodaabe peoples in **Niger**.

Current is an ocean movement that extends over long distances. The Humboldt (also **Peru**) Current and the Benguela Current (**South Africa**, **Namibia**, Angola) affect **deserts**.

Cypress is a 30- to 40-foot conifer whose twigs and branches are useful against urinary tract infection and dysentery; it can also be burned as an incense (Moore, 39–41).

Cyrenaica is Libya's eastern region.

Dades Gorge is a beautiful pass in Morocco's **Atlas Mountains**.

Daisy is a common flower of many species; it has white petals and a yellow center.

Dakar Rally (formerly Paris–Dakar Rally) is a severe vehicle race, beginning in 1978, in many divisions that formerly ran from Paris to Dakar, Senegal. It now takes place in South American **deserts** and the **Empty Quarter**, among other environments. (See **Sports**.)

Dam is an obstruction built by **animals** (beaver) or humans in order to block or control the flow of **water** for storage as in a pond or reservoir, to irrigate agricultural land, or to produce electricity. Dams are constructed out of wood (**trees**), **sand**, rock, earth, or concrete. The most famous **desert** example is **Hoover Dam** in the **Mojave**.

Damaraland is located in northwest **Namibia**.

Damaras are fat-tailed sheep in **Namibia**.

Dana Biosphere Reserve (Dana Nature Reserve), between Amman and Aqaba, **Jordan**, is replete with varied topography, **animals**, **plants**, and hiking possibilities. See https://www.aswesawit.com/dana-biosphere-reserve for an excellent series of **photographs**.

Danakil Desert (103,000 square miles) runs along the coast of the **Red Sea**, the Gulf of Aden, and then around the horn and into the heart of Somalia. According to Michael Martin, the Chalbi and

Kaisut Deserts in Kenya are part of the Danakil (as is the Afar Triangle and its **nomads** to the north), and nomadic peoples (Borana, Gabbra, Samburu, Rendille, Turkana, and Pokot) abound here (261, 262). **Salt** is mined on the surface of Ethiopia's Danikil Depression ("Creation's Hell Hole") and still transported by **camel** to market (Jake Page, 25).

Dance, indigenous. Dancing (and drumming) play an important role in many desert cultures, especially for Arabs and Native Americans who incorporate dance into their religious and cultural lives. Examples include the Hassania veil dance, the Navaho hoop dance, and the **Hopi snake dance**. (See also **Ceremonies**, **Music: indigenous**, and https://www.youtube.com/watch?v=qimlfusewec.)

Dandan-Uiliq was an ancient city in the **Taklamakan** excavated by **Sven Hedin**.

Dangers. It is often quite dangerous to travel alone or even in groups in the **desert**. It is possible to miscalculate distances; lose one's way (even with compass and GPS); run out of **water**; get caught in a **rain**, **snow**, or **sandstorm**; have mechanical or physical breakdowns; or encounter bandits, rebels, terrorists, and drug smugglers, not to mention the **heat**, which for migrants makes crossing deserts a dangerous proposition. In the spring of 1965, I hitchhiked across the **Negev**, **walking** when necessary. A Belgian student in a real Jeep stopped and loudly exclaimed, "What are you doing out here? Don't you know that Jordanian horsemen come down out of those hills and shoot people? I just turned in my guns!" Years later, crossing **Death Valley** (where the ground **temperature** was 157 degrees Fahrenheit), a fellow hiker got lost and did not return to camp until the following morning, despite my wife's enormous signal fire and continual shouting. Finally, crossing Australia's **Simpson Desert**, a bus had a flat tire and the driver was at a complete loss: He could not locate a jack and did not know how to proceed. Passengers came to his aid.

Darwin, Charles (1809–1882). The great naturalist visited the **Atacama Desert** during the five-year voyage of the *Beagle*.

Dasht (Persian or Farsi). **Desert**, plain.

Dasht-e-Kavir. See **Great Kavir**.

Dasht-e-Lut. See **Iranian Desert**.

Dasht-Margo. A gravely desert area in Afghanistan.

Date palm, Death Valley. ROBERT HAUPTMAN

Date palm is a 60- to 80-foot-tall oasis tree that has been under cultivation for 6,000 years. Dates are prized for their sweet taste and long shelf life. The trees grow in the **Arabian Desert** (https://arabian-desert1.weebly.com/plants.html) and the **Mojave** and **Negev**.

David Bellamy's Arabian Light is Bellamy's oversize, peculiarly titled but extraordinary book describing his journeys through **Saudi Arabia**, **Egypt**, **Jordan**, **Oman**, **Gilf Kebir**, and Lebanon. What makes it different from virtually all other such desert travelogues are his innumerable truly beautiful paintings that accompany the text.

Dead Sea is a large body of very salty **water** in the **Judean Desert**. It is fed by the Jordan River and is drying up. At 1,410 feet below sea level, it is the lowest point on earth. **Badwater Basin**, in **Death Valley**, at a mere 282 feet below sea level, is the lowest point in the **US**.

Dead Sea Scrolls are ancient scrolls containing biblical passages; they were found by a shepherd in clay jars in the Qumran **caves** (not far from **Masada**). Many of the scrolls are housed in the Shrine of the Book in Jerusalem. They are important because they contain some of the earliest versions of certain texts and confirm the reliability of biblical textual transmission. **Scholars**, naturally, have created unnecessary controversies.

Death. See **Desert death**.

Death adder is one of Australia's many poisonous **snakes**.

Death Valley Lake. In August 2023, Hilary, a tropical storm, produced a massive amount of **rain** in **Death Valley**; this resulted in a most unusual lake at **Badwater**, which is usually just an enormous **salt pan**.

Death Valley National Park (more than 3 million acres) is a famous area in California's and Nevada's **Mojave Desert** known for its barrenness, lack of **water**, and extreme **heat** (the old 1913 world

Desert landscape, Death Valley. ROBERT HAUPTMAN

record of 134 degrees Fahrenheit was set here, but see **Temperature**). It is considered the hottest location on earth. **Sand, salt pan**, and **dunes** are not the only topographical features; the desert is also bisected by canyons and surrounded by high **mountains. Badwater Basin**, at 282 feet below sea level, is the lowest elevation in the **US**. **Borax** was mined here, and prospectors searched for valuable ore (**gold, silver, copper**, etc.). Bill Clark, in *Death Valley*, observes that more than 600 (970, according to a park handout) types of **plants** (desert holly, **creosote bush, sagebrush**, saltgrass, **mesquite, beavertail, cholla**, Chinese lantern, desert goldpoppy) and **brittlebush** grow here (12–15); 20 plant species are unique to Death Valley. And some 700 (440, according to park handout) **animals**, including **desert bighorn, chuckwalla**, feral **burro, roadrunner**, sidewinder, and **tortoise**, live within the park (16–17). These animals are very hard to spot, and in the current author's many trips and days spent here, he has never seen an animal. Occasional **petroglyphs** can also be located.

Rolling dunes, Death Valley. ROBERT HAUPTMAN

Mesquite Sand Dunes in Death Valley, Mojave Desert.

Flower in Death Valley rubble.
ROBERT HAUPTMAN

Tracks in the sand, Death Valley.
ROBERT HAUPTMAN

Death Valley covers an enormous expanse (it is 140 miles long) and contains many **trails** and innumerable enticing areas. Richard E. Lingenfelter's massive, generously illustrated study presents a detailed history of the valley replete with **photographs**, **maps**, and documents. The outstanding, classified, 55-page bibliography contains about 1,100 entries. (See **Dehydration**, **Devil's Cornfield**, **Devil's Golf Course**, **Furnace Creek**, **Mines**, **National parks and monuments: brochures**, **Pupfish**, **The Racetrack**, **Salina**, **Scotty's Castle**, **Shoshone**, **Stovepipe Wells**, **Telescope Peak**, **Walking**, **Wildflowers**, **Zabriskie Point**, and the "Death Valley Wildflowers" sidebar. An excellent, very short video is at https://www.you tube.com/watch?v=nZrl14lXFHk.)

DEATH VALLEY WILDFLOWERS
(partial listing)

White	Yellow
• Desert-star	• Brittlebush
• Desert chicory	• Creosote bush
• Gravelghost	• Desert dandelion
• Rock daisy	• Desert gold
Orange/Red	• Golden
• Desert gold poppy	evening-primrose
• Globemallow	• Rock nettle
• Indian paintbrush	• Turtleback
Pink/Lavender	**Blue/Purple**
• Beavertail cactus	• Arizona lupine
• Desert five-spot	• Death Valley sage
• Mohave aster	• Fremont phacelia
• Purple mat	

Death Valley is a 2021 horror **film** in which rescuers are attacked by a monster. (The trailer is at https://www.youtube.com/watch?v=Wrw1erLXx1k.)

Deer. White-tailed and mule deer can be found in the southwestern US **deserts**, including the **Mojave**. (Deer are normally quite skittish, but there are exceptions: One day, coming off the Middle Teton, not in a desert, a mule deer walked calmly right alongside us for a mile or more.)

Deflation. Wind and water **erosion** lowers the desert surface by removing sand. This leaves a harder surface called desert (stone) pavement: small rocks cemented together.

Dehydration. Many people now carry water bottles when out for a short stroll on a pleasant Chicago winter day so that they can "hydrate"; this is bizarrely superfluous. But real dehydration (leading to illness or death) is a pressing problem at altitude (in the death zone; that is, above 26,000 feet) or in extremely hot environments. For years, some folks have recommended and advocated for the consumption of eight glasses (half a gallon) of **water** per day; this too seems unnecessary, but when **hiking** in the **Negev** or the extreme **heat** of **Death Valley**, a full gallon is recommended (although this author never ingested nearly that amount while walking across these **deserts**).

Demographics. Although this may be extremely simplistic, those who live in the **desert** can be divided into four major groups: **indigenous peoples** who have subsisted or even thrived for tens of thousands of years in all global desert areas; the poor who have arrived through expansion and **migration** much more recently; middle-class people who choose to live, for example, in Los Angeles or **Las Vegas**; and the wealthy who enjoy the luxury of permanent or temporary residence in locations such as Florida or **Arizona** during the winter months. (This is the very antithesis of the temporary migration of ancient Romans and others in **Israel** or wealthy 19th-century East Indians to the higher northern areas during their respective repressively hot summers.) (See "Desert Demographics" table.)

Desalination plants are expensive and the process is too, but **Saudi Arabia** gets a high percentage of its potable **water** from desalinated seawater. (See also **Iceberg**.)

Desecration. See **Desert implications**.

Desert (G: Wüste; F: désert; S: disierto; I: eyðimörk; Af: woestyn; H: מִדְבָּר [midbor/midbar];

DESERT DEMOGRAPHICS

Notes:

The size of deserts varies from source to source. At times, I take an average.

Some deserts have thousands of plant, animal, and bird species.

Only indigenous peoples are noted here, but obviously deserts are now home to innumerable other ethnic groups.

	Desert Size (square miles)	Location	Plants	Animals	People
Great Basin	ca. 175,000	W. US	Bristlecone	Bighorn sheep	Native American
Chihuahua	200,000	SW US	Agave, yucca	Snake, coyote	Native American
Sonoran	120,000	SW US, Mexico	Mesquite, saguaro	Tortoise, cougar	Native American
Mojave	25,000	SW US, CA	Joshua tree	Bighorn sheep	Native American
Sahara	3,500,000	N. Africa	Acacia	Antelope, gazelle	Tuareg
Kalahari	360,000	S. Africa	Acacia	Elephant, giraffe	San
Namib	ca. 50,000	SW Africa	Lichen, grass	Zebra, elephant	Himba
Arabian	900,000	Arabia	Acacia, date palm	Camel, lizard	Bedouin
Iranian	135,000	Iran, Pakistan	Shrubs	Oryx, lizard	Qashqai
Atacama	54,000	Chile, Peru	Cactus	Llama	Atacameño
Karakum	ca. 125,000	Turkmenistan	Shrubs	Gazelle, cat	Shepherds
Gobi	500,000	China, Mongolia	Grasses	Camel, ass	Nomads
Taklamakan	600,000	China	Shrubs	Camel, gazelle	Turkic
Thar	77,000	India, Pakistan	Acacia	Camel, jackal	Thari
Australian*	890,000	Central Australia	Acacia, eucalyptus	Kangaroo, dingo	Aborigine
Antarctic	5.5 million	Antarctica	Lichen, moss	Penguin, seal	None
Arctic	5.4 million	Arctic	Lichen, moss	Bear, wolf	Inuit

* Includes the Great Victoria, Gibson, Great Sandy, and Simpson.

(Table compiled, in part, after David Miller, 14.)

Ar: صحراء, sahra' [also badiya]; M: цөл [tsöl]; C: 沙ケ漠 [shamó]).

Etymology:

- Desert: Latin: *desertum*, "abandoned"
- *Tesert:* Egyptian: "abandoned"
- *Marustahal:* Hindi: "place of death"
- *Taklamakan:* Uyghur: "place to leave alone" (Haynes, *Desert*, 7)

An arid or semiarid area of land where **precipitation** is minimal (less than 10 inches per year). About one-third (to almost one-half) of the earth's surface is desert. Pure **white-sand deserts** such as parts of the **Sahara** have very little vegetation except in **oases**. Those with some arable land (e.g., the **Gibson**) have sparse vegetation that is barely adequate to feed **animals**, especially **livestock**. Gravel, **salt pans**, rock, canyons, and **mountains** also are part of the desert landscape. In **White Sands (New Mexico)**, the **plants** that emerge have put down roots 18 feet below the surface. **Water** is sparse and the **heat** oppressive. Nevertheless, in spring in some locations (**Death Valley**, for example) the desert blooms in extraordinary flowery profusion. Some peoples (Australian **Aborigines**, **Tuareg**, **Khoikhoi**) have adapted to desert life and thrive where others would undoubtedly perish. There exist 4 primary American deserts, 5 primary world deserts, some 23 substantial deserts, and innumerable minor deserts in most parts of the world. (See **Geography** and "The Four Major American Deserts," "The World's Most Important Deserts," and other sidebars.)

Desert (pronunciation: dih zurt') (not to be confused with dessert). Esoteric philosophical or legal term as in "just deserts," meaning something positive or negative that is deserved.

Desert (pronunciation: dih zurt'). To leave and not return (often with a negative connotation).

Desert (1996–) is a semiannual scholarly **journal** published by the University of Tehran. A typical article is titled "Early-Warning System for Desertification Based on Climatic and Hydrologic Criteria" (https://jdesert.ut.ac.ir).

Plant growing in white sand. ROBERT HAUPTMAN

THE WORLD'S MOST IMPORTANT DESERTS

- Antarctic
- Arabian (including the Rub' al Khali—the Empty Quarter)
- Arctic
- Atacama

- Australian (composed of the Great Victoria, Great Sandy, Tanami, Simpson, Gibson, and others)
- Chihuahuan
- Gobi

- Great Basin
- Kalahari
- Karakum
- Kyzylkum
- Mojave
- Namib
- Negev

- Patagonian
- Sahara
- Sechura
- Sonoran
- Taklamakan
- Thar

(See also https://www.traveller.com.au/worlds-10-most-incredible-deserts-1lzvo6 for a different take.)

Minor or Sub-Deserts

- Anza Borrego
- Black Rock
- Chalbi
- Danakil
- Darfur
- Dasht-Margo
- Dubai
- Eastern
- Erg of Chech
- Fezzan
- Gran Desierto de Altar

- Great Eastern Erg
- Great Kavir
- Great Karoo
- Great Western Erg
- Himalayan
- Hoggar
- Iranian
- Judean
- Kaisut
- Libyan
- Mauritania

- Monte
- Nubian
- Ordos
- Owyhee
- Painted
- Rann of Kachch
- Red
- Sinai
- Somalia
- Sturt Stony
- Strzelecki

- Syrian
- Tanami
- Tanezrouft
- Tassili n'Ajjer
- Ténéré
- Tirari
- Western
- Western Sahara
- White

Desert. Miranda MacQuitty has written a magnificent book, perhaps geared to younger readers but useful to and enjoyable by anyone who cares about the desert and the natural world. Each page consists of very brief texts appended to four, five, or more colorful images of everything from rocky deserts and seas of **sand** to **birds**, **reptiles**, **insects**, **dwellings**, costumes, and **jewelry**—63 pages of exquisite **photographs**.

Desert is a **novel** by the Nobelist J. M. G. Lé Clezio. Considered a masterpiece, it deals with two people in different time frames.

Le Désert (***The Desert***) is Pierre Loti's 1895 account (diary) of his camel journey from **Cairo** across the **Sinai** to Jerusalem.

The Desert is John C. Van Dyke's classic rendering of the desert and its natural environment of, for example, **rivers**, air, sky, **plants**, and **animals**. He professed both a love of the desert as well as distaste. He has been criticized for errors and distortion. (Compare **Joseph Wood Krutch**.)

The Desert: A Play in Three Acts, written by Padraic Colum (1881–1972), is considered an important cultural work.

Desert animal species, **endangered**, include the addax, Arabian tahr, bald eagle, Barbary sheep, California brown pelican, **cheetah**, **desert tortoise**, **jaguar**, **pygmy owl**, and white oryx.

Desert anomaly. In Togtoh, north **China**, on one side of the Yellow River is the Hobq Desert, on the other is a lush area of gardens. "No theory can explain the **desertification** beside a river" (Mei abstract).

Desert bedstraw is a **shrub** whose flowers have four pointed petals; it is used as a packing material and is quite fragrant. It is found in the **Sonoran Desert** (Gerald A. Rosenthal, 107).

Desert bighorn sheep is a large, wild North American sheep notable for its impressive curved horns. It inhabits various environments including the desert. Edmund C. Jaeger notes that it is a protected species but disease and poaching take their toll (37). (One day, this author was slowly descending a steep, narrow road after a Colorado climb and a bighorn, though normally quite shy, walked over to the car. The driver stopped and opened the window; he could have touched its face.)

The Desert Biome, a website of the University of California's Museum of Paleontology, in Berkeley, offers some basic desert information (https://ucmp .berkeley.edu/exhibits/biomes/deserts.php).

Desert Botanical Garden is a large park in Phoenix replete with excellent and diverse **cactuses** and **wildflowers** of the **Sonoran Desert**, the Southwest, and the world (brochure). (See also **The Arboretum at Flagstaff** and **Tucson Botanical Gardens**.)

Desert broomrape is a parasitic, clustering **plant** with purple flowers on a spike. It parasitizes the roots of bursages and **creosote bush**. It is considered a pest and is found in the **Sonoran Desert** (Gerald A. Rosenthal, 192).

A Desert Calling: Life in a Forbidding Landscape is a wonderfully informative memoir by Michael A. Mares, editor of the *Encyclopedia of Deserts*. Mares is a mammalogist and he tracked, observed, and trapped **animals**, especially small **rodents** such as the **desert kangaroo rat**, in many desert environments, including Argentina's **Monte Desert** and Iran's **Dasht-e-Kavir**. His adventures were quite frightening, once having a failed flashlight while in a precarious position deep in a dark **cave** (47) or suffering from histoplasmosis, almost always fatal even today (50ff.). He studied many genera and their countless species, at times in relation to **adaption** (114) and species convergence (convergent evolution) (125–26.), and in so doing, discovered new species (121). An appendix lists the countless species of **mammals**, **birds**, **snakes**, and **plants** he mentions in the text (287ff.).

Desert changes. There is little doubt that physical and sociological alterations in the last 100 years are greater than all of those that occurred in previous millennia, and so David Miller emphasizes these for their detrimental aspects: delimitation of the nomadic life, tourist incursions, overpopulation, **pollution**, and harm to the environment (passim) and its animal inhabitants. However, deserts have often altered over geological time because of changes in weather patterns so that **aridity** diminished and vegetation and a diversity of **animals** were more abundant, indicated by ancient **pictographs**.

Desert chicory. Small white flowers, whose undersides are lavender, adorn this **plant** (*Desert Wildflowers*, 42). It is similar to the more typical blue-colored chicory found in the northeastern US.

Desert climate (average **weather** over a period of time). In the desert climate varies from extremely hot and dry during the day to cold at night, and during the rainy season, wet, to such an extent that flooding may occur. Federico Norte presents a detailed discussion of climate generally touching on cycles, forecasts, **temperature**, humidity, and so on (120–22). (See **Climate change**, **Desert weather**, and **Paleoclimatology**.)

Desert color. The color of the **sand**, gravel, rock, **soil**, **water**, and vegetation found in deserts varies dramatically across the entire shaded color spectrum, from white, gray, brown, green, and blue to vivid red.

Desert communities. Groups of varying species interacting (ecologically).

Desert death. The desert is a dangerous place: Death occurs to **plants**, **animals**, and humans here (as it does in the **mountains**) for many reasons including poor choices, successful predators, **heat**, lack of **water**, **floods**, lightning, storms, disease, old age, and so on. Mohammed Madadin and his many colleagues explain that, surprisingly, this topic has not been fully examined: "Death in the desert seems to be obscure and little discussed in the field of forensic medicine." Their paper "aims to identify the most common causes of desert death and its medico-legal implications" (Madadin abstract).

Desert dermatoses. In the **Thar Desert**, the incidence of dermatological problems (fungal infections, leishmaniasis, skin tumors, etc.) is higher than in other regions because of **temperature variations**, lack of **water**, and poor hygienic conditions (Chatterjee abstract).

Desert development, negative aspects. The expansion of human activity, building, and industrialization in deserts stimulates the economy and increases available arable land, housing, and **resources** but it has negative consequences including depletion of **water** and decreases in desert habitat. Lihui Luo and colleagues discovered hidden costs through an assessment of meteorological and groundwater data in the desert-oasis ecotone of northwest **China** over the past six decades. The result is a depletion of groundwater, additional **pollution**, and stronger **sandstorms** (Luo abstract).

Desert Discovery Center is a facility in Barstow, **California**, run by the Bureau of Land Management. It is interested in making desert life comprehensible to visitors and is home to the 6,070-pound "Old Woman Meteorite" (https://www .blm.gov/visit/desert-discovery-center).

Desert documentaries. Innumerable desert documentaries are available on Youtube including *Faces of Africa—The Sahara* (https://www.youtube .com/watch?v=LMHfYnS65w0) and *Spectacular Namibia and Botswana* (https://www.youtube .com/watch?v=GgoGrwes4cA). See especially https://www.youtube.complaylistlist=PLm1u6z-50FrCAhcLOTBdp7tNpriV_KhM3M for a compilation of 37 **films**, some of which are quite short at 10 minutes, while others run more than an hour and a half. (See also individual entries, **National parks and monuments: videos**, and the "Desert Documentaries" sidebar.)

Desert Dream (Hyazgar) is a 2006 **film** set in the Mongolian **desert**.

Desert ecology. Here the land, **weather**, **plants**, **animals**, and human beings all interact in positive and sometimes negative ways to produce an ecology of the desert and its systems. A number of books exist that deal precisely (or exclusively) with this topic, including John Sowell's *Desert Ecology*, whose discussions are limited to the southwestern **US**; Gary A. Polis's edited *Ecology of Desert Communities*; and especially Walter Whitford's *Ecology of Desert Systems*, in which the author explains that "studies of ecological systems must focus on both pattern and process" and there exist "eight levels of organization: cell, organism, **population**, community, **ecosystem**, landscape, **biome**, **biosphere**" (2). Instead of including a comprehensive bibliography, Whitford places his references at the end of each chapter. They comprise an enormous corpus of some 800 citations.

Desert ethic. Although it is not often discussed, there is a traditional virtue in the **Arabian Desert** that one must shelter, feed, and quench the thirst of strangers (even enemies) if they show up at one's tent.

Desert Fathers and Mothers. Monks and nuns who lived in the Egyptian desert 1,800 years ago.

Desert formation. Desert etiology varies depending on many factors including location (latitude), proximity to mountain ranges and coastal **oceans**, **weather** (**wind**, **rain shadow**), **water**, barometric pressure, **temperature**, **weathering** (the heat/ cold cycle), and **erosion**. Nathaniel Harris offers an excellent technical discussion of many of these factors (18).

Desert, future. Jake Page predicts that in 100 million years there will be a single enormous desert across the African/Eurasian continent. Australia's desert lands will also be reconfigured (89).

Desert glass. The remnants of a meteorite impact, these rocks are highly prized. **Libyan Desert** glass seems to be the most important, but there are others such as **Atacama Desert** glass (**Chile**) and Edeowie glass (**Australia**). Small pieces may be had for a few dollars, but others can sell for as much as $8,000.

Desert grasses. This is an astonishingly complex subject. There are some 10,000 species of grass and some can be found in desert environments. A good discussion can be found at https://www.desertmuseum.org/books/nhsd_grasses.php.

Desert Heat is a 1999 **film**, sometimes titled *Inferno*, starring a suicidal Jean-Claude Van Damme, who changes his mind when his motorcycle is stolen in the desert and seeks revenge. *Rotten Tomatoes* saw fit to give it a 0 percent rating.

Desert implications. The desert presents vast wastelands, emptiness, nothingness, even death, but it also offers esthetically pleasing vistas, changing colors, blazing sunsets, and a plethora of **animals** and flowering **plants**—therefore, life. Unlike some other natural habitats such as **mountains**, it is anomalous, contradictory, soothing, but also frightening.

David Miller claims that the Western view of deserts as hot locations covered with **dunes**, whose nomadic peoples' lives have been unchanged for hundreds of years, is false (8). I think that this is partially hyperbolic. He also indicates throughout his panoramic overview that humans have sullied, in one way or another, the pristine aspects of the desert.

Desert Institute at Joshua Tree National Park has courses in cultural history, natural science, survival skills, and the arts for adults to explore in-depth the natural wonders of the park (https://www.youtube.com/channel/UCmyynPx64IpzOAq8V6uSxUA, www.nps.gov/jotr).

Desert iris (Negev iris) is a flower that accumulates in seas of purple that only appear in the **Negev** and **Sinai Deserts** (https://rove.me/to/israel/negev-iris-in-bloom).

Desert kangaroo rat is a small (up to 12 inches with tail) brown and white **rodent** with cheek pouches that lives underground in southwestern US deserts.

Desert kites are 9,000-year-old mega-trap stone structures found in Arabia, the Near East, and other locations. These thousands of constructions were built to trap **animals**.

Desert Laboratory on Tumamoc Hill is a University of Arizona ecological preserve located within **Tucson** that combines culture, science, and community (https://tumamoc.arizona.edu).

Desert Land Act of 1877 encourages and promotes the economic development of the arid and semiarid public lands of the western **US**.

Desert landforms. See **Desert landscape** and **Geomorphology**.

Desert landscape. Michael Martin divides desert landscape into eight types: **piedmonts** (surround higher areas); cuestas (strata of flat rock); inselbergs (isolated steep-sided hills of rock); dry valleys (wadis, washes); depressions and **pans** (of clay or **salt: sebkhas, vleis, playas**); **serir** and **reg** (gravel plains); hammadas (flat, bare-rock plateau—most frequent); and dunes and ergs (of sand) (311ff.). (See **Geomorphology**.)

Desert Life. In this small but extremely useful **guidebook**, Karen Krebbs presents an overview of the **plants** and **animals** found in the southwestern US deserts. It is divided by class (e.g., **reptiles**) and then species (e.g., **snakes**), with many examples concisely described. Bobcat, **javelina**, **bat**, skunk, raptor, **owl**, **lizard**, **spider**, **cactus**, and **shrubs** are among the many entities described here.

Desert lily is a tall, white flower with an edible bulb; it is found in the **Sonoran Desert** (Gerald A. Rosenthal, 94).

Desert Lily Sanctuary, in Desert Center, **California**, is a serene location to view an incredible profusion of **desert lilies**.

Desert locations. See **Latitudinal belts**.

Desert marigold is a common, southwestern US, large yellow flower on 18-inch stalks (*Desert Wildflowers*, 9). Inca doves enjoy their seeds (*Desert Bird Gardening*, 8).

Desert mariposa lily is an orange/vermillion, three-petaled flower that favors washes, where it spreads. It is found in the **Sonoran Desert** (Gerald A. Rosenthal, 205).

Desert Memories: Journeys Through the Chilean North is well-known author Ariel Dorfman's account of his **travels** especially in the **Atacama Desert**. He visits the famous observatory at Las Companas, a nitrate town, a copper mine, and other personal locations: "And it is in the desert, in this driest of dry dwelling places, that these and so many other secrets of the past can be uncovered because, quite simply, in a place like this they have never been truly lost" (142).

Desert National Park Jaisalmer (India) is a park in which the great Indian bustard can be observed. This endangered **bird** only lives in India. (See the "International Wildlife Refuges" sidebar.)

Desert National Wildlife Range (Nevada). Located in the **Mojave Desert**, this pristine refuge is the largest within the lower 48 states but part of it is controlled by the Nellis Air Force Base bombing range. **Plants** include cattail, **mesquite**, **willow**, **cottonwood**, **Joshua tree**, larkspur, aster, **Indian paintbrush**, and fleabane. Among the 240 bird species are grebe, bittern, heron, osprey, hermit thrush, grosbeak, and bunting. Fifty-two **mammals** include **pronghorn**, **coyote**, bobcat, **bat**, **fox**, and skunk, plus **tortoise**, **lizard**, and speckled **rattlesnake**. Most important are the imperiled **desert bighorn sheep**, which succumb to legally limited **hunting** (Wall, 32–41). (See https://www.fws.gov/refuge/ desert and an enticing **film** at https://vimeo.com/ 86424950.)

Desert Notebooks: A Roadmap for the End of Time is Ben Ehrenreich's highly praised 2002 memoir/ travelogue concerning climate science, mythologies, nature writing, and personal experiences.

Desert of Maine is an anomalous 47-acre tract of **sand** and **dunes** in Freeport, Maine. This most unusual "desert" of glacial silt was farmland in the 18th and 19th centuries but the **soil** eroded and exposed the underlying sand. Pine tree trunks are buried in some 50 feet of sand and only their tops are visible (Maura J. Casey). (See https://www .desertofmaine.com.)

The Desert of the Tartars. Based on a **novel** by Dino Buzzati, this 1976 **film** depicts a garrison waiting for an attack that may never arrive. It stars a host of well-known actors such as Vittorio Gassman, Jean-Louis Trintignant, and Max von Sydow, among others.

Desert Oracle: Strange True Tales from the American Southwest is a collection of brief vignettes, sometimes just a page or two, concerning various occurrences that take place in the desert. In one, the author, Ken Layne, describes some of the survival techniques outlined elsewhere in this encyclopedia; then, in another, a hiker disappears and barely manages to survive. Many deal with the supernatural: Yucca Man, Pahranagat Man, space aliens, and the white stag ghost. William S. Burroughs, Marty Robbins, and **Edward Abbey** make cameo appearances.

Desert origins. See **Desert formation**.

Desert pavement. See **Deflation**.

Desert peoples. Innumerable tribal and ethnic peoples have inhabited and about one billion continue to live in the world's deserts. They have foraged, hunted, raided, farmed, traded, and transported as **nomads** or sedentary urbanites. Richard A. Pailes, in the *Encyclopedia of Deserts*, has an extensive, 11-page essay on peoples in which he discusses a broad array of divergent (peripheral) topics (barley, animal domestication) (156–67). (See **Ancient desert civilizations**, **Ancient desert peoples**, and the "Indigenous Desert Peoples (International)" sidebar.)

Desert Plants (1979–) is a **journal** that publishes articles on indigenous and adapted arid land **plants**.

Desert Research Center is an Egyptian agency devoted to the desert. Material is also available in Arabic. It publishes *The Egyptian Journal of Desert Research* (https://drc.gov.eg/en/home).

Desert Research Institute is a research campus and branch of the University of Nevada, Reno. Faculty and students deal with environmental issues related to deserts (https://www.dri.edu).

Desert research institutions. Examples include the Desert Laboratory of the Carnegie Institution of Washington, Southwestern Research Station, **Desert Research Institute**, Institute of Arid Lands Research, and **Philip L. Boyd Deep Canyon Desert Research Center**.

Desert rheumatism (valley fever or coccidioidomycosis) is a fungal infection prevalent especially in Arizona's **soil**. It is spread through inhalation.

Desert rose-mallow is a medium-size **shrub** with pale yellow to cream-white flowers, whose petals overlap; it is found in the **Sonoran Desert** (Gerald A. Rosenthal, 95).

Desert Shield/Desert Storm. The Gulf War (1990–1991) was fought in the deserts of **Iraq** and Kuwait. In addition to the normal offensive armaments used, there were toxic materials in evidence.

Desert Solitaire is Edward Abbey's excellent personal overview of the American desert. Inexplicably, it is confused with fiction and it sometimes rubs people the wrong way.

The Desert Song is a 1926 operetta by Romberg and Hammerstein. It is based on a Moroccan revolt against French colonial rule.

The Desert Speaks is a long-running PBS program that presents lovely landscapes and stories concerning people and natural wonders. It ran for at least 19 seasons and each presents many episodes, some of which are available on Youtube.

Desert Storm syndrome (Gulf War syndrome) is an ailment that is characterized by fatigue, headaches, cognitive dysfunction, musculoskeletal pain, insomnia, and respiratory, gastrointestinal, and dermatologic complaints.

Desert studies. Many institutions offer courses and programs that emphasize desert environments. (See **Copper Mountain College**, **California Desert Studies Consortium**, **Desert Research Institute**, and **Desert Studies Institute**.)

Desert Studies Institute (Boise State University) provides educational programs and scholarly presentations concerning the prehistory, history, **ecology**, and politics of Idaho's desert environments and deserts worldwide (https://www.boisestate.edu/anthropology/research/desert-studies-institute).

Desert Sunrise is a play (reminiscent of *Waiting for Godot*) in which an Israeli soldier and a Palestinian spend time together in a **wadi** (Hunka).

Desert Time. See **Travel**.

Desert tortoise is a threatened species that lives in the **Mojave** and **Sonoran Deserts**. They spend most of their lives in underground burrows and hibernate and estivate. It is against the law to even touch these

Desert tortoise. ROBERT HAUPTMAN

creatures in the wild and, amazingly, to release pet tortoises into the desert. Some members of this species can be viewed at the **Arizona-Sonora Desert Museum** or in the wild at the **Desert Tortoise Natural Area**. It is California's state **reptile**. The Desert Tortoise Preserve Committee (founded in 1974) helps to protect these creatures. Sadly, the Las Vegas Desert Tortoise Conservation Center closed in 2013. (See https://www.youtube.com/watch ?v=2m2H6GHW7Qo.)

Desert Tortoise Natural Area (25,000 acres), located northeast of Los Angeles in the **Mojave Desert**, is a ca. 40-square-mile preserve or refuge that protects these 15-inch **reptiles**, which have a 60- to 100-year lifespan.

Desert types. There are a number of ways of designating or describing deserts: (1) **hot** and dry (arid), semiarid, coastal, **cold**; (2) tropical (arid), relief (in shadow of **mountains**), continental (inland), coastal (near **oceans**), polar; (3) high and low; (4) **erg**, hammada, **reg**, **serir**.

Desert varnish is the polished surface of some rocks.

Desert weather is extremely divergent. During the day it may be 100 degrees Fahrenheit, but when the sun retires the **temperature** can drop so low that **water** will freeze. **Wind** can pick up **sand** and make life very uncomfortable as it flings the stinging particles into one's face. It is for this reason that desert dwellers often wear face coverings. (See **Climate**, **Climate change**, **Drought**, **Global warming**, **Precipitation**, **Snow**, **Clouds**, and Christopher C. Burt's extraordinary *Extreme Weather* in the bibliography.)

Desert willow (mimbre). A **shrub** or diminutive **tree**, desert willow indicates that there is **water** below the surface. Its pink/lavender/white flower can be used to make a tea or poultice against coughing, and it is effective as an antifungal and anti-candida for infections (Moore, 44–45). It can be found in all **four major American deserts**. (See "The Four Major American Deserts" sidebar.)

Desert Wind is a **documentary** concerning 13 men who wander in the Tunisian Sahara for two weeks in order to discuss their anxieties and fears.

"The Desert World" is Alonzo W. Pond's expression (and his book title) described as a land where contrasts, contradictions, and spectacular extremes dominate the landscape (23).

Desert X is a biennial exhibition in the desert in Coachella Valley, **California**. The artworks consist of large installations of immersive works that span **sculpture**, painting, photography, writing, **architecture**, design, **film**, **music**, performance and choreography, education, and environmental activism (https://desertx.org). (See https://www.youtube .com/watch?v=opq5sayGrjA.)

Desert zinnia. Quail, finches, and sparrows enjoy this plant's white flower petals (*Desert Bird Gardening*, 27).

The Deserted Station is a 2004 Iranian **film** about a couple whose car breaks down in the desert and who seeks help from a man who cares for other people's children.

Desertification. North Africa and parts of **China**, Russia, **Spain**, and the southwestern **US** are all losing ground (and **soil**) to desertification, which derives from **drought** but also from overgrazing, overfarming, and tree removal. An increase in loss of nutrients and parched land means that crops and **livestock** cannot be maintained (Nicas, A1). Jiri Chlachula describes the worsening situation in the **Thar Desert**: Decreased **precipitation** and increased **wind** produce **aridity**, **erosion**, salinity, and groundwater depletion (Chlachula abstract). In 1980, so quite some time ago and perhaps from a rather eccentric position, although based on case studies, James Walls claimed that **deserts** exist because of **climate**; climate varies but there are no grounds for indicating that climate is now more arid; **climate change** is not responsible for desertification, which derives from human abuse of land; manifestations include **dust storms**, wind and water erosion, and loss of fertility; and desertification can be halted everywhere, because its causes are always social (247ff.).

From an external perspective, an increase in the Sahara's boundaries, for example, does not appear to be of crucial importance, but the dwellers and **nomads** who live in affected areas face increased hardships, are harmed, and may die. The United Nations Convention to Combat Desertification (UNCCD), adopted in 1994, is the sole legally binding international agreement linking environment and development to sustainable land management. (See **Desertification: myth**, **Desertification: restoration**, **Dust Bowl**, **Global warming**, **Great Green Wall**, **Oasification**, https://wad.jrc.ec.europa.eu/atlas, https://www.unccd.int/land-and-life/desertification/overview, and https://www.youtube.com/watch?v=KTpaJn22w4I.)

Desertification: myth. In 1977, Michael H. Glantz edited a collection of 12 papers that maintained that **desertification** is a pressing natural and human-induced problem, and this has long been the univocal consensus. What follows challenges this. In 2016, that is, almost 40 years later, Diana K. Davis published *The Arid Lands*, an iconoclastic study that offers a historical overview of the **desert** and then claims that basically desertification is a myth ("estimates of desertification [are] questionable at best, . . . have been significantly exaggerated") and restoration projects have had little effect. That "there is insufficient scientific evidence of large-scale permanent desertification" (1–2) is reminiscent of climate change denial. Davis objects that despite this new perspective, policies continue to conform to the older "desertification dogma" (19), for example, the Arid Zone Program (that ended in 1964) (145–47). Confirming *The Arid Lands*' contentions, Roy Behnke and his colleagues' edited volume, *The End of Desertification? Disputing Environmental Change in the Drylands*, also appeared in 2016 and the message of its many authors is perhaps even more extreme: The concept of desertification is meaningless and harmful! It is possible that both of these books offer evidentiary proof that their contentions are valid, but it is equally possible that much of what they offer is politically (conservatively) motivated and hyperbolically misleading.

Desertification: restoration. David A. Bainbridge, in *A Guide to Desert and Dryland Restoration*, insists that it is possible to restore damaged **deserts** (1) and serious efforts to this end began in 1980, but also as early as 1900 (8). He points out that about 50 percent of rangeland areas on six continents suffer from *severe* **desertification** (7). Restorative actions include soil modification, seeding, planting, and **irrigation** (96). Bainbridge's volume is useful and his ca. 600-item bibliography is invaluable. Michael Allaby adds dry farming, new crops, vetiver (a **grass**), and other **plants** (176–77), and Jiri Chlachula indicates that tree plantations, hydrology regulation, and farming based on new xerophytic cultigens can help in the **Thar Desert** (Chlachula abstract). (See **Conservation** and *Ecological Management & Restoration* and Robert H. Webb in the bibliography.)

Deserts is Michael Allaby's excellent, generously illustrated reference tool, which explicates a host of technical matters such as plate tectonics, **Hadley cells**, and Milankovich cycles.

Deserts is also the title of James A. MacMahon's Audubon Society Nature Guide, a superb and expansive **guidebook** to the US deserts and their wildlife: **birds**, butterflies, **fish**, **insects** and **spiders**, **mammals**, mushrooms, **reptiles** and **amphibians**, **trees**, and **wildflowers**. Extensive lists of these **plants** and **animals** are followed by more than 600 color plates and species descriptions.

Déserts (1950–1954) is Edgard Varèse's extremely esoteric musical composition for 14 winds, other instruments, and electronic tape. It is not for everyone.

Deserts: American. There are four overarching American deserts: the **Chihuahuan**, **Great Basin**, **Mojave**, and **Sonoran**. Some add a fifth: the **Painted Desert**. Philip Hyde, in *Drylands: The Deserts of North America*, presents a series of texts and extremely gorgeous **photographs**, some of which are spread across two pages with an approximate width of 26 inches.

THE FOUR MAJOR AMERICAN DESERTS
1	Chihuahuan	**3**	Mojave
2	Great Basin	**4**	Sonoran

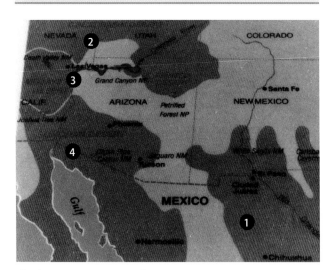

The four major American deserts. ROBERT HAUPTMAN

Deserts Are Not Empty is Samia Henni's edited collection of poems and articles from **architecture**, curatorial studies, comparative literature, film studies, and photography that question the empty desert myth, which allowed for exploitation and **pollution** of desert lands.

Deserts: Miracle of Life is Jim Flegg's generously illustrated overview of the desert but with detailed expositions on **mammals**, **birds**, **amphibians**, **reptiles**, invertebrates, and people rather than the usual lengthy discussions of etiology, **geomorphology**, **climate**, **weather**, and remarks on specific deserts such as the **Gobi**, **Thar**, or **Patagonian**.

Deserts of America is Peggy Larson's detailed survey of US desert land, its constituent parts, **plants**, **animals**, and people.

Deserts of the World is M. P. Petrov's 1973/1976 substantial comparative survey of the physical and environmental features as well as the **resources** of the world's deserts. Unique to this book is a pocket containing six large folded **maps**.

DesertUSA is a website for those who love the desert. It contains information on **national parks**, **state parks**, BLM land, and the **Colorado River** and its lakes and articles on the cities and towns located in or near the desert regions (https://www.desertusa.com/dusablog).

Desiccation theory postulates that deforestation resulted in desiccation that, in turn, turned land into **desert**.

Devil's claw. A large yellow flower adorns this **plant**, and its fruit produces fibers used for basket weaving. It is found in the **Sonoran Desert** (Gerald A. Rosenthal, 163).

Devil's Cornfield is located in **Death Valley** east of **Stovepipe Wells**. The **plant** found here is arrowweed and it looks like a haystack (https://www.usgs.gov/media/images/devils-cornfield).

Devil's Garden (**California**) is an area below Mount San Jacinto. In a black-and-white **photograph** from the 1930s, the **desert** is replete with **barrel** and other **cactuses**; in a comparable color image taken about three-quarters of a century later, the desert is gone, replaced by weeds, highways, and windmills, a sad ecological commentary (Pavlik, 62).

Devil's Golf Course is located between **Badwater Basin** and **Furnace Creek** in **Death Valley**. Here one finds stalagmites and spiky mounds rolling along for 30 miles.

Devil's Hole, near **Death Valley**, is a cavern filled with **water** in which the rare and endangered **pupfish** can be found.

Devil's Marbles (**Australia**). In the Tanami Desert, which lies to the east of the **Great Sandy**, an area is covered with large granite **boulders**, which figure in Aboriginal cosmology (Nathaniel Harris, 161).

Devil's Postpile National Monument is found near Mammoth Mountain in **California**; it is here that one can see the strange, parallel rock formations created out of columnar basalt.

Dhahran is near the site of the 1938 discovery of **oil** in **Saudi Arabia**. (See **Petroleum**.)

Diamond Sutra is the most famous of the tens of thousands of manuscripts found by Aurel Stein in the Cave of a Thousand Buddhas. It is considered the earliest, dated, printed book.

Digital-Desert. See Mojave.

Dingo is a wild Australian dog, despised by many, hunted, and killed, but it still thrives in the **desert**. It is sometimes kept as a pet by the **Aborigines**, although it is the largest land predator here (Moffitt, 70, 22).

Dingo fence is a 6,000-mile wire barrier built to keep **dingoes** in the Australian **outback** and away from sheep and cattle **stations** (Moffitt, 70).

Dinosaur National Monument is located in Colorado and **Utah** and preserves 1,500 dinosaur **fossils** embedded in a cliff, protected by a building, as well as **petroglyphs**.

Dinosaurs such as velociraptors, protoceratops, tarbosaurus, and saurolophus lived in **deserts**. **Fossils** have been found in deserts in **Arizona**, **Mongolia** (along with eggs), and **Morocco**.

Distances in the **desert** are extremely deceptive. What might appear to be a quick 1-mile jog may turn into a 10-mile, waterless marathon. One can become dehydrated and the result may be a tragedy.

Diyafa means hospitality. (See **Desert ethic**.)

Djenné (**Mali**) is an extremely old city within the **Sahara**. It was a link in the trans-Saharan gold

Canis dingo (dingo) on the road. JARROD AMOORE, 2009. CREATIVE COMMONS ATTRIBUTION 2.0

trade and a center for the spread of **Islam**. From the air, it looks like a crossword puzzle.

Documentaries. See **Desert documentaries**.

Donkey (**ass**, **burro**) is a horselike **animal** used for various tasks such as pulling wagons. In **Death Valley**, there currently are some 4,000 feral donkeys wandering around harming wetlands. The National Park Service wants them eliminated; **mountain lions** stalk and kill them (Elbein). A mule is a sterile offspring of a horse and a donkey. (See **Twenty-mule team**.)

Door to Hell. In Derweze, Turkmenistan, there is a large crater that exudes gas. Somehow, it caught **fire** and it continues to burn. It is a major tourist attraction.

Dot paintings. Australian Aborigine artists create intricate geometric designs that reflect their **dreamtime** though symbolic representations. Occasionally, an **animal** such as a **snake**, turtle, or **kangaroo** is included. The dot method resembles the way in which the Pointillists such as Seurat worked. These paintings can sell for thousands of dollars. (Also see **Sand paintings**.)

Doughty, Charles Montagu (1843–1926) (**Arabian**), was a great British adventurer and author of the 1888 *Travels in Arabia Deserta*. Goudie notes that he visited **Petra**, copied the inscriptions at **Mada'in Saleh**, and spent about two years wandering in Arabia and living with **nomads**. (See also **T. E. Lawrence**, **Wilfred Thesiger**, **Explorers**, and the "Desert Explorers" sidebar.)

Doum palm grows in the Dungul Oasis in Egypt's **Western Desert** (Zahran, 110).

Draa (Arm) is a large sand dune in the **Sahara**.

Dreamtime. In the Australian Aborigine cosmology, this is the mythic time before time began. It is when the spirits created the world. (See **Myth** and **Songlines**.)

Drought. Although **deserts** are arid, some areas, especially cities located in deserts, can suffer from drought; that is, **rain** fails to fall. In 2006 Phoenix underwent an extreme drought: 142 days with no **precipitation** (the record is 160). Even **cactuses** can die when deprived of **water**, and **animals** and **birds** fail to reproduce (Wilson). In 2022 a megadrought in the southwestern **US** produced the driest 20-year period in the past 1,200 years. The cause is laid primarily at the feet of human-induced **climate change** (**global warming**) (Fountain). (See **Sahel** and https://www.youtube.com/watch?v=r-wTq0EuqKCM, a short video on the looming water crisis in the southwestern US.)

Drumming. Indigenous peoples drum, and drumming can become part of a broader musical expression as well as cultural and religious rituals. (See **Music: indigenous**.)

Charles Montagu Doughty, 1908. PUBLIC DOMAIN

Druze are a people in **Israel**, Lebanon, and Syria who follow the somewhat secretive Druze **religion**, which differs from the three major **Abrahamic religions**. Note that one day my sister-in-law was passing by a Druze wedding in Israel and the dancers invited her to dance along with them, which she did.

Dryland archeology. Graeme Barker and David Gilbertson lay out the long-term concerns and themes of this sub-discipline. They include **drought**, **desiccation**, **desertification**, **irrigation**, and especially the resilience of farmers, similarity of tactics, variability of systems, and non-commonsense results—all related to farming and irrigation in the Pleistocene and then the Holocene (6–8). They conclude their introductory chapter with the following caution: "As dryland peoples face the uncertainties of the twenty-first century, understanding the richness, diversity and, above all, the complexity of the **archeology** of their antecedents has never been more urgent" (16). (See **Negev Desert**.)

Drylands. About 40 percent of the earth's land surface area is dry; that is, arid to one degree or another. (See **Dryland archeology**.)

Dubai Desert Conservation Reserve is an 87-square-mile natural reserve and the UAE's first **national park**. Here one may see how **Bedouins** live.

Dumont Sand Dunes, in Baker, **California**, is an off-highway vehicle (OHV) recreation area. The dunes range in height from 700 to 1,200 feet, and many different types of vehicles are allowed here. (See **Off-road adventure** and **Sports**.)

Dune (G: Düne; F: dune; S: duna; I: sandöldu; Af: duin; H: חוֹלִית [kholis]; Ar: الكثبان الرملية [alkuthban alramlia]; M: манхан; C: 沙ケ丘 [Shāqiū]). **Wind** rolls through **deserts** with powerful gusts that pick up the sand particles and pile them up in ever-changing, rolling dunes, some quite tall. They range from a small mound to more than 4,000 feet in height; they also come in many different shapes and colors, and they occur all over the world. The types are created by varying wind direction and speed.

Although most people associate dunes with the desert, they only comprise about 20 percent of the desert environment. An excellent example of an esoteric if enlightening study is Siegmar-W. Breckle and colleagues' *Arid Dune Ecosystems: The Nizzana Sands in the Negev Desert*, which contains an astounding 30 papers on every aspect of the Nizzana dunes, which are located southwest of **Beersheba** and stretch into **Sinai**. Topics include **geomorphology**, **climate**, **soil**, **flora**, **fauna**, **erosion**, and **desertification** inter alia. (See **Dune types**, **Erg**, and https://www.muchbetteradventures.com/magazine/tallest-sand-dunes-world.)

Dune is Frank Herbert's well-known 1965 science fiction **novel** of desert life on a distant planet. It has been turned into **films** a number of times, most recently in 2021, starring Zendaya and Javier Bardem. (See the trailer at https://www.youtube.com/watch?v=8g18jFHCLXk. The complete film is available on Netflix.)

Dune grass (American dune grass, Marram dune grass) is a species of grass that can take root and grow on sandy **dunes**.

Dune types include linear (seif), crescentic (sickle, barchan), star (radial), parabolic (U-shaped), and sand sheets (stringers) (Nathaniel Harris, 54–56); also nabkha (around vegetation).

Dunes, petrified, are a series of rock formations located in **Arches National Park**. (See also **Pinnacles Desert**.)

Dunes, rippled. The **wind** creates long, parallel, rippled dunes that can be extremely symmetrical.

Dung (animal excrement) is used as a potent fuel, in the **Gobi** for example, especially when wood (or coal, peat, or **petroleum**) is unavailable.

Dunhuang, "ancient Shazhou—City of Sands" (Sherman, 52), is an important urban center in **China** along the **Silk Road** near the Kumtag Desert. (See also **Hexi Corridor**, **Gobi**, and **Mogao Caves**.)

Dust. The **wind** picks up tiny **sand** particles in the **desert** and carries them to distant locations where they settle on streets and buildings. They can be physically harmful. Andrew Goudie discusses how dust particles from the **Sahara** and the Middle East, among other locations, cause health issues (asthma, pneumonia, allergic rhinitis, silicosis, cardiovascular disorders, conjunctivitis, skin irritations, meningococcal meningitis, valley fever, algal bloom ailments, and mortality) in many cities including Phoenix, Athens, Dubai, Beijing, and Sidney. Pollutants include heavy metals, pesticides, spores, fungi, and **bacteria** (Goudie, "Desert"). Hundreds of papers and articles covering most parts of the world claim that this is an increasing problem. Ritesh Gautam and his colleagues explain that satellite observations of desert dust (an "optically thick aerosol layer") indicate that it is responsible for darkening Himalayan **snow**. This is important because of "potential implications to accelerated seasonal snowmelt and regional snow albedo feedbacks" on "summer monsoon and regional climate forcing" (Gautam abstract). (See also **Pollution**.)

Dust Bowl. During the 1930s, a **drought** and subsequent soil **erosion** caused a major alteration in Colorado, Kansas, Texas, and Oklahoma farmland, which turned to **dust**, making **agriculture** difficult or impossible. The wind-whipped dust was so intense that people sealed their door frames with tape. In 2021 Ken Burns created a four-part series on the Dust Bowl. (See **Desertification**, **Drought**, and a Burns short trailer at https://www.youtube.com/watch?v=ELCArPSAST8.)

Dust devil is a whirlwind, much smaller and far less destructive than a tornado.

Dust storm. See **Dust** and **Sandstorm**.

Dwellings. The world's **deserts** are now filled with modern buildings—houses, apartment complexes, casinos, skyscrapers—in **Las Vegas** and the **United Arab Emirates**, for example, but traditional dwellings are still in use and include the Mongolian **ger**, or yurt; the Himba round, mud-covered hut; the North American Indian tepee and adobe constructions; the **Inuit** igloo; and various shacks (I have even seen a nontraditional **Aborigine** overturned car in use as a protective enclosure). In the past, indigenous new-world peoples built stone structures such as **pueblos** and buildings within rock (cliff) enclosures. (See **Mesa Verde National Park**, **Petra**, and **White House Ruins**.)

Dzungaria. This valley is located between the Asian Altai and Tien Shan mountains not far from the **Gobi** and is covered with **reg**, **salt**, clay, and **dunes** (Michael Martin, 95).

Earthships is a community near **Taos** in the **desert** that contains architecturally unusual, self-sustaining homes, built in part out of old tires. The people here are doing things similar to what is being done in parts of the third world (https://www.youtube.com/watch?v=wgUkjbMhF18). (See also **Arcosanti**.)

East Is West is one of Freya Stark's many works. It is not merely a travel account but also includes historical as well as current commentary on the situations that obtained in the mid-20th-century Near East. She discusses **Yemen**, **Egypt**, Palestine, **Iraq**, and other locations in 28 chapters, replete with 85 incisive black-and-white **photographs**.

Marsa Alam Desert near Marsa Mubarak, Egypt. MARC RYCKAERT, 2008. CREATIVE COMMONS ATTRIBUTION 3.0

Eastern Desert. See **Western Desert**.

Eberhardt, **Isabelle** (1877–1904), was a desert **explorer** who traveled in the **Sahara** disguised as an Arab. *The Oblivion Seekers* is one of her many books.

Echidna (spiny anteater) is a small, egg-laying **mammal** found in **Australia**. Because it has spines, it resembles a porcupine.

Echinacea (purple coneflower) is a perennial that can grow to a height of 1½ feet. Medically useful for immunological issues, tissue repair, and the inhibition of swelling (Moore, 45, 47–49).

Ecology. See **Desert ecology**.

Ecosystem is an area in which an environment and its organisms interact.

Egypt is a North African country to the east of **Libya**. Although parts are well irrigated by the flooding **Nile**, the **Sahara** covers some of its southern area. **Sinai**, to the east and near **Israel**, is a different type of **desert** where some scrub exists. Egypt is the single most important country in relation to a historical perspective involving a desert. It stretches back 5,000 years to the biblical account of Israelite life in Egypt, the **pyramids**, the

Egyptian nomad. WEX MAJOR 98, 2008. CREATIVE COMMONS ATTRIBUTION-SHARE ALIKE 4.0

Exodus, and the discovery of treasures including Howard Carter's opening of Tutankhamun's tomb in 1922. M. A. Zahran and A. J. Willis's *Vegetation of Egypt* devotes more than 400 pages to a comprehensive overview of Egypt's diverse desert **flora**. (See **Antiquities**, **Anthropogenic influence**, **Nile Valley**, **Population**, **Thebian Desert Road Survey**, *Traveling Through Egypt*, and individual plant listings.)

Egypt: new capital. President Abdel Fattah el-Sisi is building a new, $59 billion capital city in the **desert** outside of **Cairo**. This and other new projects are draining the Nile's **water** (Walsh).

The Egyptian is a 1945 **novel** by Mika Waltari and a subsequent 1954 **film** starring Jean Simmons, Victor Mature, and Peter Ustinov. It is a tale of a physician in the pharaoh's court.

Egyptian art. The art of ancient **Egypt**, which covers some 3,000 years, is often of and within the **desert**. Monumental architecture (**pyramids**, temples,

obelisks) plus **sculpture** (see directly below), paintings, statues, reliefs, vases, and decorated sarcophagi are usually found in the tombs or pyramids of the elite. (See Abbate in the bibliography.)

Egyptian sculpture. In ancient **Egypt** both small and monumental sculpture played an important role in the cultural and religious lives of the people. Enormous examples adorn tombs, and **obelisks** were spread around the country. They are so impressive that many of them were stolen and now reside in other countries. (See **Sculpture** and **Theft**.)

Eid is the large meal at the conclusion of **Ramadan**.

Eilat is the southernmost city in **Israel**. It lies at the confluence of the **Negev** and **Sinai Deserts** and the Gulf of Aqaba on the **Red Sea**.

Ein Gedi is an **oasis**, a 7,000-acre nature preserve, in Israel's **Judean Desert**. Here one will find a waterfall along with hyrax, **ibex**, **goat**, grackle, **raven**, and kingfisher (Shulevitz).

Ein Gedi Botanical Gardens is located on the Ein Gedi **kibbutz** near the **Dead Sea** and features many desert **plants**. (See https://deadsea.com/explore/out-doors-recreation/nature-hiking/ein-gedi-botanical-garden, **Negev Desert Botanical Garden**, and individual plants.)

El Camino del Diablo (the Devil's Highway) is a difficult, 130-mile dirt **road** between Ajo and Yuma, **Arizona**, in the **Sonoran Desert**. It is apparently safe to travel, although it is located within the Barry M. Goldwater Bombing Range and one is warned against explosions, bombs, missiles, and other hazards. But it also bisects a **wildlife refuge** and **Organ Pipe Cactus National Monument** and so the scenery, 275 animal species, and ca. 400 plant types are overwhelming. The road runs along the Mexican border (Benanav, "A Drive"). (See **El Gran Desierto del Altar**, **Immigration**, and *Sunshot*.)

El Gran Desierto del Altar is in the western portion of the **Sonoran Desert** through which **El Camino del Diablo** passes. Bill Broyles informs readers that it is also called the Stinking Hot Desert and the Cactus Coast, among other sobriquets (ix). (See *Sunshot*.)

El Malpais National Monument (west of **Albuquerque**). Lava flows and tubes, cinder cones, and **petroglyphs** are what one will discover here (brochure).

El Moro National Monument (west of **Albuquerque**). Here there are 2,000 **petroglyphs** on a cliffside called Inscription Rock. Two ruined **pueblos** can be found on top of the mesa (brochures).

Eland. Some species of this large, horned **antelope** live in African **deserts**.

Electrical production. Electricity is generated in many ways including through the application of combustion (of coal and other fossil fuels), flowing

water (at **Hoover Dam**, for example), nuclear reactions, geothermal energy, and **wind** and solar power. In the past solar panels harnessed the *light* of the sun, but in 2007 an array of 47 miles of trough-shaped mirrors, emplaced in the Nevada desert, began to produce electricity using the sun's *heat* (Wald). There is a plan to emplace almost 2,000 enormous wind turbines and 10 million solar panels in the remote **Australian Desert**. The resulting electricity will be used to produce green hydrogen that will power the many iron ore mines there. Max Bearak's article is long, detailed, and colorfully illustrated (Bearak, "High-Tech").

Elephant (desert elephant, African bush elephant) is the largest land **animal**. Some herds manage to survive in the **Namib Desert**.

Empty Quarter (Rub' al Khali الربع الخالي) is a vast (250,000 square miles) part of the **Arabian Desert** that is known for its desolate nature. It can be found in **Saudi Arabia**, **Oman**, **Yemen**, and the **United Arab Emirates**. It is the largest continuous sand sea in the world, and few people, **plants**, or **animals** live here. Vast quantities of **oil**, gas, and some minerals are in evidence.

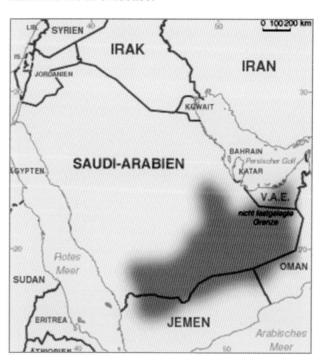

The Empty Quarter in the Arabian Desert. FLOMINATOR. CREATIVE COMMONS ATTRIBUTION-SHARE ALIKE

Emu is a very large, fast-running **bird** that can be found in Australia's **deserts**.

Encyclopedia of the Biosphere: Humans in the World's Ecosystem. The fourth volume of this massive, 11-volume encyclopedia, edited by Ramon Folch, project director, with the help of many others, is titled *Deserts*. Originally written in Catalan and translated into English, it covers the world's **deserts** under a number of broad headings, e.g., **Desertification**, Geologic Life, and **Biosphere Reserves**. The detailed overviews are beautifully illustrated with intercalated color **photographs** and cover various aspects of the desert including plant and animal life. Its 11-page species index contains some 2,700 entries. (See also **Reference books**.)

Encyclopedia of Deserts, edited by Michael A. Mares, is an enormous (654 pages) volume with contributions from 38 **scholars**. It has sometimes extremely lengthy entries on concepts, processes, creatures (full columns on the aardvark and **roadrunner**, for example; this is because the editor heavily favors individual **animals**), people (four columns on **Cochise**, two on **Geronimo**, but none on **T. E. Lawrence**), locations (almost two columns on pampas), body parts (two full columns on kidneys), and other apposite topics, but it contains only 122 diminutive black-and-white **photographs**, 11 **maps**, and fairly limited references to papers and articles and none to the internet. It is an excellent collection of materials directly related to the physical desert but other possibilities such as **book** entries, **ceremonies**, **clothing**, festivals, organizations, **wildlife refuges**, ad infinitum, are elided.

Some peculiarities include choice of topics ("renal papilla"); the often enormous number of *see also* references (25 in the case of "heat stress"—very few readers would be willing to follow up on so many additional terms); and the way in which "peoples" are treated: There are no entries for individual groups such as the **Aborigine** or Bushmen (now **San**) but instead they are all gathered together under "Desert Peoples" in a 12-page essay, essentially a little booklet so that one would have to fight one's way through to locate the **Tuareg** or **Druze**, who are not there, or the Mongols (through the index), who are. A lack of balance is also in evidence: There are three columns on the **fox** but less than two on the **Atacama** and only three on the deserts of **Australia**; and of various adventurers, explorers, and scholars are excluded, e.g., neither **Shackleton** nor **Thesiger** is found in the 45-page index. (See also **Reference books**.)

Encyclopedia of Global Warming is a three-volume set that covers various aspects of **climate change** in the desert environment. Sections cover **desertification** and **deserts** generally in some detail (Dutch, 304–9).

Endurance: Shackleton's Incredible Voyage. Alfred Lansing's classic account tells the amazing story of Shackleton's adventure and rescue. (See also ***The Endurance*** [directly below], **Books**, and the "Desert Explorers" and books sidebars.)

The Endurance: Shackleton's Legendary Antarctic Expedition is Caroline Alexander's superbly illustrated account of the Shackleton debacle and rescue. (See also ***Endurance*** [above], **Books**, and the "Desert Explorers" and books sidebars.)

The English Patient is Michael Ondaatje's superb 1992 **novel** and adapted 1996 **film** that presents a complex story and love affair in the **Sahara**. Of its 12 Oscar nominations, it won 9. (See also **Almásy**.)

Eremophobia is fear of solitude or the **desert**.

Erg is an unchanging, flat, windswept, sand-covered area devoid of **plants**. There are many of these in the world's **deserts**. (Not to be confused with its technical meaning: a unit of energy.) (See **Desert types**, **Great Nafud**, and **Kum**.)

Erg Murzuk (Murzuq) is an **erg** in Libya's southwest in the **Sahara**.

Erg of Admer (Erg Admer) is located in Algeria's southeast in the **Sahara**.

Erg of Bilma is in the **Ténéré Desert** (part of the **Sahara**) in northeastern **Niger**.

Erg of Chech Desert

46,000 square miles	Algeria	Hot desert	Almost no precipitation	Few if any indigenous peoples

Located in the central **Sahara**, this small **desert**, separated from both the **Great Eastern** and **Great Western Ergs**, is covered with enormous, red sand dunes (Stoppato, 116–17).

Erosion is a natural process through which **wind** and **water** (as well as **glaciers** and ocean waves) wear down the earth's surface (the ground, **mountains**, canyons, gorges). Over geological time, erosion creates **Grand Canyons** and decreases mountain heights from, for example, ca. 15,000 feet to 5,000. This occurs in all desert environments. Erosion and **weathering** are why rock that is of different hardnesses wears out at different rates and therefore takes such diverse form. The **freeze/thaw cycle** also causes rock to split.

Erta Ale is a live **volcano** in the northern **Danikil Desert**.

Estivation is the ability of certain creatures to sleep deeply or suspend their metabolism in order to avoid the **heat**. It is similar to **hibernation**, which takes place in colder **climates** among larger **mammals**, **bears** for example.

Ethics. Acting ethically in **deserts** is complex and often difficult. Countries, corporations, and individuals sully the land through **mining**, oil exploration and extraction, military exercises and testing, **hunting** and abusive poaching, robbery, **theft** from archeological sites, graffitiing desecration, and murder. In 1991 the Iraqis set fire to ca. 600 oil wells, **polluting** the desert in many ways (Nathaniel Harris, 171). Doing the right thing in desert environments is extremely challenging. In the past and even today, settlers and colonialists dramatically harmed desert dwellers in the western **US** and **Australia**. (Also see **Anthropology, Cave, Desert ethic, Leave No Trace, Rock art, Walking,** and **Uluru**.)

Ethnobotany is the study of the relationship between **plants** and the **indigenous peoples** who use them. It is a critical discipline since the peoples' lives often depend on foraging and **agriculture**; that is, plant usage. As an example, consider James P. Mandaville's work in *Bedouin Ethnobotany*, a magnificent study (with an accompanying CD-ROM that provides hundreds of color **photographs** of people and plants). Mandaville is fluent in the Najdi dialect of the **Arabian Desert** area (just west of the Persian Gulf) where his many nomadic tribes live. All plants are indicated by their Arabic names. He discusses different types of shrubland, the many tribes, and usages (**food**, medicine, fuel, etc.) and offers a 90-page annotated list of the plants. There is a lot more here but this, I think, will suffice. (Also see *Native Plants of Southern Nevada*.)

Ethnographic (anthropological) studies of the customs and cultures of **desert peoples**, including the **Mardu**, abound.

Etosha National Park (Namibia) has a partial desert landscape and is a wonderful location to observe **elephant**, giraffe, **hyena**, leopard, **lion**, rhino, springbok, and zebra, among many others. (See the "International Wildlife Refuges" sidebar.)

Euphorbias are African succulent trees such as the candelabra (**saguaro**).

European Southern Observatory is in the **Atacama Desert**. The telescopes give astronomers some of the best optical results in the world (https://www.eso.org/public).

Evaporation ponds are created in order to extract **salt** from ocean water in the **Atacama** and in **Utah** and also to dispose of wastewater.

Exploration. Deserts have lured **explorers** ever since Herodotus visited **Libya** and **Egypt**. Unlike the many other types of desert visitors and **scholars**, the true explorer is primarily interested in discovery, rather than trade or cultural exchange, animal observation, or **hunting**. Digital-Desert (http://mojavedesert.net/people) lists hundreds of historical individuals, many of them explorers, who had an influence on the **Mojave**, including Kit Carson, Padro Fages, John Charles Frémont, Joseph Christmas Ives, Jedediah Strong Smith, Ewing Young, and C. B. Zabriskie.

Explorers. Many explorers with many diverse motives have visited desert environments. Some were mere adventurers, a few cared about knowledge, and many were interested in trade as well as riches and therefore wanted to claim lands for their sovereigns. Others wished to convert heathens, a handful cared about locating the source of **rivers** or discovering what lay inside tombs and **pyramids**, and, finally, the purest were obsessed with setting records, for example, by reaching the poles first. Many of these adventurers died at an early age, their unpleasant deaths caused by disease, fever, stroke, starvation, madness, murder, or killing (Goudie, *Great*, 3–4). These men (and women) often composed reports, accounts, and memoirs, some of which are noted in their respective locations in this listing. (See, among others, **Roald Amundsen (Antarctica)**, **Ibn Battuta (China)**, **Johann Ludwig Burckhardt (Arabian)**, **Richard E. Byrd (Antarctica)**, **Sir Richard Burton (Mecca)**, **T. E. Lawrence (Arabian)**, **Marco Polo (Silk Road)**, **Sir Ernest Shackleton (Antarctica)**, and **Sir Wilfred Thesiger (Arabian)**; **Travel**; the "Desert Explorers" sidebar; and in the bibliography, Andrew Goudie's superb *Great Desert Explorers*, in which the author offers succinct biographies of 64 pertinent people, some of whom are quite well known and others who are rather obscure, and Robin Hanbury-Tenison's *Great Explorers*, which allocates sections to 10 polar and desert explorers including **Nansen**, Amundsen, **Barth**, **Bell**, and Thesiger. Naturally, full-length biographies and even Goudie's summaries are more informationally replete than the capsule entries presented on those individuals who appear in these pages.)

DESERT EXPLORERS

- Roald Amundsen (Antarctica)
- Roy Chapman Andrews (Gobi)
- Ralph Alger Bagnold (Sahara-Libyan)
- Gertrude Margaret Lowthian Bell (Syrian)
- Fabian Gottlieb von Bellingshausen (Antarctica)
- Johann Ludwig Burckhardt (Arabian)
- Robert Burke (Simpson)
- Sir Richard Burton (Arabian)
- Richard E. Byrd (Antarctica)
- René-Auguste Caillié (Sahara)
- Hugh Clapperton (Sahara)
- Sir Bede Edmund Hugh Clifford (Kalahari)
- Charles Montagu Doughty (Arabian)
- John Charles Frémont (Great Basin)
- Sven Anders Hedin (Gobi, Taklamakan)
- Friedrich Konrad Hornemann (Sahara)
- T. E. Lawrence (Arabian)
- David Livingstone (Kalahari)
- Douglas Mawson (Antarctica)
- Théodore André Monod (Sahara)
- Gustav Nachtigal (North Africa, Sahara)
- Fridtjof Nansen (Arctic)
- William Palgrave (Arabian)
- Robert Edwin Peary (Arctic)
- Marco Polo (Silk Road)
- John Wesley Powell (Grand Canyon)
- James Clark Ross (Antarctica)
- Robert F. Scott (Antarctica)
- Sir Ernest Shackleton (Antarctica)
- Jedediah Smith (Great Basin)
- Freya Stark (Arabian)
- Charles Sturt (Australian, Simpson)
- Sir Marc Aural Stein (Thar, Taklamakan)
- John McDouall Stuart (Simpson)
- Sir Wilfred Thesiger (Arabian)
- Peter Warburton (Great Sandy)
- Friedrich Martin Josef Welwitsch (Namib)
- Francis Edward Younghusband (Gobi)

Extraterrestrial deserts. Because **water** is sparse or nonexistent (except as ice, vapor, or underground) on most of the solar system's eight planets and ca. 200 moons, one must conclude that there are many extraterrestrial deserts. **Mars** and Venus are excellent examples: They have desertlike surfaces, and even Jupiter and Saturn contain extremely arid land. Titan, a Saturn moon, has Sahara-like sand dunes composed of ice or organic solids rather than silicates (Chang).

The Extraterrestrial Highway. State Route 375, not far from **Las Vegas**, **Nevada**, is extremely desolate and has more reported UFO sightings than any other **road** (Regenold).

Eye of the Sahara (*Guelb er Richat*, Richat Structure) is a large circular impression in Mauritania's **Sahara** that some people think may be the site of Atlantis. (See **Théodore André Monod**.)

Fairy circles. See **Circles, mysterious**.

Far from Men. Inspired by "The Guest," a Camus short story, this is a 2015 **film** about two disparately different, isolated, and nomadic men walking through the Algerian **desert** (Dargis).

Farthest North: The Incredible Three-Year Voyage to the Frozen Latitudes of the North is Fridtjof Nansen's account of his 1893 trip to the north country in the *Fram*, a ship designed to be frozen into the pack ice. He then traveled by sled toward the pole. (See also **Books** and the books sidebars.)

Fata morgana. See **Mirage**.

Fauna. See **animals** and individual species.

Fauna & Flora International is the world's oldest international wildlife conservation organization. It has been shaping and influencing conservation practice since its founding in 1903 (https://www.fauna-flora.org/).

Festival au Desert (2001–2015 in exile) was a music festival held in Essakane and then **Timbuktu, Mali**; well-known musicians such as Bono and Jimmy Buffett as well as **Tuaregs** have performed here. It was cancelled because of terrorism and civil war. (See **Music: indigenous** and https://www.youtube.com/watch?v=PvPOjV_HTss.)

Fezzan (**Fazzan**) **Desert** (212,000 square miles). In Libya's **Sahara**, this section in the southwest is called the Fezzan. Sandstone, limestone, **dunes**, **oases**, **fossils**, and Neolithic images of sheep, **goats**, and so on are all found here (Stoppato, 124–27).

The Fighting Dervishes of the Desert (1912) is a very early **film** (made just 17 years after the first commercial film showing). Here, an Arab sheik falls in love with a priest's daughter. (See also **Whirling Dervishes**.)

Film. **Desert documentaries** abound, but Hollywood and others also produce countless desert extravaganzas. Examples include *The Sheik* (1921), *Son of the Sheik* (1926), *The Thing* trilogy (1951, 1982, 2011), *Lawrence of Arabia* (1962), *Woman in the Dunes* (1964), *Raiders of the Lost Ark* (1981), *The Mummy* (1932, 1999), *Mad Max* (1979, 1981–1985), *Dune* (1985, 2021), *Stargate* (1994), *Ice Station* (1999), and countless others. (See **Film: Westerns** and the "Desert Films: Fiction" and "Desert Documentaries" sidebars, which will lead to individual entries.)

DESERT DOCUMENTARIES
(partial listing)

- Bushmen of the Kalahari
- Desert Wind
- !Kung People of the Kalahari
- Lawrence of Arabia: The Battle for the Arab World
- Lion Pride Documentary—The Realm of the Desert Lion
- March of the Penguins
- Mojados: Through the Night
- Mojave Mystery: Vanished in the Desert
- Mongolian Ping Pong (fictional documentary)
- N!ai: Story of a !Kung Woman
- Nostalgia for the Light
- Skeletons on the Sahara
- The Story of the Weeping Camel

Film: Westerns. This requires a separate entry because US Western films starring cowboys and Indians are often shot in the **desert**, although the desert as such may play an insignificant role. Just west of Lone Pine, **California**, on the Mount Whitney Portal access road, there is a magnificent stretch of complex desert landscape that has frequently served as a backdrop in Westerns.

DESERT FILMS: FICTION
(partial listing)

- The Adventures of Priscilla, Queen of the Desert
- Antonio das Mortes
- Bab'Aziz: The Prince Who Contemplated His Soul
- Beau Travail
- Black God, White Devil
- Black Stallion
- Breakdown
- Casablanca
- Death Valley
- Desert
- Desert Heart
- Desert Heat
- The Desert of the Tartars

(continued)

- The Desert Rats
- The Double Steps
- Duel at Diabolo
- Dune
- The Egyptian
- Escape from Zahrain
- Far From Men
- Fata Morgana
- Fighting Dervishes of the Desert
- Flight of the Phoenix
- Gerry
- The Gods Must Be Crazy
- A Hologram for the King
- Hidalgo
- The Hitch-Hiker
- House of Sand
- Japón
- Lawrence of Arabia
- Lightning Strikes Twice
- Lilies of the Field
- Lost Patrol
- Marco Polo
- Nine Men
- Oedipus Rex
- Passion in the Desert
- The Power of the Dog
- Queen of the Desert
- The Red Desert
- Sahara
- The Searchers
- The Sheltering Sky
- Simon of the Desert
- Stagecoach
- The Swiss Family Robinson
- Thelma and Louise
- Theorem
- The Thirteen
- Treasure of the Sierra Madre
- Until the End of the World
- Valley of Love
- Walkabout
- The Wild Bunch
- Woman in the Dunes
- The Young Black Stallion
- Zabriskie Point

(https://www.imdb.com/list/ls004453539 and other sources)

Fire. See **Wildfire.**

Firewheel (Indian blanket). Found in the semi-desert grasslands of the southwestern **US**, this unusual large flower's petal edges are yellow; the flower is red with a deeper red in the center (*Desert Wildflowers*, 14).

First People are the **indigenous people** who were in North America and especially in **Australia** before the white man arrived. See **Aborigine.**

FiSahara is an international film festival that takes place in Sahrawi refugee camps in southwestern **Algeria**. The camps have been inhabited since 1975. The festival empowers, entertains, and instructs (https://fisahara.es/?lang=en).

Fish might seem anomalous in the **desert** but there are **rivers**, lakes, **wadis**, **oases**, and peripheral **oceans** and so fish do play a role here, in reality a fairly large role as one can see in Robert J. Naiman and David L Soltz's *Fishes in North American Deserts*, a 552-page, edited collection of 16 papers. The **pupfish** (genetic differentiation, breeding systems, thermal tolerance), naturally, come in for careful scrutiny but so too do Cenozoic fish, habitat size, and threatened species. Where there is constant flowing or standing **water**, fish and other aquatic creatures somehow find their way there and

often prosper. Surprisingly, nonnative fish manage to get into reservoirs in the Moroccan desert; of 13 species found, 8 were not native (Clavero abstract).

Fish River Canyon in **Namibia** is a sparsely beautiful canyon and the second largest in the world. (See https://africafreak.com/fish-river-canyon, where a video is also available.)

Fish Springs National Wildlife Refuge. Deep within the desolate **Great Salt Lake Desert** are substantial marsh wetlands that are a lure for waterbirds. Migrating visitors include heron, egret, cormorant, pelican, ibis, osprey, and eagle, plus **pronghorn**, cougar, and preying **raven** and **coyote** as well as **lizard**, **rattlesnake**, and **frog**. **Hunting** is allowed (Wall, 72–78). (See the "National Wildlife Refuges (NWR): USA" sidebar.)

The Five Civilized Tribes. The Cherokee, Chickasaw, Choctaw, Creek, and Seminole were removed from their homes in the South and sent to Indian Territory, now Oklahoma. During the **Dust Bowl** (1930–1936), Oklahoma's land was tantamount to a **desert.**

Flamingo. Different species of (pink) flamingo inhabit the Salar de Atacama, the Atacama saltflats.

Flamingos in an Algerian sebkha. MEKDAS ZOUHIR 2019. CREATIVE COMMONS ATTRIBUTION-SHARE ALIKE 4.0

Flash flood is an enormous surge of **water** that occurs after a deluge (monsoon) or dam burst, which might occur many miles away from the flooding area. If one occurs in an **arroyo**, it will become impassable; in a low-lying **desert** (in **Australia**, for example), one may be inundated; and if in a gorge, one must exit immediately or drown. At Vermont's Quechee Gorge (not in a desert), if one has descended 165 feet, is wandering around, and hears a loud horn blast, one is adjured to race up; otherwise, he or she will be swept away. Sadly, more than 100 people in a single year may die in flash floods, though not in a desert and not all at once. (See **Negev Desert** and an incredible flash flood in **India** at https://www.youtube.com/watch?v=cpfUQH5N9gY or a **river** reborn in the Israeli desert at https://www.youtube.com/watch?v=S02RRTlWDPM.)

Flood. See **Flash flood**.

Flora. See **plants**, **wildflowers**, and individual species.

Fly. See **Uluru**.

Fog. Low-lying **clouds**. Natural fog deposition of **water** is crucial in very dry **deserts** such as the **Namib** or **Atacama** for plant and animal life. It is now possible to set up screens and catch the fog that then leaves water that can be used for additional purposes. (See **Fog basking** and **Ocean**.)

Fog basking. "Fog Basking by Namib Desert Weevils" is a superb example of an abstruse paper that would seem to have little relevance to virtually anything, but as it turns out, the way in which these diminutive **insects** access **water** from fog has potential applications for people through "biomimetics (the study of biology-inspired technology, which can be utilized to solve complex human problems). In short, synthetic surfaces that mimic the external texture of fog-dependent organisms can be used to extract water from the atmosphere" (Lovegrove abstract). (See a very short video at https://www.youtube.com/watch?v=TmyfqjXOf7M.)

Foggara is a system in which the bottoms of wells are connected via a sloping tunnel; the **water** runs downhill and is thereby collected.

Food. Animals prey on other species or eat **plants**. The food of those peoples who live naturally in the **desert**—indigenous **Aborigines**, **Berbers**, and **San**, among many others—often eat the foods available to them in their natural habitat, including some plant life and creatures such as **kangaroo**, **antelope**, wolf, **gazelle**, and, in some cases, **fish**, as well as their domesticated animals such as **camels** and sheep, but also now foodstuffs imported from farmed areas. When urban environments develop in the desert, the people eat the products that are normally eaten by their countrymen in urban settlements, so whatever is consumed in **Timbuktu** may also be the typical food for those who thrive in the surrounding **Sahel** (or vice versa).

Tess Mallos's wonderfully illustrated *Complete Middle East Cookbook* presents the food of its many lands and it varies dramatically. The food of the desert Gulf states is different from what one will find in Iran and very different from the fare in Greece or Cyprus. Typical foods for those who have always made the Near Eastern deserts their home are dates, dried fruit, nuts, **milk**, yogurt, cereals, barley rice, beans, chickpeas, dolma, okra, kibbeh, tabouleh, tahini, hummus, meats, and various combinations of things. Multiple spices such as cardamom, cinnamon, cumin, and fenugreek are used by the diverse peoples of the Near East.

For those who live in the **Gobi**, the food consists of **goat**, camel, and mare's milk, yogurt, cheese, vegetables, fruits, and mushrooms. In the **Kalahari**, one finds **Hoodia cactus**, gemsbok cucumber, and tsamma melons. The **San** eat **insects**, meat from hunted animals, and even white hot coals! (See *Adventures in the Kalahari*, https://www.travelchannel.com/videos/adventures-in-the-kalahari-0163115.) Australian **Aborigines** eat insect larvae, bees, **ants**, and **termites**. Additionally, they have Bush tomatoes, wattle (**acacia**) seeds, desert limes, quandongs, sandalwood nuts, and bush bananas (https://www.terrain.org/articles/16/cribb_latham_ryder.htm). In the **Atacama**, the Inca enjoyed potatoes and corn.

Hopi cultivating corn, 1911. PUBLIC DOMAIN

In American deserts, we have the **Hopi** who eat corn, beans, and squash, as well as domesticated turkeys. They hunt deer, **antelope**, and small game and gather nuts, fruits, and herbs. And the **Navaho**, whose diet consists of corn, beans, and squash, as well as pumpkins, mutton, acorns, and **yucca**, also hunt elk and **rabbit**. (See **Food gathering**.)

Food desert. This is a metaphoric usage of the word *desert*: It means that a rural or urban area does not have affordable, healthful food for its inhabitants.

Food gathering. Gary Paul Nabhan, in *Gathering the Desert*, presents 12 detailed chapters on 12 distinct Sonoran Desert food products (e.g., **agave**, amaranth greens, **devil's claw**, and gourds), of the 425 available edible **plants**, that one may easily collect as **indigenous peoples** have for millennia.

Additionally, anyone can gather **cactuses** (**saguaro**, **barrel**, **prickly pear**) or their fruit, when legal, or purchase them for consumption—either raw or cooked.

Fool's Paradise is one of the finest accounts of life in a 20th-century **desert**. It takes place in **Saudi Arabia**, its well-known areas as well as some very distant, hard to reach, and primitive villages. The book presents a harsh picture of this desert kingdom, where the extremely strict **Wahhabism** version of **Islam** controls life. A noteworthy feature of this account is the intercalation of Arabic terms and their translations thus: "Are those *gadeem* [old], Abdullah?" (66). Interestingly, the author, Dale Walker, informs readers that Charles Montague Doughty's ***Travels in Arabia Deserta*** is the finest book ever written on Arabia. (See also **Books** and the books sidebars.)

Fort Defiance (**Arizona**),within the Navaho Nation, is an 1851 fort built to allow the military to inhabit Navaho territory.

Fort Sumner State Monument (New Mexico). The US Army forced the Mescalero **Apache** and the **Navaho** to march hundreds of miles from their homes to the Bosque Redondo Indian Reservation, which surrounds this fort. The Apache left in 1865 and the Navaho in 1868. This Long Walk for the Navaho was a great tragedy (brochure).

Fossil Falls (Mojave). The name is a misnomer, since neither **fossils** nor falls are in evidence. Instead, this is what remains of a now-dry waterfall as the Owens River flowed across an eroding lava bed; some archeological **artifact**s can be observed (handout).

Fossils are often found in **deserts**, the **Gobi**, for example, where dinosaur eggs (and bones), **lizards**, and unknown **mammals** have cropped up (Wilford, "For Fossil"). (See also https://www.amnh.org/exhibitions/fighting-dinos/fossil-preservation-in-the-gobi and https://www.youtube.com/watch?v=H_W9pwT0B3Y.)

Gobi Desert, Mongolia dinosaur eggs. CHRISTOPHER MICHEL, 2017.
CREATIVE COMMONS ATTRIBUTION 2.0

The four major American deserts. See the sidebar.

Fox. Various species (Andean, bat-eared, fennec), sometimes quite different from each other, inhabit different desert environments.

Freeze/thaw cycle. During the hot day, the sun heats rocks; at night the **heat** dissipates and the cold may freeze. This constant temperature differential causes the rock to fracture, and at times this produces an explosive sound. Eventually, the rock crumbles and produces **sand**.

Frémont, John Charles (1813–1890), was one of the most important explorers of western America. He married Thomas Hart Benton's daughter and employed Kit Carson. He was court-martialed twice but was also an excellent scientist, specimen collector, and cartographer. He crossed the **Great Basin** and visited the **Great Salt Lake** (Goudie, *Great*, 269–74). (See **John Wesley Powell**, **Explorers**, and the "Desert Explorers" sidebar.)

French Foreign Legion. Founded in 1831, this famous military contingent was often stationed in the **Sahara**.

Frog. See **Amphibian**.

From Heaven Lake is an enticing account of Vikram Seth's **travels** across Sinkiang and **Tibet** (at least part of which is in the torrid **desert** that one finds in northwestern **China**), beginning in Turfan and scrolling through **Dunhuang**, Lhasa, to Kathmandu and ultimately Delhi, much of which is done by catching rides in extremely uncomfortable and unreliable trucks. (See also **Books** and the books sidebars.)

Frozen in Time: The Fate of the Franklin Expedition. Owen Beattie and John Geiger have produced a most unusual study: Sir John Franklin's 1845–1848 expedition nearly reached the Northwest Passage but then met with the usual horrors of scurvy, starvation, and death. The ships, *Erebus* and *Terror*, were discovered, corpses exhumed, and Margaret Atwood wrote an introduction to a later edition. (See also **Books** and the books sidebars.)

Fulani (Fula, Fulba) people live in the **Sahel**. At 20 million, they are considered the largest group of **nomads**.

Furnace Creek is a "town" in **Death Valley** that is 190 feet below sea level. Very few people reside there.

Gahwa (gahhwa, qahwa) is an Arabic coffee and by extension a desert rest stop/gas station in **Saudi Arabia**.

Gambel's oak is a small **tree** up to 20 feet tall that prefers watery areas. Native Americans ate the acorns and made bows and spears from the wood, which they also used for fuel (Rhode, 38ff.).

Gambel's quail, similar to a bobwhite, is a ground **bird** that is found in the southwestern US **deserts**. (See **Buenos Aires National Wildlife Refuge**.)

Gander's cholla is a **cactus** found in the **Sonoran Desert**. (See **Cholla**.)

Gaochang is an ancient city in ruins near **Turpan**, along the **Silk Road**, on the border of the **Taklamakan**.

Garua is an Atacama coastal **fog**, so named by Peruvians; in **Chile** it is called chamanchaca (Michael Martin, 199).

Gazelle. There are various species of this attractive, small **antelope** that has horns and colorful stripes.

Gebel (جَبَل) means "hill" or "mountain" in Arabic.

Gebel Kamil Crater. This desert site in southern **Egypt** was struck long ago by a **meteorite** and looters swooped in and stole the fragments so that now scientists cannot study them. They are sold illegally to collectors (Broad). (See also **Ocucaje**, **Cactus trade**, and **Wildlife trade**.)

Gecko is a small, multi-species, omnipresent **lizard** in desert environments but also in urban areas.

Gems are found in America's southwestern **deserts**. They include turquoise, opal, quartz, topaz, amethyst, jade, chalcedony, and petrified wood (https://bizfluent.com/list-6834001-resources-do -deserts-.html).

Gemsbok (South African oryx) is an enormous, long-horned **antelope** that can weigh more than 600 pounds. It lives in the **Namib** and **Kalahari Deserts** and was introduced into **New Mexico**, where **hunting** it is legal.

Gemsbok National Park is located in South Africa's **Kalahari** and protects the gemsbok antelope. A connecting park with the same name is located in **Botswana**. (See the "International Wildlife Refuges" sidebar.)

Geodiode: *Biomes*, **"Deserts: The Bones of the Earth Exposed"** presents an exceptionally attractive, wonderfully illustrated, informative text on the **desert** (https://geodiode.com/biomes/deserts).

Geoglyphs are often-extensive large lines or figures incised on the desert's surface, in the **Atacama Desert** (the 390-foot "Atacama Giant" and some 5,000 others), in the **Thar Desert**, in the southwestern **US**, and in **Peru**. Harry Casey and Anne Morgan point out that some are ca. 1,000 feet in length and represent abstractions, humans, and **animals** (2, 3–4). They then present 161 photos of southwestern US geoglyphs and rock alignments taken from the air. The abstractions are only notable for their sometimes enormous size but the humans, **mountain lion**, horse, stick figure, **snakes**, and **fish** are astonishing, especially when seen from high above. The **Arabian Desert** is home to thousands of ancient (geometric) stone structures (rock alignments). (See **Nazca Lines**, **Petroglyphs**, and https://www.youtube.com/watch?v=z8A0LpX7_yM.)

Geographers. American and European geographers view (define) **deserts** differently.

Geography is the discipline that takes as its domain the physical aspects of the world and the ways in which humans have divided it into its constituent national parts as well as its populations' influence. See the superb Geography website; it contains a wealth of information on **deserts** (https://geography.name/the-extreme-earth-deserts.html).

Geology is the discipline that studies the earth's natural surface and undergirdings including its structure, workings, activities, **mountains**, waterways, elements, magma, **lava**, minerals, **gems**, rocks, and so on.

Geomorphology is the discipline that covers the etiology, physical landforms, and topography of the earth. J. A. Mabbutt, in his 1977 *Desert Landforms*, presents a technical discussion of the subject including lack of vegetation (1), **climate** (5ff.), zonal (low latitudinal) aspects (3ff.), shield and platform and **mountain** and basin (13ff.), mechanical and chemical **weathering** (21ff.), hillslopes (39ff.), drainage (61ff.), **piedmonts** (81), stony **deserts** (119ff.), lake basins (180ff.), and **sand** (215ff.). A more recent general overview can be found in editor Andrew Goudie's enormous 2003 *Encyclopedia of Geomorphology*, which presents a detailed historical framework of desert geomorphology and remarks on **desert varnish** and **desertification** (248–57). For all **desert peoples**, the morphological aspects of the desert are physically and spiritually important, but for Australian **Aborigines**, the land and its features are crucial and they have names for most of the features in their immediate vicinity, all of which are memorized. (See **Desert landscape** and **Songlines**.)

Ger is a smaller or large, round Mongolian tent or yurt. It can be deconstructed and moved to another location because it is built on a basic wooden frame and covered with felt. Amazingly, it contains a stove that has a chimney. The interior can be quite beautiful. Similar permanent abodes in other

Gers (yurts), Gobi Desert, Mongolia. CHRISTOPHER MICHEL, 2017. CREATIVE COMMONS ATTRIBUTION 2.0

countries—Nepal or the **US**, for example—often do not contain a chimney and the open fire pollutes the interior, although the smoke is supposed to escape through a hole in the roof.

Germa (Garama) is a Libyan archeological site.

Geronimo (1829–1909) was an Apache shaman who lived and fought in the **Sonoran Desert**.

Geronimo (on horse at left). LIBRARY OF CONGRESS

Ghaf tree, which does not require a lot of **water**, originated in the **Arabian Desert** (https://arabian-desert1.weebly.com/plants.html).

Ghan Railway is a famous, luxurious Australian train that runs through the **desert**.

Ghat (khat, qat) is a narcotic **plant**, a stimulant, that is chewed in **Yemen** and **Saudi Arabia**.

Ghazu are raids by Arabs against other tribes.

Ghost gum tree, a eucalyptus, is found in central **Australia** and can grow as much as 10 feet a year. Its name derives from its luminous bark. It is a symbol of the Australian bush and spirit.

Ghost towns. Many abandoned desert towns exist and not just in the **US**; e.g., Gleeson (**AZ**), Goldfield (**AZ**), Rhyolite (**NV**), Ballarat (**CA**), Calico (**CA**), Skidoo (**CA**), Farina (South Australia), Humberstone and Santa Laura (**Chile**),

Kolmanskop (**Namibia**), and Old Town Al'Ula (**Saudi Arabia**) (https://www.loveexploring.com/gallerylist/114482/ghost-towns-hiding-in-the-worlds-deserts). Digital-Desert lists more than 100 ghost towns and gold mines in the **Mojave** (http://digital-desert.com/ghost-towns). (Also see the video *Top Ten Ghost Towns in America*, https://www.youtube.com/watch?v=4DSrGFx_vG4.)

Ghutra is an Arab head covering. It differs from the keffiyeh and the shemagh in its pattern and cloth. (Also see **Thobe**.)

Gibber. In **Australia**, arid land covered with rocks and pebbles.

Gibson Desert lies just south of Australia's **Great Sandy Desert** and north of its **Great Victoria Desert** (Nathaniel Harris, 158). It is replete with feral **camels**, **wildflowers**, **grass**, **spinifex**, and the mulga tree (Michael Martin, 137).
   Extent: 60,000 square miles
   Environment: Hot
   Surface: Sand plains, dunes, red pebbles, and gravel
   People: Very few
   Animals: Birds, bustard, dingo, emu, feral camel, lizard, parrot, red kangaroo, and skink
   Plants: Acacia, mulga, shrubs, and spinifex
   Resources: Petroleum
   (See **Australian Desert**.)

Gila Cliff Dwellings National Monument, northwest of **Alamogordo**, **New Mexico**, is a Mogollon culture site; its cliff houses are built into **caves** (brochure). (See the "Southwest US Indian Ruins" sidebar.)

Gila monster. The **Sonoran Desert** is where one will find this venomous **lizard**, but underground, for it stays there most of the time. It is up to 2 feet long and has black and yellow spots.

Gilf Kebir is a limestone plateau located where the southern portions of **Egypt** and **Libya** meet. It is a **national park**.

Giza (not to be confused with Gaza). Located here are Egyptian **pyramids** including the Great Pyramid, which is where Khufu (Cheops) is entombed; the **Sphinx** is nearby.

Glacier is a permanent accumulation of **snow** and ice that increases and shrinks in area depending on the **weather**. It slowly moves downward under the force of gravity. Surprisingly, there are glaciers in the **mountains** in the **Gobi**.

Global warming. Records indicate that the earth and its atmosphere are warming—due to human activity (**human-induced global warming**), the result of which is **drought**, which causes **desertification**. (See also **Climate change** and **Weather**.)

Goat is a domesticated bovid that is bred by many peoples and used for its **milk** and meat. But there are also wild goats, and one day high above Independence Pass in Colorado (but not in a **desert**), we stumbled on a herd of 60 wild white goats; they were extremely skittish and moved away as soon as we approached.

Gobekli Tepe is a 10,000-year-old **archeological** site consisting of large circular structures supported by massive stone pillars—the world's oldest known megaliths. It is located in a desertlike environment in Anatolia, Turkey, and may have been a temple. As with many ancient sites and **artifacts**, conspiracists and mythologists assign it an extraterrestrial origin.

Gobero is a Saharan site in the **Ténéré Desert** of **Niger**. It contains a large Stone Age cemetery.

Gobi means "waterless place" or "**desert**" in Mongolian, but Michael Miller also observes that the term can describe a landscape of rock and stone, whereas shamos are sand deserts (97).

Gobi bear is a smallish brown bear subspecies. There are only about 50 of these **animals** left. The Gobi Bear Project is attempting to save this endangered creature.

Gobi Desert

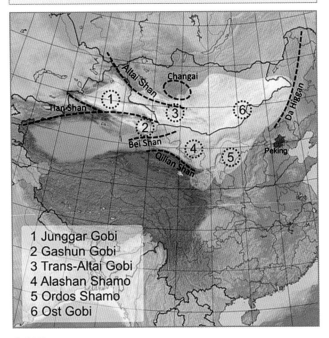

500,000 (374,000) square miles | China, Mongolia | Cold desert | Little precipitation | Nomad: Mongols, Uyghurs, and Kazakhs

1 Junggar Gobi
2 Gashun Gobi
3 Trans-Altai Gobi
4 Alashan Shamo
5 Ordos Shamo
6 Ost Gobi

Gobi Desert map. NIEK VAN SON, 2019. CREATIVE COMMONS ATTRIBUTION 2.0

The Gobi is an enormous and well-known **desert** in northern **China** and **Mongolia**. The southern area is called Badain Jaran (along with the Tengger), with 1,640-foot **dunes**. The Trans-Mongolian Railroad crosses the Gobi from Ulaanbaatar to Russia in the north and China in the south. Its dark **soil**, rock, and **dunes** compete with some **grass** and bushes; there are also wild **camels** and **asses** plus a few Przewalski's horses as well as nomadic peoples (Nathaniel Harris, 70–73). David Miller claims that the Gobi is one of the world's most disaster-prone areas, subject as it is to blizzard, **dust**, **zud**, flooding, earthquake, **wildfire**, **drought**, and **desertification** (107). The Gobi's **sand** and dust particles, picked up by the **wind**, sometimes inundate distant cities such as Beijing in a yellow smog that is harmful to the health of their inhabitants. Note that some claim that the **Taklamakan** is merely part of the Gobi, and a few others think that it is shrinking in size (Michael Martin, 97).

Paleontological remains have been found here. Lisa Janz and her colleagues claim that "the palaeoenvironmental and faunal record of Zaraa Uul show

Gobi Desert, Mongolia. NIEK VAN SON, 2019. CREATIVE COMMONS ATTRIBUTION 2.0

that Early-Middle Holocene hydrology and species distributions were distinct from all other periods of human occupation. Holocene hunter-gatherers inhabited an **ecosystem** characterized by extensive marshes, riparian **shrub** and arboreal vegetation"(Janz abstract). And according to Christopher McCarthy and his six coauthors, the Gobi is an iconic desert, "a functioning, healthy ecosystem home to spectacular landscapes that support an impressive variety of biological diversity, including many rare and endangered species." However, "despite a wealth of natural and cultural heritage the Gobi Desert in Mongolia lacks any recognition as UNESCO World Heritage." They argue that with "sites with exceptional geological, ecological, and ethnological features," it meets UNESCO's criteria and therefore should be designated a World Heritage Site (McCarthy abstract).

Extent: 500,000 square miles

Environment: Cold desert

Surfaces: Soil, rock, sand dunes, and grass

People: Nomadic herders: Mongols, Uyghurs, and Kazakhs

Animals: Bactrian camel, black-tailed gazelle, Gobi bear, Gobi ibex, Gobi viper, jerboa, marmot, snow leopard, and reptiles

Plants: Grasses, Mongolian chive, pearshrub, salt cedar, saxaul tree, Siberian elm, sophora, and wild onion

Resources: Coal, copper, gold, and oil

(See **Glacier**, **Pastoralists**, and **Yelyn Valley National Park** and https://www.scmp.com/video/china/3142543/heavy-sandstorm-engulfs-chinas-northwest-dunhuang-causing-chaos-ancient-silk for a cinematic example of an enormous and engulfing sandstorm in **Dunhuang, China**.)

Gobi Gurvansaikhan National Park is a park in southern **Mongolia**. (See the "International Wildlife Refuges" sidebar.)

The Gods Must Be Crazy is a 1980 comedic **film** about a San tribesman who encounters modern civilization. The full movie is available on Youtube.

(See the trailer at https://www.youtube.com/watch ?v=hvgFqdqPIuE.)

Gold is the most valuable **ore**, relished by all peoples for its malleability and lustrous beauty. Additionally, like **silver**, it is a superb electrical conductor. (See **Mongolia** and **Salt**.)

Goldeneye covers fields with 3-foot-tall, yellow-blossomed flowers in southwestern US **deserts** and **arroyos** (*Desert Wildflowers*, 29).

Grand Canyon National Park. Arizona's enormous and staggering Grand Canyon (277 miles long, 10 miles wide, and 1 mile deep) is the only place in the world where all five climatic zones—tropical, dry, temperate, polar, and highland—are represented in one location (the designating terminology often varies). The park is in the **Sonoran Desert**, and some Indian tribes and many **mammals** reside there. Stanford University Press has created *Enchanting the Desert*, a 2016 digital project by Nicholas Bauch (https://www.sup.org/books/title/ ?id=25726). Descending on an old, abandoned, and extremely dangerous mining trail in 1972, it took me four days to hike the 17 miles to the **Colorado River** and back. As I approached the top, I encountered a descending young college student reading a book. I warned her that if she did not put the book away and pay attention to every step, she might very well fall into the abyss as someone else did that Easter. (See **Yavapai Museum** and https:// www.youtube.com/watch?v=hNcG76DqOiI.)

Grand Egyptian Museum is a new (2023), enormous museum in **Giza** devoted to Egyptian **antiquities**, including more than 5,000 items extracted from Tutankhamen's tomb.

Grand Erg Occidental (Algeria) is in the northwestern **Sahara**. See **Great Western Erg**.

Grand Erg Oriental (Algeria) is in the northeastern **Sahara**. See **Great Eastern Erg**.

The Grapes of Wrath is John Steinbeck's great **novel**; it describes the trip as the Joad family makes its way across the **desert** in order to escape the **Dust Bowl** and reach **California**.

Grass. There are 10,000 species of grass and some do thrive in the **desert**. **Desert grass** is a coarse, tussocky, tropical, and subtropical variety suited to dry areas. Others include California brome, fluffgrass, bull grass, and curly mesquite grass. (A very detailed discussion can be found at https://www .desertmuseum.org/books/nhsd_grasses.php.)

Grasshopper is a small jumping **insect**. The desert **locust** is an especially germane example. Another is the large western horse lubber grasshopper.

Great Artesian Basin, near Australia's **Simpson Desert**, is a groundwater reservoir, the largest such basin in the world. It supplies much of **Australia** with **water**, and in 120 years of exploitation, it has used up less than 0.1 percent of the water stored. (See **Mound Springs**.)

Great Basin Desert

158,000 (175,00 to 208,000) square miles | Western US | Cold desert | Minimal precipitation | Native Americans

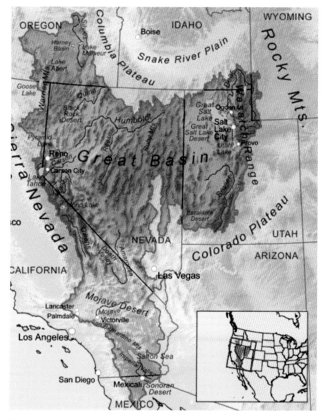

Great Basin map. KMUSSER, 2022. CREATIVE COMMONS ATTRIBUTION-SHARE ALIKE 3.0

One of the **four major American deserts**, the Great Basin is quite large and stretches into five states—Oregon, Idaho, **Utah**, **California**, and **Nevada**—and its cities include **Reno**, Carson City, and **Salt Lake City**. The **Great Salt Lake** is located here. **Plants** include **bristlecone pine**, **sagebrush**, and wormwood, and **animals** (105 mammal and 204 bird species) include **desert bighorn sheep**.

Extent: 175,000 square miles
Environment: Hot in summer, cold and snowy in winter
Surfaces: Sand dunes and vegetation
People: Native American tribes including Paiute, Ute, Washoe, and Western Shoshone
Animals: Beaver, bighorn sheep, mountain lion, porcupine, yellow-bellied marmot, and water shrew

Plants: Sagebrush, saltbush, and horsebrush
Resources: Gems, gold, lithium, magnesite, mercury, and silver
(See "The Four Major American Deserts" sidebar.)

Great Basin National Park is located in eastern **Nevada** and offers a summative experience of the enormous **Great Basin Desert**. It contains the oldest **bristlecone pines**, the Lehman Caves, Wheeler Peak, and a **glacier**. For some reason, perhaps its extraordinary beauty, there exist a plethora of videos on the park. Here is a brief but stunning example: https://www.youtube.com/watch?v=fXIruBaCezo.

The Great Cacti. In this oversize book, replete with one to four magnificent color photos on almost every page, David Yetman manifests his fascination with (and love of) the world's largest (columnar) **cactuses**. These amazing **succulents** thrive in the southwestern US and Mexican **deserts**, but they also can be found in **South America** (Bolivia, Brazil, **Chile**, **Peru**) and the Caribbean (Bahamas) (20). The **saguaro** is often thought of as the most iconic cactus, but others such as the **organ pipe** can be impressively broad and tall. Still others such as sahuesos can reach almost 60 feet (102). Galloping cactus is shrublike (98), and totem pole grows straight and tall (143). Note that there exist hundreds of species of just columnar cacti. **Indigenous peoples** use these extraordinarily diverse **plants** for building as well as for **food** and medicinal purposes (12ff.).

The Great California Deserts is W. Storrs Lee's paean to the **Anza-Borrego**, **Colorado**, and **Mojave Deserts** and **Death Valley**. It is a history of the search for **borax** and **gold** and the later rush to purchase plots of government land.

Great desert skink. Among the 5,000 **lizard** species, this skink, which lives exclusively in northern **Australia**, is the only one that builds and maintains an elaborate, underground home (Bakalar).

Great Dividing Range (East Australian Cordillera) is 2,300 miles of plateaus and **mountains** that separate the eastern, more arable part of **Australia** from the **deserts**.

Great Eastern Erg (Grand Erg Oriental)

74,000 square miles	Algeria	Hot desert	Very little precipitation	Few if any inhabitants

This sub-desert, located southwest of **Tunisia**, is the central part of the **Sahara** and consists of a sea of sand dunes formed and altered by complex wind patterns. A 60-mile-long rocky plateau intervenes between this **erg** and its western neighbor (Stoppato, 106–7). (See the **Great Western Erg** and the "Sahara Sub-Deserts" sidebar.)

Great Gobi B Strictly Protected Area is a nature reserve in the **Gobi Desert**, situated in the southwestern part of **Mongolia** at the border with **China**. A similar reserve in a drier part of the Gobi, the Great Gobi A Strictly Protected Area, exists farther to the east. (See the "International Wildlife Refuges" sidebar.)

Great Green Wall. Eleven Sahel countries just south of the **Sahara** are building a 5,000-mile wall of **trees** to combat **desertification** and increase arable land (https://www.youtube.com/watch?v=y-FN6TlO0z6w). (See **Taklamakan Desert**.)

Great Karoo is an enormous semi-desert area in **South Africa**. The Little Karoo lies to its south.

Great Kavir (Dasht-e-Kavir) is a barren **salt desert** in Iran.

Great Man-Made River Project. Employing enormous 13-foot-diameter pipes, the project brings ice age **water** from the Nubian sandstone aquifer system below the **Sahara** to Libya's urban areas. It is claimed that the water supply will last more than 100 years (Tyler). By 2022 the project was still not completed. (See https://www.youtube.com/watch?v=oIQNnF7mwWE.)

Great Nafud (Al Nafud) is an **erg**, a large, red, sandy area in the **Arabian Desert**.

Great Salt Lake (Utah) is a lake in the **Great Basin Desert** that is much saltier than the **ocean**. It has been drying up during the long 21st-century, western **US**, HIGW **drought**, and by mid-2022 things had reached a crisis, since aquatic animal life could soon die off and, therefore, migratory **birds** would have little on which to feed.

Great Salt Lake Desert (19,000 square miles) is a dry lake in **Utah** with white evaporite salt deposits, including the Bonneville Salt Flats. Its **dust** is harmful. A military installation is here as well.

Great Sand Dunes National Park and Preserve, southwest of Pueblo, Colorado, contains the tallest North American **dunes** (at 750 feet). Only some **beetles** and a cricket manage to survive out here (Kappel-Smith, 7). It is a popular location for **sandboarding**.

Great Sand Sea (28,000 square miles) lies between western **Egypt** and eastern **Libya**. This **erg** is one of the world's largest **dune** fields.

Great Sandy Desert

154,000 (139,000) square miles	Australia	Hot desert	Little precipitation	Aborigines

Located in northwestern **Australia**, the Great Sandy is a large, empty wasteland of salt marshes and sand dunes (Nathaniel Harris, 158). A few Aborigine settlements and a gold mine are all that can be found here (Michael Martin, 135).

Extent: 154,000 square miles
Environment: Hot desert
Surfaces: Rock, red sand plains, and dunes
People: Aborigines and miners
Animals: Birds, camel, dingo, donkey, horse, lizard, and pig
Plants: Grass, woods, and shrubs
Resources: Petroleum
(See **Australian Desert**.)

Great Victoria Desert

> 250,000 (163,000) square miles | Australia | Hot desert | Little precipitation (but with many thunderstorms) | Aborigines

The Great Victoria stretches from the western through the central **Australian Deserts** and is considered a sandy wasteland. Much of it is Aboriginal land (Nathaniel Harris, 160–61). Plant-covered **dunes** are in evidence.

Extent: 250,000 square miles
Environment: Hot desert
Surfaces: Sand dunes, rock, and dry salt lakes
People: Aborigines, and others for testing weapons
Animals: Copperhead, dingo, fox, lizard, sandhill dunnart, skink, southern marsupial mole, and malleefowl
Plants: Acacia, eucalyptus, grass, gum, mulga, and shrubs
Resources: Copper, diamond, gold, nickel, and uranium
(See **Australian Desert** and https://a-z-animals.com/blog/great-victoria-desert, which will also lead to other deserts.)

Great Western Erg (Grand Erg Occidental)

> 31,000 square miles | Algeria | Hot desert | Very little precipitation | No people except on the periphery

Located southeast of the **Atlas Mountains**, this **erg** contains few **plants**, **reptiles**, and **insects** but much **sand** in the form of **dunes**. The **wind** is fierce and sandstorm particles can reach Europe. **Oases** can be found along the erg's edge (Stoppato, 108–9). (See the **Great Eastern Erg** and the "Sahara Sub-Deserts" sidebar.)

Greenland Ice Sheet (700,000 square miles). This ice cap covers more than three-quarters of Greenland and it is melting, losing an average of around 250 billion metric tons of ice per year. The result will be a catastrophic rise in sea level.

Grey, Zane (1872–1939), was a novelist of US western **deserts**. His books include *Desert Gold*, *The Heritage of the Desert*, and *Riders of the Purple Sage.*

Guadalupe Mountains National Park. The Texas high point, Guadalupe Peak, is located in the **Chihuahuan Desert** near an ancient marine fossil reef that hovers over McKittrick Canyon. As one ascends the peak, it is possible to encounter both long-dead marine **fossils** and living (petrifying) **mountain lions**.

Guano is the detritus (excrement) left when large numbers of **bats** inhabit an area, often a **cave**, such as **Carlsbad Caverns**. Over the years it builds up, and the truly enormous amounts that accumulate defy conception. (It is too bad that bats have to reside in their bathrooms.) It also derives from seabirds and is a very valuable fertilizer. Francisca Santana-Sagredo and 13 of her colleagues claim that beginning in AD 1000, seabird guano helped to intensify **agriculture** in the **Atacama**. This white gold, so named because it appears white on its surface, helped to support "a substantial **population** in an otherwise extreme environment" (Santana-Sagredo abstract).

Guayule. Rubber can be extracted from this **plant**, which grows in the southwestern **US**.

Guban Desert. This narrow "burnt land" runs 150 miles to the easternmost tip of the northern Somalian coast and is hot and dry, with only a tiny amount of **precipitation**. The Dir and Isaaq clans live here (https://www.travelawaits.com/2663066/best-experiences-in-africas-ten-unique-deserts).

Guelta is a collection of **water** that forms in a Saharan canal or **wadi**.

Guelta d'Archei is a famous permanent **oasis** in **Chad**.

Guidebooks are invaluable for those people who wish to identify species of **birds**, other **animals**, and **plants** as well as for tourists and adventurers. Ian Sinclair's *Birds of Southern Africa* depicts almost 1,000 examples, many of which can be found in the **deserts** there; Iain Campbell's *Birds of Australia* covers ca. 2,000 species, some of which are found in the desert; Hadoram Shirihai's *Complete Guide to Antarctic Wildlife* depicts 500 species of birds and marine **mammals**; Pinau Merlin's *Guide to Southern Arizona Bird Nests & Eggs* is an unusual compilation. James A. MacMahon's *Audubon Society Nature Guides: Deserts*; the Sierra Club's *Naturalist's Guide to the Deserts of the Southwest*; David Rhode's ***Native Plants of Southern Nevada***; Karen Krebbs's *Desert Life: A Guide to the Southwest's Iconic Animals and Plants and How They Survive*; Nora and Rick Bowers's *Cactus of Arizona Field Guide*; and ***Sahara Overland: A Route and Planning Guide*** all give readers an idea of how extensive and diverse this group of helpful aids can be. (See ***Sonoran Desert Life***.)

Gully is a ravine, gorge, or cleft or a couloir in the **mountains**. (See **Water**.)

Gullywasher is a short, heavy rainstorm.

Hadley cell is a low-latitude overturning circulation wherein air rises at the equator and falls at about 30 degrees latitude. The cells produce trade winds and **weather**.

Hajj (haj, hadj) is the annual pilgrimage to **Mecca** that Muslims make at least once in a lifetime.

Halophyte is a **plant** that thrives in a high-saline environment.

The Halt in the Desert is an 1845 painting by the mad artist Richard Dadd, who is famous for *The Fairy Feller's Masterstroke* and infamous for killing his father. This is a darkening image of people sitting around a campfire.

Hami (Kumul, Qumul) is an **oasis** town on the **Silk Road**. The Hami Desert is part of the **Gobi**.

Hamada (hammada) is a sandless, rocky plateau. (See **Desert types**.)

Hamada du Draa (Dra Hamada) is an Algerian plateau.

Han Desert Ruins is in the **Gobi Desert** in Gansu. (See https://www.youtube.com/watch?v=cTs5EedaMuc.)

Hanson, Erin (1981–), is a contemporary Impressionist painter of desert scenes (https://www.erinhanson.com/portfolio).

Harm. Harm and destruction occur in **deserts** for many reasons and through countless activities. **Heat** and **dehydration** kill, and humans desecrate, litter, graffiti, steal, poach, wage **war**, pillage, and plunder. Craig C. Billington and his coauthors discus the results of illegal border crossing in the **Sonoran Desert**. In 1987, two million migrants were apprehended (109). The extraordinary increase in people crossing from **Mexico** has now increased and the problems therefore exacerbated: garbage dumping, unsafe target shooting, harm to vegetation and **petroglyphs**, damage to **livestock**, and so on (116). (**See Immigration**, **Resources**, and **Theft**.)

Harmattan is a fall/winter dry **wind** in the **Sahara** and the **Sahel**. It makes people ill.

Hartani (pl. Haratin) is a freed slave or descendent in the **Sahara**.

Hassanein, Ahmed (Bey) (1889–1946), crossed the **Libyan Desert** and continued to **Sudan** (Michael Martin, 358). He is the coauthor of *The Lost Oasis*.

Hassani is a **Saharan tribe** that lives by the sword, whereas the Zawaya live through **books**.

Havasu Falls is a famous waterfall in the **Grand Canyon** on the Havasupai Indian Reservation. To visit requires an expensive permit (at least $100) and a hike of some 20 difficult round-trip miles, but for serious hikers it is worth the effort.

Hayden, Julian D. (1911–1998), was an archeologist whose *Field Man* is an overview of his work in the southwestern **US**. He "camped and argued with a who's who of the discipline, including Emil Haury, Malcolm Rogers, Paul Ezell, and Norman

Tindale." He worked in Pima, Papago, and Seri territory (43ff., 103ff.).

Hayduke Trail is an extremely difficult, dangerous, and hard-to-follow route through 812 miles of **desert**; it runs from **Arches National Park** in **Utah** along the **Grand Canyon** in **Arizona** and back to Zion Canyon in Utah. The rocky canyons, striated cliffs, and desert scenery are extraordinary. The elevation gain/loss is an astonishing 125,092 feet. Naturally, it takes many days to complete it, and hikers have recorded videos of their trials and triumphs. (See http://www.hayduketrail.org, one of the finest and most useful websites this author has ever seen, and an impressive video at https://www.youtube.com/watch?v=BkB2Abzu31A.)

Healing. The **desert** is often considered a place where physical and spiritual healing through its "restorative power" can occur. People have often moved to **Arizona** because the air there improves breathing—for asthma sufferers. Marilyn Berlin Snell notes that in Wickenburg, Arizona, one finds dude ranches and treatment centers for the addicted, bulimics, and anorexics.

Heard Museum (Phoenix). It is not unusual for a pointed collection to contain countless Native American **artifacts** but the Heard, according to Don and Betty Martin, is "probably the world's greatest monument to Native Americans." It presents baskets, **jewelry**, **pottery**, textiles, **kachinas**, and an artist-in-residence. Equally appealing is the shop and bookstore, which sells original traditional and contemporary works; the book selection is also impressive (brochure). Surprisingly, the traditional kachinas are reasonably priced (ca. $325), but contemporary models can be quite expensive (up to $17,000). Baskets run $300 to $500; Navaho **rugs**, under $1,000 to $12,000 (https://www.heardmuseumshop.com and see https://heard.org).

Heat. See **Temperature**.

Hedgehog cactus (calico) is a colorful cactus that produces magenta flowers. It can be found in Joshua Tree National Park's Cholla Cactus

Garden (brochure). It is 2 feet tall with long spines and provided **food** for indigenous tribes (Rhode, 112–13.).

Hedin, Sven Anders (1865–1952) (**Gobi**, **Taklamakan**), traveled widely, mapped, and used motorized vehicles. His willingness to sacrifice colleagues and **animals** (indeed, he almost died in the **desert**) as well as ties to the Nazis sully his reputation. Among his ca. 50 books are *My Life as an Explorer* (Goudie, *Great*, 145–50), *Across the Gobi Desert*, and *The Silk Road*. (See **Explorers** and the "Desert Explorers" sidebar.)

Hejaz is a region around **Mecca**, Medina, and other Saudi cities.

Hejaz Railway was finished in 1908 and ran 800 miles from Damascus to Medina, but soon fell into desuetude. Now it only allows passengers to travel 50 miles from Amman to Al Jizah in **Jordan**. It is the railroad one observes in the **film** *The Mummy Returns* (Hubbard).

Hemotoxic. In the past, a useful distinction between hemotoxic and neurotoxic **snakes** was made. The former type's venom attacks through the blood, which allows more time to fight off the poison or reach a hospital; neurotoxic creatures' venom delimits physiological processes such as breathing and one has less time administer antivenin before succumbing. Note that today, this distinction is out of favor.

Hemp is grown in southwestern US **deserts**.

Herb. See **Medicinal plants**.

Hermit is an individual who chooses to live apart from society, often for religious reasons. Hermits sometimes live in desert environments, as the Desert Fathers in third-century **Egypt** did.

Hexi Corridor (Gansu Corridor), a 600-mile branch of the **Silk Road**, runs southeast from the **Taklamakan** and **Gobi**. Along the way are oasis towns every 50 to 100 miles (Sherman, 36).

Dunhuang is at its entrance. (See the *Yanni Hexi Corridor—China* video at https://www.youtube.com/watch?v=r9oQsv1RNG0.)

Hibernation. See **Estivation**.

Hidalgo. This 2004 **film** presents an emotionally wounded half–Native American who enters a horse race in the **Arabian Desert**. It stars Viggo Mortensen and Omar Sharif.

Hi-Desert Nature Museum is located in Yucca Valley, **California**. It specializes in the **plants**, **animals**, and geological aspects of the desert.

High Desert Museum, located in Bend, Oregon, presents wildlife, culture, **art**, and the natural **resources** of the high desert.

Hijab is a scarf or shawl that covers the hair and neck of Muslim women; but more germane in this context, it is a protecting amulet, a pouch containing Koranic verses, that is worn by Sahel desert dwellers from Senegal to **Sudan**.

Hiking. See **Walking**.

Himalayan desert is a little-used locution that describes the area running east to west across some 600 miles of the Karakorum range, **China**, and **Tibet** from **Ladakh** in **India** to **Mustang** in **Nepal** and south to Kinnaur and the Satluj River in India (Rao, 8–9). It is a barren land of steep **mountains**, valleys, little vegetation, and Buddhist *gompas* (fortified monasteries) built in inaccessible locations (12ff.). Villages can be found at 12,000 to 16,000 feet (40), and Leh, Spiti, and Dunai are population centers. **Plants** and **animals** include poplar, dog rose, wormwood, and marsh marigold (52, 44) and **camels**, wild **asses**, **goats**, and geese (57, passim).

Himba is a nomadic people who inhabit part of **Namibia**. David Miller delineates a nasty history of colonialism, apartheid, and **war** that harmed these people (19). (See "Ceremonies" sidebar.)

Historical overview. There exists an extensive human, natural, and ecological history associated with the world's **deserts**. It would be impossible in a work devoted exclusively to the desert to cover Egyptian, Tuareg, San, Chinese, Mongolian, Mesopotamian, Persian, Arabian, Aboriginal, and South and North American indigenous and colonial historical events and personages as well as the evolution of the natural world—its **oceans** and waterways, forests and plains, **mountains** and valleys, **glaciers** and ice caps—that once were deserts or evolved into desert environments. To exemplify this latter point, that deserts come and go, consider Texas's Guadalupe Mountains that were once an ocean: The reef still exists, and this high area is littered with marine **fossils**. And recall that pharaonic and imperial dynastic shifts, Near Eastern and Indian astronomical and mathematical discoveries and knowledge acquisition, and the Aztec and Anasazi eradications that led to contemporary ethnic groups all could be discussed in great detail—but elsewhere.

History and ecology. The **desert** influences history, and history conversely has a potent impact on the desert, especially environmentally: Nomadism, **hunting**, settlement, farming, **exploration**, **travel**, **mining**, and petroleum extraction alter the natural balance sometimes quite detrimentally. An excellent example, as Jennifer Keating observes, is Imperial Russia's 19th-century engagement in Turkestan (which at the time encompassed Turkmenistan, Uzbekistan, Kyrgyzstan, Tajikistan, and part of Kazakhstan) and the role played by cotton, **water**, **animals**, **plants**, **sand**, silt, **salt**, and especially the railroad as well as sand inundations, what she calls "a political ecology of **aridity**" (9, jacket, 29, 69, 1).

Hogan is a low, round building in which the **Navaho** live. It has ceremonial and religious significance.

Hoggar Desert (Ahagger)

| 31,000 square miles | Algeria | Hot desert | Virtually no precipitation | Tuareg only in oasis |

This is another small area in the central **Sahara** in the Hoggar Mountains. Three hundred unusual monoliths (such as Ben Amira) are found here, as is the Tamanrasset oasis (TamTam), where some 45,000 **Tuareg** reside. In other locations, **rock art** (**animals** and symbols) have been discovered (Stoppato, 118–21).

Hohokam is a southwestern US tribe that flourished long ago by building irrigation canals in the **desert**. (See **Casa Grande Ruins National Monument** and https://www.desertusa.com/ind1/du_peo_hoh.html.)

A Hologram for the King is a 2016 **film**, based on a Dave Eggers **novel**, starring Tom Hanks, in which the protagonist attempts to sell a holographic teleconferencing system to a Saudi king. (See the trailer at https://www.youtube.com/watch?v=UW4OE1eg bHs and the "Desert Films: Fiction" sidebar.)

Hoodia gordonii in the **Kalahari** is also called Hoodia cactus, though it is not a real **cactus**. It has clusters of pinkish flowers and is useful against digestive disorders (https://kalaharidesert12.weebly.com/plants.html).

Hoover Dam is an enormous structure near **Las Vegas**. It was built between 1931 and 1936 and required the displacement of the **Colorado River**. It protects against flooding, produces electricity, and sends **water** to Los Angeles. Its concrete is still curing.

Hopi is a Native American tribe that lives in desert environments including the **Grand Canyon**.

Hopi children in a photo from 1921. PUBLIC DOMAIN

A 1911 photo of a Hopi pueblo. PUBLIC DOMAIN

A 1911 photo of a Hopi Snake Dance. PUBLIC DOMAIN

Hopi Snake Dance. In this religious ritual, participants dance with **snakes** in their mouths. (See **Dance, indigenous**.)

Horned melon tree (gemsbok cucumber, hedged gourd) grows in the **Kalahari** and produces a melon, the kiwano, that is both delicious and nutritious (https://kalaharidesert12.weebly.com/plants.html).

Hot desert. A distinction is made between hot deserts such as the **Sahara** or **Mohave** and colder locations such as the **Antarctic** but also the **Gobi** or **Karakum**. This, naturally, does not mean that these latter are not extremely hot during summer days.

Hotan is an important **oasis** on the **Silk Road** in the **Taklamakan Desert**. (See **Kashgar**.)

Hottentot (now derogatory). See **Khoikhoi**.

House of Sand is a stunning 2005 Brazilian **film** about two women who must survive in a harsh desert environment from which escape seems impossible. (It should not be confused with *House of Sand and Fog*.) It is reminiscent of the 1964 existential Japanese film ***Woman in the Dunes***. (See the trailer at https://www.youtube.com/watch?v=algdilTBfRs.)

Housing development. A lack of **water** is causing officials to rethink new housing developments in desert environments such as **Arizona** and **New Mexico**. Nevertheless, officials sometimes sanction enormous projects such as Teravalis, a 37,000-acre plot in the **Sonoran Desert** that will contain 100,000 houses, depending on assuring adequate water **resources** (Schneider).

Hovenweep National Monument. Located on the Cajon Mesa, northwest of Durango, Colorado, Hovenweep ("deserted valley") contains **kivas** and mysterious, desolate, ruined towers.

Hueco Tanks State Historic Site, near El Paso, Texas, contains some 2,000 cave wall **pictographs**.

Human-induced global warming (HIGW). The earth's **climate** has changed quite radically during the past century, and humans, through fossil fuel combustion (and **pollution**), are responsible. **Desertification**, among many other horrors, is the result. Philipp Lehmann observes that shrinking **glaciers** and expanding **deserts** are the alarming consequences of **climate change**; this became evident as early as the 19th century, and the **Sahara** played a crucial role (13–14). Things were so bad that scientists came up with ideas such as flooding the Sahara for reversing matters (38ff.), just one of many such projects in **climate engineering** (53ff.). The decrease in Sahel rainfall (155–56) and the expanding Sahara, as well as desertification, generally have continued to stimulate an ongoing and now politically motivated debate that began more than a century ago (169).

Humboldt Current. See Current.

Humpy (gunyah, wurley) is a temporary shelter for Australian **Aborigines**.

Hunter-gatherers. See Anthropology.

Hunting. Many indigenous desert dwellers hunt game of one sort or another. (Others are shepherds with herds of **goat**, sheep, cattle, **camel**, or **yak**.) Native Americans took buffalo, but only in small numbers; Europeans, who arrived fairly recently, decimated most of the big herds. Miller notes that the **San** use arrows tipped with poison and must follow quarry for many miles until the **animal** finally succumbs (187).

Huntington Botanical Gardens. The Desert Garden at the Huntington is renowned for its extraordinary wealth of international **cactuses** and **succulents**. In *Desert Plants*, Gary Lyons presents a history of the garden and an overview of some of the thousands of botanical specimens preserved here in the world's largest collection. Many of the hundreds of specimens are presented in gorgeous, colorful illustrations that are awe-inspiring. (See https://huntington.org/desert-garden and an amateur video at https://www.youtube.com/watch?v=iriBnLcnmzM.)

Hyena, including the **jackal**, can be found in southern African **deserts** such as the **Kalahari**.

Hyperthermia is a dangerous increase in **body temperature** that is harmful or fatal. It results from activity in extreme **heat**.

A lagoon in the desert (Huacachina, Peru). LAUREN DAUPHIN, NASA

Ibex is a wild goat of many species found in **Oman**, **Egypt**, **Israel**, **Jordan**, and **Saudi Arabia**. They have enormous horns that curve back upon themselves. An imported herd exists in **New Mexico**.

Ica Desert is a coastal Peruvian desert whose ocean **fossils**, including whale, are among the best in the world.

Ice cave. See **Perpetual ice cave**.

Ica Desert. (Credit unknown.)

Iceberg. Innovative methods of bringing potable **water** to **deserts** and drought-stricken areas include towing icebergs from the **Arctic** or **Antarctica**. Although smaller bergs have been towed for short distances, no large one has ever been hauled to a water-starved country because of a variety of technical and financial problems. (See also **Desalination**.)

Iceland deserts such as Odadahraun cover parts of this country with **sand** and **volcanic** glass.

Iguana. The desert iguana (one of 35 species) is a well-known **lizard** that can be found the **Mojave** and **Sonaran Deserts**. Some people keep them as pets.

Illegal activities. See **Cactus trade**, **Gebel Kamil Crater**, **Immigration**, and **Ocucaje**.

The Immeasurable World is William Atkins's detailed account of his **travels** in some of the most potent **deserts** (**Sahara**, **Arabian**, **Gobi**). He intercalates historical events within his experiences, which makes this an unusual memoir.

Immigration. In late 2022, illegal immigration into the southwestern US **desert** across the Mexican border reached crisis proportions with cities such as El Paso inundated with thousands of refugees every day. Some locations simply shipped these people to other places such as New York City. But this is hardly a recent development. In 1999 John Annerino published the "gut-wrenching" *Dead in Their Tracks* (reissued in 2009 in a new format from a new publisher). The story he tells is horrific enough with death as a constant companion: From 1994 to 1998, more than 5,000 people perished in "America's killing ground . . . the U.S.-Mexico borderlands" (vi–vii). The book concludes with "In Memorium," the historic death toll through 2000 in a 177-numbered listing many noting multiple deaths; this is followed by more recent catastrophes and rescues (126 ff.). An extensive 25-page bibliography rounds out this appalling account. (See **El Camino del Diablo** and **Harm**.)

Imperial Valley Desert Museum, in Ocotillo, **California**, contains Harry Casey's ca. 10,000 aerial geoglyph **photographs** (Harry Casey, 97).

In the Kingdom of Ice: The Grand and Terrible Polar Voyage of the USS **Jeannette**. Hampton Sides retells the story of this ship, which was breached and sank in the ice as it attempted to reach the North Pole. The crew had to walk hundreds of miles to reach safety. (See also **Books** and the books sidebars.)

In the Land of White Death: An Epic Story of Survival in the Siberian Arctic is Valerian Ivanovich Albanov's account of his 1912 attempt to locate new Arctic hunting areas. His ship was frozen in and the crew got scurvy. Some of the men attempted to cross the ice on a 90-day, 235-mile journey that was horrifically painful. Only two survived. (See also **Books** and the books sidebars.)

India is an enormous country (and subcontinent) in which the **Thar Desert** is located.

Indian paintbrush. Tiny flowers are overshadowed by bright red floral bracts (*Desert Wildflowers*, 31) in this well-known but smallish **plant**. Indigenous tribes used it against venereal disease and to purify blood; a tea made love medicine (Rhode, 144ff.).

Indian Petroglyph State Park. Located in **Albuquerque**, this park has many prehistoric rock carvings.

Indian Pueblo Cultural Center. This **Albuquerque** cultural center represents the 19 Pueblos in a **museum**, archives, and library and presents events, celebrations, and **dances** (brochure). (See **Pueblo** and "The 19 Pueblos" sidebar.)

Indian Wild Ass Sanctuary (Gujarat, **India**). Here is India's largest sanctuary; it offers protection to wild asses as well as other **mammals** and **reptiles** (https://www.holidify.com/places/kutch/indian-wild-ass-sanctuary-sightseeing-2560.html). (See the "International Wildlife Refuges" sidebar.)

Women harvesting millet in the Thar Desert. JI-ELLE, 2019. CREATIVE COMMONS ATTRIBUTION-SHARE ALIKE 4.0

Indigenous peoples inhabit desert environments. As the modern world has encroached, their traditional ways of living have altered: Mechanized conveyances, tourists, and governmental strictures all influence the lives of these people, especially the **nomads** who are settling down in specific locations. External influences such as the Spanish invaders, missionaries, and corporations seeking **oil** or **gold** have taken advantage of these people. (See the "American Desert Indian Tribes" and "Indigenous Desert Peoples (International)" sidebars as well as individual entries.)

AMERICAN DESERT INDIAN TRIBES
(partial listing)

• Acoma	• Navaho	• Shoshone
• Apache	• Pai	• Tanoan
• Cocopah	• Paiute	• Tohono
• Hopi	• Papago	O'odham
• Keresan	• Pascua Yaqui	• Ute
• Laguna	(Yoeme)	• Zuni
• Mojave	• Seri	

INDIGENOUS DESERT PEOPLES (INTERNATIONAL)
(partial listing)

- Aborigine (many subgroups) (Australia)
- Atacama (Atacameño) (Chile, Argentina, Bolivia)
- Aymara (Chile)
- Bedouin (Saudi Arabia)
- Berber (North Africa)
- Gadaria Lohar (India)
- Houthi (Yemen)
- Kalbelia (India)
- Mongols (Mongolia, China)
- Nama (South Africa)
- Obatjimba Herero (Namibia)
- Ovahimba (Namibia)
- Raika (India)
- Sa'ar (Yemen)
- Sahrawi (Saharawi) (Morocco, Mauritania)
- San (Botswana, Namibia, South Africa)
- Tehuelche (Argentina)
- Topnaar Nama (Namibia)
- Tuareg (North Africa)
- Uyghur (China)

Indio is a city 127 miles to the east of Los Angeles, **California**. It is the home of the **Coachella Valley Music and Arts Festival**.

Insect is a small **arthropod** of many species with a three-part body, three pairs of jointed legs, and compound eyes. The **ant**, **termite**, **centipede**, and **beetle** are among the multitude of insects that inhabit desert environments.

Inspiration Peak is a **mountain** in **Joshua Tree National Park**.

Institute of American Indian Arts (Santa Fe). Founded in 1972, this college's **museum** displays contemporary Indian **art**.

International Year of Deserts and Desertification took place in 2006.

Inuit (formerly Eskimo). The Inuits inhabit Alaska, northern Canada, and Greenland.

The Invention of the American Desert: Art, Land, and the Politics of Environment is a collection of 10 essays edited by Lyle Massey and James Nisbet. It has a postmodern slant, and the scope, viability, and esthetic appeal of the **art** and **architectural** works discussed may be measured by Walter de Maria's 1968 *Mile Long Drawing*, which entails two parallel chalk lines about 10 feet apart drawn on the surface of the **Mojave**. It rolls along for a mile and the artist is pictured lying on the ground perpendicular to the lines (215).

Iranian Desert

135,000 (150,000) square miles | Iran, Pakistan | Hot desert | Minimal precipitation | Qashqai

Much of Iran is covered by **deserts**, including the **Dasht-e-Kavir** (Great Salt) and the **Dasht-e-Lut** (Nathaniel Harris, 64). (See also **Great Kavir**.)

Iranian Desert Masileh Qom (Desert Ruins). MOSTAFAMERAJI, 2009.

Iraq. The **Arabian** and **Syrian Deserts** extend into this Near Eastern county which is rich in **oil**. They are called Al-Ḥajarah and Al-Dibdibah here.

Ironwood (desert ironwood, palo fierro) is a **tree** (a legume) that grows to a height of 45 feet in the **Sonoran Desert**, where it offers shade to small creatures.

Irrigation. Of course, the **desert** is irrigated naturally, when it **rains** or **snows** or from artesian **aquifers** and springs at **oases** or **wadis**, but this has never sufficed for peoples who make permanent homes here, so they have built walled ditches, underground aqueducts, and in modern times small concrete canals that line enormous melon fields, for example. On a larger scale, **dams** (**Hoover**, **Aswan**) and lakes (**Mead**, **Powell**) are constructed and large amounts of **water** are diverted, some of which heads for urban areas but much of which irrigates farmland. Bad irrigation techniques cause much harm, which is perhaps why dams are being removed. (See **Drought**, **Puquio**, **Qanat**, and **Wet desert**.)

Islam is one of the three **Abrahamic religions**. It was founded by Mohammad in AD 622 and traces its beginnings to both **Christianity** and **Judaism**, which is why the biblical patriarchs are honored here. Its holiest sites are **Mecca** and Medina, both located in the **Arabian Desert**. Its three main sects are Sunni (**Saudi Arabia**), Shiite (Iran), and Wahabi (Saudi Arabia).

Israel is a small desert country bordered by **Egypt** to the west, **Jordan** to the east, Syria to the north, and the **Red Sea** in the south. It is the "land of milk and honey" promised by God to the Israelites in the **Bible**. After it gained its independence in 1948, the people began to eke out viable farmland and made the desert bloom. (See **Negev Desert** and **Judean Desert**.)

Israel: solar tower. In 2019 an enormous, 800-foot solar tower was constructed in the **Negev Desert**. Fifty-thousand ground mirrors reflect the sun's rays toward the top of the tower where **water** is heated, which then turns to steam; this drives some turbines that produce electricity. Unfortunately, photovoltaic panels are now more efficient (Kingsley).

Jabès, Edmond (1912–1991). Born in **Egypt**, Jabès was a pseudo-mystical writer and poet who emphasized the **desert**.

Jackal is a small canine that inhabits the **Kalahari** and **Sahara Deserts**.

Jacob Blaustein Institutes for Desert Research is located at Ben-Gurion University of the **Negev**. It is interested in delimiting **desertification** by working on food security, water scarcity, and clean energy (https://in.bgu.ac.il/en/bidr/Pages/default.aspx). (See **Research**.)

Jaguar. This large cat has been absent from US soil for years, ever since ranchers decimated its population. Now it is returning to **Arizona** and **New Mexico**. Weighing as much as 350 pounds and resembling a leopard, it is a fearsome foe that attacks the head of its victim. In this author's extensive wilderness experiences, **animals** he fears most are **mountain lion**, wolverine, and jaguar. Black bear, moose, **coyote**, and even grizzly leave him alone, although an aggressive bear destroyed the car in pursuit of our Fig Newtons!

Jaisalmer is a city in Rajasthan, **India**, in the **Thar Desert**. Its "Golden Fort" is one of the world's oldest inhabited forts (Juskalian). It was a Silk Road caravan halting place. (See also **Meherengarth Fort**.)

Jarhead is an unconventional 2005 **film** based on a true account concerning the First Gulf War.

Javelina ("skunk of the **desert**") resembles a smelly pig but is a collared peccary that lives in the **Sonoran Desert**.

Jemez Mountain Trail, north of **Albuquerque**, is a series of automobile **roads** that pass various sites including the Jemez Pueblo, **Los Alamos**, and **Bandelier National Monument** (brochure).

Jerboa is a small, burrowing **mammal** that can jump long distances. It makes its home in the **Gobi** among other locations (https://www.desertusa.com/du_gobi_life.html).

Jericho is the famous biblical city whose walls Joshua caused to crumble.

Jewelry. If **clothing**, including face coverings, is often though not always necessary for protection against **heat** and cold, sun and **wind**, as well as to accommodate certain religious demands, jewelry in many desert locations, especially in the **Sahel**, is mandatory. Naturally, precious metals such as **gold**, **silver**, and even **copper** are valued, but beading is preferred. The intricate designs of enormous body coverings made entirely of small beads are extraordinary. Both males and females of some tribes cover their midriffs with a beaded girdle, but not their upper or lower torsos at all (Beckwith and Fisher, passim). In *Africa Adorned*'s chapter on the **Sahara**, Angela Fisher indicates that the clothing and jewelry derive from the desert's dictates (191). The ornate jewelry is overwhelming, as in the case of Tuareg amulets, rings, and bracelets (208ff.), and this is even more the case in the **Maghreb**, where the protective hand is fashioned into pendants probably of silver, and a pictured **Berber** is so over-ladened with jewelry from head to torso that it is a miracle that she can walk around (227ff., 241, 252). (See **Books**, **Ceremonies**, and **Clothing**.)

Jojoba (goat or deer nut, pignut, wild hazel) (pronunciation: hohoba) is a **shrub** with large leaves speckled with hairs. Its nut produces an excellent lubricating oil. It can be found in Joshua Tree National Park's Cholla Cactus Garden (brochure).

Jordan is a country located to the east of **Israel** bordering on the **Negev Desert**. Much of it is in the **Syrian Desert** along with **Wadi Rum**. Jordan is one of the driest countries in the world, and by late 2022 its water shortage had become acute: The Jordan River was at about 10 percent of its average and its outlet, the **Dead Sea**, was disappearing. It is therefore draining its **aquifers** (Zraick).

Jornada del Muerto is a wilderness study area (WSA) south of Socorro, **New Mexico**. It has a 90-mile **trail** through lava flows, is uninhabited, and is the site of the first atomic bomb test. It is also a Linkin Park song. (See https://www.blm.gov/visit/jornada-del-muerto-wsa.)

Joshua tree. See **Yucca**.

The Joshua Tree is a 1987 album by the band U2.

Joshua Tree National Park (794,000 acres), located in southern **California**, contains many of these **cactuses** as well as **wildflowers**. It is also very popular with rock climbers, who come here to boulder; that is, climb large rock formations that are only 10 or 15 feet high but present technical problems that one might not encounter on a much higher cliff. Very sadly, the Dome Fire, in Mojave National Preserve, burned 43,000 acres

and over a million Joshua trees in the summer of 2020 (https://www.nps.gov/jotr/learn/photosmultimedia/mojavefire.htm). There are five **oases** in the park, including the Oasis of Mara, Forty-Nine Palms, and Lost Palms. **Plants** in evidence are **creosote bush** and various cacti; **animals** include **coyote**, bobcat, **roadrunner**, **lizard**, **rattlesnake**, and tarantula; and the Lost Horse Gold Mine is located within the park's boundaries (brochure). The Cholla Cactus Garden (with **teddy bear**, **silver**, **hedgehog**, and **jojoba**) is also inside the park. A bird checklist includes 240 species including shorebirds (killdeer), **owls** (barn, elf), woodpeckers (hairy), flycatchers (kingbirds), wood warblers (yellow, ovenbird), and many more (brochure). (See **Inspiration Peak** and **Art**.)

An unpleasant development occurred in the nearby town of Joshua Tree after the pandemic struck in late 2019: People began to turn desert properties into expensive bed-and-breakfasts and investors began to build at an astonishing rate in order to create more Airbnbs, thereby sullying the desert environment (Murphy).

Climbing (bouldering) in Joshua Tree National Park. ROBERT HAUPTMAN

Journal of Arid Environments (1978–) is a monthly scholarly publication that offers abstruse material such as "Arbuscular mycorrhizal fungi improve nutrient status of *Commiphora myrrha* seedlings under **drought**." Some articles are available under an open-access agreement. The journal's author instructions consists of 14 dense pages. (See also ***Arid Land Research and Management*, *Journal of Arid Land*, and Journals.**)

Journal of Arid Land (2009–) is similar to the ***Journal of Arid Environments***. A typical arcane article is entitled "Physiological and biochemical appraisal for mulching and partial rhizosphere drying of cotton." (See also ***Arid Land Research and Management* and Journals.**)

Journal of Desert Research (1981–) is a Chinese publication that presents esoteric scientific papers. The links do not function.

Journals. There are many germane scholarly periodicals that focus on esoteric scientific papers, **aridity**, and **deserts**, e.g., ***Arid Land Research and Management***, *The Egyptian Journal of Desert Research*, ***Journal of Arid Environments***, ***Journal of Arid Land***, *Journal of Desert Research*, and *Journal of Geographical Research on Desert Areas*.

Journeys. See **Travel**.

Juan Bautista de Anza National Historic Trail runs 1,210 miles from Nogales, **Arizona**, to San Francisco.

Judaism. The oldest of the three **Abrahamic religions**, it traces its beginnings back some 4,000 years to Adam and Noah but especially to the patriarch Abraham. The **desert** plays an enormous role here because the Israeli tribes wandered in the **Negev** for 40 years before reaching **Israel**, the promised land, which is located in the desert. Later, these people moved to **Egypt** where, as **slaves**, they helped to build the **pyramids**. **Christianity** and **Islam** both evolved from Judaism.

Judean Desert is a small desert in Israel's West Bank. Here one will find the **Dead Sea** and Massada. It figures prominently in biblical history. (See **Ein Gedi**.)

Juniper is a **tree** of many species. It can be found in the **Arabian Desert**, where it is eaten by **animals** and used to make utensils (https://arabiandesert1.weebly.com/plants.html).

Kachina (katsina, tihu) is a small doll carved out of cottonwood root and then painted and decorated by Hopi artists. A kachina represents a *katsina*, or spirit (Teiwes). They vary in size, shape, and complexity and have cultural and religious significance for the **Hopi**, who also sell them commercially to tourists and collectors for as much as $1,700; an older model sold for hundreds of thousands of dollars. Note that although katsina is the correct spelling, kachina is ubiquitously used.

Kaffir ("to cover" or "infidel" in Arabic; highly insulting and offensive). Black African or non-Muslim person; the term is used in **Saudi Arabia**.

Kailash, Mount. A holy mountain in **Tibet**, sacred to Hindus and Buddhists.

Kais: Or, Love in the Deserts is an 1808 **opera** by Isaac Brandon.

Kaisut Desert. See Danakil Desert.

Kalahari Desert

> 347,000 (275,000, 193,000, 100,000) square miles | South Africa and see below | Hot desert | Minimal precipitation | The San people

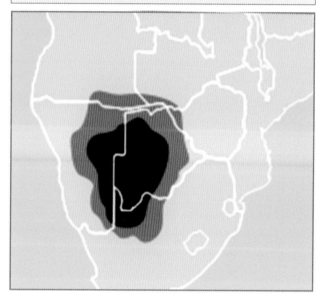

Kalahari Basin map showing distribution of three major language families. THE KALAHARI BASIN PROJECT, 2019. CREATIVE COMMONS ATTRIBUTION-SHARE ALIKE 4.0

This **desert** is not especially large (depending on the source), but it covers parts of **Botswana**, **South Africa**, and **Namibia** and is home to the San people as well as **animals** such as **antelope**, **elephant**, giraffe, **birds**, and predators that are found in preserves. Surprisingly, it **rains** here and the **pans**, at times with **water**, include the Nxai and the Deception (Michael Martin, 275). Because the Kalahari may receive more than 10 inches of rain each year, it is not, by definition, a real desert, but a desert nonetheless. (See *The Lost World of the Kalahari*.)

A southern yellow-billed hornbill in the Kalahari Desert. MATHIAS, 2009. CREATIVE COMMONS ATTRIBUTION 2.0

Children of the Kalahari. SARA ATKINS, 2005. CREATIVE COMMONS ATTRIBUTION 2.0

Alice Springs Desert Park kangaroo. PRITCHARD 16000, 2016. CREATIVE COMMONS ATTRIBUTION-SHARE ALIKE 4.0

Extent: 347,000 square miles
Environment: Hot desert
Surfaces: Plain covered with sand
People: San
Animals: Antelope, birds, elephant, and giraffe
Plants: Acacia, grass, and shrubs
Resources: Coal, copper, diamond, and nickel
(See **Gemsbok National Park, Central Kalahari Game Reserve,** and **Okavango Delta.**)

Kalut (Persian). A **yardang**: wavelike parallel ridges.

Kaluta. An Australian semelparous **marsupial** whose male dies after mating.

Kangaroo. A well-known, large, Australian **marsupial**. The newborn joey is still a fetus and must crawl up to the mother's pouch. If it falls, it dies. I recall a tribute that noted that its climb is like a blind man's ascent of Everest. Tourists like kangaroos, but farmers consider them pests.

Kangaroo rat. See **desert kangaroo rat.**

Kaokoveld. Located in the northeastern **Namib**, it is inhabited by nomadic Himba cattlemen.

Karaburan (black hurricane). In the **Taklamakan Desert**, this is an enormous **dust storm**.

Karakorum (Nubra Shyok) Wildlife Sanctuary is located in the eastern Karakorum (in Leh, **Ladakh, India**); it protects the Tibetan **antelope** (Chiru) as well as wild **yaks, snow leopards,** red foxes, wolves, and otters (https://www.rampfesthudson .com/which-animal-is-protected-in-karakoram -wildlife-sanctuary). There are also some endangered medicinal **plants** here (https://en.bharatpedia .org.in/wiki/Karakoram_Wildlife_Sanctuary). (See the "International Wildlife Refuges" sidebar.)

Karakum Canal. Built to irrigate parts of Turkmenistan, it is draining the Aral Sea (Miller, 206), which now is also polluted.

Karakum Desert (Kara Kum) (Turkestan Desert)

> 135,000 (115,800, 189,000) square miles | Turkmenistan | Cold desert | Little precipitation | Turkmen, shepherds

The Karakum ("black sand") covers most of Turkmenistan, where **oases** make life possible. **Animals** include **gazelle**, cat, **fox**, **lizard**, and **snake**. Domestically raised Karakul lamb's wool is highly prized (Nathaniel Harris, 62–63).

Karez (well) is a tunnel that brings **water** to desert areas in **China**. (See also **Qanat**.)

Karkom, Mount. The mountain that some now think is **Mount Sinai**. First, it is in **Israel** rather than in Egypt's **Sinai Peninsula**, but more importantly, at the time of the winter solstice, the sun lights up some rocks so that it appears as if they are on fire: This could be the burning bush that Moses encountered. Additionally, there are many rock carvings, some of which appear to be the tablets that Moses received, but chronological discontinuities delimit this possibility (Kershner). (See **Sinai, Mount**.)

Karoo Desert. Located northeast of Cape Town, **South Africa**, it is divided into the separated Great Karoo and Little Karoo, both of which are minuscule in contrast to the **Kalahari**, their northern neighbor.

Karst is a topographical feature characterized by dissolved limestone, dolomite, and gypsum, which results in sinkholes, **caves**, and underground **water**. (See **Nullarbor Plain**.)

Kartchner Caverns State Park. Near **Tucson**, under the **Sonoran Desert**, lies the Kartchner cave with more than 2 miles of interior **trails**. Here one will find one of the world's longest soda straw stalactites at 21 feet, 3 inches. Outside are hiking trails and wildlife.

Kasha-Katuwe Tent Rocks National Monument (northeast of **Albuquerque**). The monument's rock formations, as tall as 90 feet, resemble large tents, hoodoos with boulder caps. At Kasha-Katuwe

("white cliffs") one may see manzinita, **Indian paintbrush**, marigold, hawk, kestrel, eagle, elk, turkey, and **coyote** (brochure).

Kashgar is an **oasis** on the edge of the **Taklamakan**. (See **Hotan**.)

Katsina. See **Kachina**.

Kavir, in Iran, is a salt marsh. (See also **Sebkha**.)

Kelso Dunes is located near Baker, **California**, and is a dune field protected by the Mojave National Preserve. The dunes boom and within them one finds two endemic cricket species.

Kern National Wildlife Refuge is located near Delano, **California**, and contains desert uplands and marsh. There are **willow**, cedar, and **cottonwood** in evidence and creatures include 211 bird species including great horned and long-eared **owls**, mourning dove, ring-necked pheasant, curlew, sandpiper, raptors including golden eagle, and tens of thousands of Pacific Flyway waterbirds, plus **lizard**, kit fox, badger, and **coyote**. Flooding, pesticide incursions, and alien seeds are problems. **Hunting** is allowed (Wall, 8–14). (See https://www.fws.gov/refuge/kern.)

Khamsin is a Saharan wind in **Egypt**.

Kharga Oasis is located a few hours from **Luxor** and is the largest **oasis** in the **Libyan Desert**. Ancient **ruins** have been found, and 70,000 people live here.

Khartoum is the capital of the **Sudan**. I once met a young man who nonchalantly remarked, "When I was in Khrartoum . . ." What he failed to note is that a single woman cannot rent a hotel room there.

Khoikhoi (Khoekhoe) (formerly Hottentot, now derogatory) are a people who inhabit South Africa and Namib Desert areas. Sadly, as Russel Viljoen points out in his microhistorical account (an account that considers insignificant individuals and events rather than Khans, Churchills, and Stalins or battles at Waterloo and Gettysburg), Khoikhoi society fragmented and during the 17th, 18th, and 19th centuries some Khoikhoi left their traditional homes and became Cape-Dutch colonial society farmers' servants (1). Bad things were bound to occur, e.g., **theft** of land (79) and murder of a Khoikhoi wagon driver (45ff.) recounted in protracted detail befitting microhistorical scholarship. Note that Viljoen's study appeared in 2023, but microhistory is not a new subdiscipline. Carlo Ginzburg published *The Cheese and the Worms: The Cosmos of a Sixteenth-Century Miller* in 1976, almost half a century ago. (See also **San** and the "Indigenous Desert Peoples (International)" sidebar.)

Kibbutz is a communal village in **Israel** in which all of those who live there work for the common good. Often they are in the **Negev Desert**, and the members have managed to make the desert bloom. Kibbutzim prosper because young people from all over the world (America, Denmark, Japan) volunteer their services.

Kimberley Plateau is a large volcanic area in **Australia** prone to raging **floods** and bushfires (Nathaniel Harris, 158).

King Solomon's Mines lies about 15 miles north of **Eilat**, **Israel**. Copper ore is still mined here.

Kipuka is an island of high ground not covered by a lava flow.

Kitt Peak National Observatory is located in Arizona's **Sonoran Desert** on the Tohono O'odham Nation Reservation. It has 23 optical and 2 radio telescopes, the largest collection of optical telescopes in one location in the world. (See **Atacama Desert**.)

Kiva is an underground room used by the Puebloans for ceremonial purposes. (See https://www.youtube.com/watch?v=pqfNGmYgFxE and https://www.youtube.com/watch?v=YbbiXtCgZ-4.)

Kiva, New Mexico. ROBERT HAUPTMAN

Kofa National Wildlife Refuge. Home to about 1,000 **desert bighorn sheep** that frolic on cliffs above the **Sonoran Desert** floor, this refuge is wild and beautiful but part of it is littered with the remnants of **mining** and dangerous vertical shafts. Its 289 **plants** include **mesquite, ocotillo, ironwood, saguaro,** desert marigold, **prickly pear, teddy bear, cholla, jojoba,** and **willow.** There are 185 bird species including hawk, eagle, cactus wren, warbler, and shrike; bighorn, **mountain lion, fox, coyote, bat, rattlesnake, gecko, Gila monster,** and **desert tortoise** also make Kofa NWR their home. As with most US refuges, **hunting** is allowed (Wall, 130–38).

Koko Nur is a large, holy lake in **China.**

Kokopelli is a **kachina,** a flute-playing spirit, fertility god, prankster, and healer for the **Hopi** but also the **Zuni** and other Puebloans.

Kopit-Dag Oasis is found near the Kopit Dag mountain range in Turkmenistan.

Kraal is a South African village usually with a stockade and a location for cattle.

Krutch, Joseph Wood (1893–1970), author of the critically acclaimed *Modern Temper*, wrote two influential books on the **desert**, *The Desert Year* and *The Voice of the Desert*, both of which recount his personal, impressionistic, Thoreauvian experiences in the **Sonoran Desert.** From his *Desert Year*, we learn that "these toads who surprised me by coming from nowhere after our first big **rain** and who sang their hallelujah chorus on every side have surprised me again" (101). (Also see *The Desert* and *When the Rains Come.*)

Kufra is an **oasis** in southeastern **Libya.**

!Kung are a San people who reside in the **Kalahari Desert.** Their language includes the click sound that often precedes a word and is indicated orthographically by an exclamation mark. (See **Language.**)

Kunoth-Monks, Rosalie (1937–2022), was reputed to be the first Australian Aborigine woman to star in a **film** and an Aboriginal rights activist.

Kutch Desert Wildlife Sanctuary (Gujurat, **India**) is a saline wetland that dries up annually and produces desert. **Flamingos** breed here, and **fox** and **hyena** can be spotted. Crocodile and whale as well as invertebrate **fossils** are also in evidence (https://www.gujarattourism.com/kutch-zone/kutch/kutch-desert-sanctuary.html). (See the "International Wildlife Refuges" sidebar.)

Kutse Game Reserve adjoins the **Central Kalahari Game Reserve** and protects **lion,** leopard, **cheetah, hyena,** and **eland.** (See the "International Wildlife Refuges" sidebar.)

Kyzylkum (Kyzyl Kum) ("red sand") (116,000 [77,000] square miles) is a **cold desert** located in Uzbekistan and Kazakhstan between the Syr Darya and the Amu Darya (Arax or Oxus) **rivers** whose **waters** irrigate, which is draining the Aral Sea.

Ladakh is a contested area in **India**, Pakistan, and **China** in the **Himalayan desert**. **Water** is in short supply and so during winter, engineers create mini-glaciers that then supply towns in the spring, when water is needed to irrigate crops (https://www.youtube.com/watch?v=kptgonELj00).

Lake Chad was an enormous lake in **Chad** and Nigeria, but **climate change** has resulted in a reduction of 90 percent of its former size.

Lake Eyre is an important body of **water** about 400 miles north of Adelaide, South Australia.

Lake Mead is a large man-made lake, created by **Hoover Dam**, in southern **Nevada** that feeds **Arizona**, **California**, and Nevada. In 2007 it suffered a dramatic loss of **water** because of a nine-year **drought**, which has continued through 2023. As the water level has fallen, human bodies have been discovered.

Lake Nasser, formed when the **Aswan High Dam** was constructed, is an enormous man-made lake in **Egypt** and **Sudan**.

Lake Powell is located in **Arizona** and **Utah** and was formed by the Glen Canyon Dam. Because of the southwestern US **drought**, it continues to shrink quite dramatically.

Language. Different **desert peoples**, naturally, speak diverse and often complex languages from Aborigine Wati, **Berber**, and Arabic to Chinese and Mongolian as well as **Navaho** and **Hopi**, but there is one linguistic group that requires special mention: the click languages spoken in southern **Africa** by the Xhosa and Khoisan peoples. There are five different clicking sounds in Xhosa, and for some non-speakers they are sometimes very difficult to produce (https://www.youtube.com/watch?v=lrK-XVCwGnI). (See **!Kung** and "The 19 Pueblos' Languages" sidebar.)

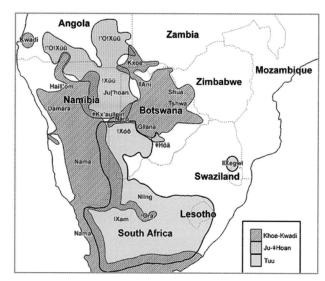

The distribution of the three major language families in the Kalahari Basin area. THE KALAHARI BASIN AREA PROJECT, 2019. CREATIVE COMMONS ATTRIBUTION-SHARE ALIKE 4.0

Lanzarote is one of the Canary Islands, dominated by **volcanoes**. Its surface is often desertlike and replete with **camels**. (See **Maspalomas Dunes**.)

Larapinta Trail runs 139 miles west from **Alice Springs**, **Australia**, through harsh, hot **desert** to Ellery Creek. It takes two to three weeks to **hike** the entire **trail**. **Water** is provided in tanks along the way; if replenishing **rain** fails to materialize, rangers refill the containers (Hutchinson).

Las Vegas is a city in **Nevada**, surrounded by **desert** and known for its casinos and high life. It is very popular and at one time was the fastest-growing city in the **US**. It consumes enormous amounts of **water**.

The Last Place on Earth: Scott and Amundsen's Race to the South Pole, Roland Huntford's biography of **Robert Scott** and **Roald Amundsen** as they attempt to reach the pole, is a gripping tale; they both achieved their goal but Scott perished.

Latitudinal belts. Most of the world's (hot) **deserts** exist between two belts located at 15 and 40 degrees: The first stretches from North Africa's **Sahara** across **Arabia** to the **Gobi** and then around to the **four major American deserts**. The second lies farther south and encompasses the **Kalahari**, **Atacama**, and Australian **outback** (Jake Page, 21).

Lava is the material ejected from a volcano's interior magma.

Lawrence of Arabia is an epic 1962 **film** directed by David Lean starring Peter O'Toole and Omar Sharif. It is highly regarded and won seven Oscars.

Lawrence of Arabia: The Battle for the Arab World is a narrated, dialogue-less 2003 **documentary** shown on PBS. It is revelatory and probably

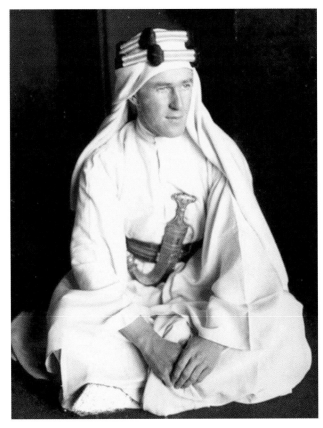

T. E. Lawrence. HARRY CHASE, 1918. PUBLIC DOMAIN

presents facts not in evidence in the David Lean extravaganza (see previous entry).

Lawrence, T. E. (Lawrence of Arabia) (1888–1935) (**Arabian**), was a British military officer and subsequently an enlisted man who fought alongside Arab tribes against the Turks. It was said that he spoke Arabic dialects like a native. His ***Seven Pillars of Wisdom*** is highly regarded. John E. Mack's *Prince of Our Disorder: A Life of T. E. Lawrence* is a massively expansive, Pulitzer Prize–winning biography. (See also **C. M. Doughty** and **Wilfred Thesiger**, among many others, and the "Desert Explorers" sidebar.)

Leave No Trace is an increasingly relevant philosophy (with seven principles) that demands that visitors to wild places including **deserts** leave them as they find them, not disturbing or removing anything and not littering or polluting. Extremists also require eliminating noise and loud colors, on tents for example. (Also see **Ethics**.)

Lava landscape in the Tularosa Valley (Basin). ROBERT HAUPTMAN

Lençóis Maranhenses National Park contains Brazil's **Sahara Desert** with its many miles of sand dunes, which, amazingly, are transformed into blue lagoons after the rainy season arrives.

Libraries. Manuscripts and **books** play a pivotal role in **deserts** across the world. The **Dead Sea Scrolls** were found in the **Negev**, and some 50,000 manuscripts of Buddhist sutras and other items were found in the **Mogao Caves** in **Dunhuang, China**. **Chinguetti**, Mauritania (in the past, a holy Islamic city), once had 30 libraries but most have disappeared in the harsh desert; the 10 remaining are usually small collections (up to 1,000 books) and have been maintained by the same families for hundreds of years. The most heartening and extraordinary account concerns the librarians and people in **Timbuktu, Mali**, who in 2012 rescued, secreted, and transported some of the 300,000 (perhaps as many as 700,000) manuscripts from al Qaeda's destructive psychosis. (See **Manuscripts, rescued**; https://newrepublic.com/article/112898/timbuktu-librarians-duped-al-qaeda-save-books; and also **Sinai**.)

Libya is a large North African country to the west of **Egypt**. The **Libyan Desert**, part of the **Sahara**, has enormous petroleum reserves.

Libyan Desert (Red Hammada)

463,000 square miles | Libya | Hot desert | Very little precipitation | Few people

The Libyan Desert is a substantial area within the northeastern **Sahara** in **Libya**. Few people live here and only at **oases**. Marco C. Stoppato points out that it is a plateau covered with reddish marl, sandstone, and limestone. Cereal, olives, dates, sheep, and **goat** can be found here as well as **rodents**, **fox**, **insects**, and **snakes** (122–23).

Libyan Sands is Ralph A. Bagnold's stunning account of innumerable trips he took into the **deserts** of northern **Africa** and across Suez to Transjordan, **Petra**, and other locations. These journeys were done in the 1920s and used **automobiles** (Ford Model Ts) for the first time. Amazingly, some 60 years later, Bagnold wrote an epilogue to his famous book.

Lichen is a symbiotic entity composed of fungi and algae. It grows on rocks and **trees** and can be green, yellow, orange, red, or brown; takes many forms; and is found in virtually all terrestrial environments. Most people ignore lichens but they are always worth admiring when wandering, especially in the **desert**.

Lilies of the Field is a 1963 **film** starring Sidney Poitier (for which he won an Oscar) in which he helps some nuns build a chapel in the **desert**.

Lima (**Peru**) was built in the **Sechura Desert**.

Literature. The enormous literature of the **desert** is bifurcate. On the one hand, there exists a scholarly corpus that covers every aspect from **climate** to **succulents**, from genomics to species ethology, textually as well as cinematically articulated in **documentaries**; on the other, there is also a creative **library** of thousands of essays, poems, and fictional works (**novels** and **films**), including *Dune*, *Holes*, and *The English Patient*, that happen to be both novels and films. (See also individual entries and the "Desert Novels" and "Desert Films" sidebars.)

DESERT NOVELS

(partial listing)

- *Blood Meridian* (McCarthy)
- *Ceremony* (Slko)
- *Dune* (Heinlein)
- *Dune Messiah* (Heinlein)
- *La femme aux trois déserts* (Bédard)
- *Holes* (Sachar)
- *In Desert and Wilderness* (Sienkiewicz)
- *Le Désert* (Loti)
- *Painted Desert* (Barthelme)
- *Quarantine* (Crace)
- *The Sheltering Sky* (Bowles)
- *Song of Slaves in the Desert* (Cheuse)
- *The Swiss Family Robinson* (Wyss)
- *Through the Desert* (Sienkiewicz)

(See https://www.goodreads.com/shelf/show/desert for an extensive list, but be wary because some of these titles are not **novels**.)

Lion is a majestic **mammal** that can be seen in **Gemsbok National Park** in South Africa's **Kalahari**. Lions travel in groups called prides and are ferocious predators.

Lion documentary. There exist innumerable **desert documentaries** and many on **lions**. *The Realm of the Desert Lion* deals with these creatures in the beautiful **Namib Desert**. (View at https://www.youtube.com/watch?v=-L9pxZxXztA.)

Lithop is a most unusual **succulent** that resembles a stone.

The Little Prince. Antoine de Saint-Exupery's children's book takes place in the **Sahara Desert**.

Little Sahara is a 1,600-acre Oklahoma **desert** resembling the **Sahara**. There is also a similarly named area in **Utah**.

Livestock. Different **animals** have differing feeding patterns and some, such as sheep, will eat **grasses** down to the roots, whereas cows, goats, and **yaks** will not. None of these creatures normally pull the roots out of the ground. But **camels** do harm the

environment in various ways, e.g., by increasing **desertification**.

The Living Desert is a 1953 Disney **film** that won an Academy Award. It presents a single day in the lives of the **animals** that inhabit southwestern US **deserts**.

Living Desert Zoo and Gardens is located in **Palm Desert**, **California**, and has more than 500 desert **animals** and 50 immersive gardens. It is considered one of the best zoos in the **US**.

Lizard is a **reptile** of many diverse species including the **Gila monster**, the exotic thorny devil, and the very peculiar blue-tongued skink. Hundreds of species can be found in the **Australian Desert**. Héctor Tejero-Cicuéndez and his colleagues observe that the diversity of desert species can be traced to the "effects of environment, time and evolutionary rate." But "the extreme species richness of the Australian deserts cannot be explained by greater evolutionary time, because species began accumulating more recently there than in more species-poor arid regions. We found limited support for relationships between regional lizard richness and environmental variables" (Tejero-Cicuéndez abstract). (See the "Reptiles" sidebar.)

Desert spiny lizard. ALAN SCHMIERER, 2008. CREATIVE COMMONS ZERO, PUBLIC DOMAIN DEDICATION

Desert locust swarm, Imililik, Western Sahara, Morocco.
MAGNUS ULLMAN, 2004. CREATIVE COMMONS ATTRIBUTION-SHARE ALIKE 3.0

Schistocerca gregaria, desert locust. MICHA L. RIESER, 2014.
CC ATTRIBUTION

Llama. One of the four new-world camels (the others are alpaca, guanaco, and vicuña), llamas can be found in the **Atacama Desert**.

Locust. The old-world desert locust swarms, darkens the sky, and destroys crops. **Precipitation**, in the form of **rain**, stimulates breeding, as occurred in the late 1980s and again in 2004 in the **Sahara** (Craig S. Smith). It is difficult for someone who has never experienced such an invasion to fully comprehend the extent and the agricultural impact it would make. Consider Viktor, an African doctoral student at the University of Pittsburgh in 1978. One day, he opened his front door, looked out, and screamed hysterically for his wife because the city was under attack by a swarm of locusts, something with which he was intimately familiar. Belatedly, he discovered that locusts do not swarm in winter in Pittsburgh. He was caught in a very powerful snowstorm that put down

more than 10 inches of **snow**! The flakes resembled locusts.

Loess is silt (sand and clay sediment) composed of quartz, mica, and other substances. This **soil** is extremely fertile.

Loma is a fog-watered **oasis** in **Peru**.

Lop Desert covers 19,000 square miles of sand at the eastern end of the **Taklamakan Desert**.

Lop Nor (Lop Nur) is a dried-up salt lake located between the Kumtang and **Taklamakan Deserts**.

Los Alamos (at ca. 7,000 feet) is located in New Mexico's **desert**. It is the birthplace of the atomic bomb (whose development also took place in New York City and Chicago).

Desert delight: grape soda lupine and bee. TERRY LUCAS. CREATIVE COMMONS ATTRIBUTION 3.0

The Lost World of the Kalahari is Laurens van der Post's classic 1958 account of his **exploration** of the **Kalahari Desert** in order to locate the Bushmen (now **San**) who were exterminated (35ff.), but eventually he located a remnant: "We had made contact at last! I was so overwhelmed by the fact that for a moment I barely knew what to do" (225). This is considered an edenic depiction of the !Kung people, now, according to Roslynn Haynes, discredited (Haynes, *Desert*, 73). A **film** of the book also exists.

Lupine. Various species (including Mojave and Arizona) of this tall purple or blue flower are found in Arizona **deserts**. The flowers grow along a stalk and can be toxic, especially to sheep (*Desert Wildflowers*, 18).

Lut. See **Iranian Desert**.

Luxor (Thebes) is a pharaonic capital city. Near it are the Luxor and Karnak Temples and the royal tombs of the Valley of the Kings and the Valley of the Queens.

1913 photo of the temple of Amen-Ra in Karnak. PUBLIC DOMAIN

Madagascar. There are some desert areas in Madagascar, but recently **climate change**, **drought**, and deforestation have transformed exceptional arable land into a dust bowl desert, the result of which is a famine (Katanich).

Mada'in Saleh (Hegra, al-Hijr) is a 2,000-year-old Saudi Arabian group of 111 tombs carved by the ancient **Nabataeans** into a rock face and very similar to **Petra**, an analogue in southwestern **Jordan**. The two sites are only about 300 miles apart (https://whc.unesco.org/en/list/1293).

Maghreb (Maghrib) ("west" in Arabic) is an area in northwestern **Africa** that includes **Algeria**, **Tunisia**, **Libya**, and Mauritania.

Mali is a country south of **Algeria**, with the **Sahara** in the north and the **Sahel** in the south. (See **Manuscripts, rescued**; **Tellem Burial Caves**; **Salt**; and **Timbuktu**.)

Mallee is an Australian eucalyptus **tree** or **shrub**.

Malpais–Valley of Fires. The 160-foot-thick Malpais ("bad land") lava flow, north of **Alamogordo**, **New Mexico**, runs along for 44 miles and is inhabited by diverse **plants**, **animals** (deer, sheep, **coyote**, bobcat), and **birds** (warbler, **owl**, eagle) (brochure).

Mammal is a class of **animals** characterized by live birth and suckling young on **milk**. All of the

Alluaudia montagnacii (devil's dagger), a rare plant found in Madagascar.

world's large land creatures as well as whales are mammals. Examples that can be found in the desert environment include mouse, **rabbit**, vicuña, **gazelle**, **oryx**, **ibex**, and **baboon**.

DESERT MAMMALS

(partial listing)

• Armadillo	• Kangaroo
• Badger	• Kangaroo rat
• Bat	• Kit fox
• Bobcat	• Mountain lion
• Camel	• Mouse (many
• Chipmunk	species)
• Cottontail	• Mule deer
• Coyote	• Pack rat
• Desert bighorn	• Peccary
• Donkey	• Ringtailed cat
• Fennec fox	• Skunk
• Grasshopper mouse	• White-footed mouse
• Jackrabbit	• Yak
• Javelina	

Mano and metate (quern) are stones used for grinding grain and seeds. The mano is held in the hand, while the metate rests on the ground.

Manuscripts, rescued. In the early 21st century, Malian librarians (especially Abdel Kader Haidara) rescued innumerable valuable manuscripts from zealous al Qaeda purifiers who were bent on destroying them. They were secreted in trunks and shipped from **Timbuktu** to Bamako in southern **Mali**. See Joshua Hammer's *Bad-Ass Librarians of Timbuktu*, which tells the story in glorious detail. (See also **Libraries**.)

Manzanar National Historic Site is the most infamous of the 10 relocation camps built to house about 120,000 Japanese Americans during World War II. It is located in the **Mojave Desert** north of Los Angeles and contained apartments, schools, a Buddhist temple, gardens, orchards, a cemetery, and so on. Similar desert camps were Heart Mountain (Wyoming), Poston (**Arizona**), and Gila River (Arizona) (handout). A visit would show that the layout is unchanged, but almost all of the 800 buildings are gone.

Manzanita. These 3-foot bushes to 30-foot **trees** with red bark and twisted branches are useful against uterine-tract infections and for sitz baths (Moore, 67–69).

Maps. Cartographic depictions of **deserts** take different forms. Some are incorporated in maps of larger areas; others cover only the desert itself. Some are **guidebooks** or trail maps, while others are general overviews. Contour maps indicate elevation alterations; road maps guide one along highways. For a person unfamiliar with a particular desert environment, a map can be life-saving. Nathaniel Harris's *Atlas of the World's Deserts* contains 8 general maps and 21 cartographic representations of specific desert environments. The National Geographic Society produces germane maps on an ongoing basis, some of which are extremely specific such as the 10-by-31-inch, vertical, encyclopedic, illustrated overview of Egypt's **Nile Valley**, which includes every city, lake, **dam**, and **pyramid**. Special mention should be made for the US government's Surface Management Status Desert Access Guides. A superb example is *California Desert District Palm Springs*, an enormous and colorful 1:100,000-scale topographic foldout map, one of 22 similar maps covering California deserts. It is divided into many hundreds of square boxes so that any location can be pinpointed. The backside is replete with localized information. The Bureau of Land Management and the Forest and National Park Services cooperate in producing these exceptional maps that are also available in digital form (e.g., http://npshistory .com/brochures/blm/ca/dag/palm-springs-1998 .pdf). (See also **Atlases**, **Books**, and **Conservation**.)

In 2007, NASA and other entities created a digital map of **Antarctica** from 1,100 Landsat satellite images. It is extremely accurate and the high-resolution images indicate both texture and color. See http://lima.usgs.gov (Leary).

Maps: history. The history of cartographic representations of the **desert** is fraught with danger because the early attempts are, naturally, inaccurate and therefore misleading. Indeed, it is well known that early travelogues (that may contain maps), such as *The Travels of Sir John Mandeville*, for example, contain descriptions that are purely imaginative (and no less bizarre than those portrayed in *Gulliver's Travels*) and therefore false. This, however, does not negate their historical relevance. Early examples include Battuta's *Rihlah* and the *Catalan Atlas*. (See https://www.oldmapsonline.org.)

Marabout is a holy man in North Africa.

Marco Polo is a Netflix series concerning the explorer's **travels** and later adventures. (See the first-season trailer at https://www.youtube.com/watch?v=TjxmgKuL7ZM and **Polo, Marco**.)

Mardu (Mardujara) are an Aboriginal people who live in western Australia's **Gibson Desert**. Like many of Australia's **indigenous peoples**, they have their own **language** and ways of doing things but there are, naturally, cultural, physical (**hunting**, gathering), religious, and mythic similarities with the many other groups despite the size of this enormous continent. The lives they led in the (distant) past are both different from and similar to the way things play out today for those who remain nomadic rather than living in permanent enclaves (with television access!). Robert Tonkinson's **ethnographic study** of the Mardu can serve as an exemplar for other groups. In brief, physical and material comforts are less important than **dreamtime**, **songlines**, and other cosmological/mythic demands. (I once saw an Aborigine sleeping in the main road that crossed the central **desert** near **Alice Springs**; although he was almost run over, I doubt he really cared.) **Food** is supplied by many **plants** and more than 50 types of **animals** (31); magic and sorcery play a role (128ff.), as does the influence of the white settlers, especially recently (160ff.). "The Mardu have lost their battle to keep the white world at bay, though for many older people 'Whitefella business' remains less important than the need to keep the Law strong" (179). In 1965 and 1967, *Peoples of the Australian Western Desert*, a series of 19 black-and-white short **films**, were shot here (197). (See also Siegfried Passarge in the bibliography for another ethnography.)

Marree Man (Stuart's Giant) is an oversize carving (a **geoglyph**) of an indigenous hunter in the **Australian Desert**. It is 1.7 miles tall within a perimeter of 17 miles. It is a mystery because no one knows how it was created or how it got there, although it is suspected that Bardius Goldberg, an artist, may be responsible. It did not exist prior to 1998 and it is visible from above. Although it is not a particularly well-known manifestation, Wikipedia nevertheless devotes an extremely long and scrupulously documented article to it (https://en.wikipedia.org/wiki/Marree_Man).

Mars. The surface of Mars is barren, desolate, and arid with some salt deposits (though there is a south polar ice cap)   a perfect **desert**.

Marshall, John (1932–2005), was a documentary filmmaker whose 30-plus **films** include *A Kalahari Family* and *The Hunters*. Some are available in part or in their entirety on Youtube. See, for example, *Bushman of the Kalahari* (https://www.youtube.com/watch?v=ocR6f1ItshA) or *The !Kung People of the Kalahari* (https://www.youtube.com/watch?v=AAJCtSKzN-E). His sister is the anthropologist Elizabeth Marshall Thomas. (See *N!ai: Story of a !Kung Woman*.)

Marsupial. Marsupials are **mammals** with pouches and include the **kangaroo**, rat-kangaroo, wallaby, and wombat, all of which live in **Australia**.

Marusthali Desert. See **Thar Desert**.

Masada ("fortress") is located in central **Israel** in the **Judean Desert**. It is a high mesa redoubt that was occupied by Jewish patriots who refused to surrender to the Roman army, which began to build an enormous ramp, still visible today, in order to reach the summit, which lies 1,500 feet above the **Dead Sea**. Eventually the ca. 1,000 people committed suicide. One may now ride a gondola to the summit.

Masdar City is a sustainable city built in the **Arabian Desert** in Abu Dhabi, **UAE**. It has not been a booming success.

Masonry. The ancient Egyptians and the Incas are famous for their stonework, the former for the massive **pyramids** and the latter for walls and Machu Pichu. Both of these peoples used finely cut stone and did not apply mortar between the impossibly tight joints. Their stone blocks were consistently symmetrical. A brochure informs readers that the Chaco Canyon Anasazi Puebloans' stonework varies. There is a jumble of quite thin mortared stones as well as symmetrical and nonsymmetrical blocks of varying sizes and shapes, at times with inner rubble. The end result, however, is always a smooth-finished facing unlike New England's stone walls that are composed of arbitrary, differently sized rocks piled upon each other to a height of about 2 feet. In all of the former cases, the walls were the goal. In New England they were merely a way of getting rid of the rock that came up (and continues to appear) when plowing, although they often served as boundaries between properties. Many other **desert peoples** built with stone (**ziggurats**) or carved **caves**, buildings, and facings out of cliff walls (**Petra**).

Maspalomas Dunes can be found on the south coast of Gran Canaria. (See **Lanzarote**.)

Mawson's Will: The Greatest Polar Survival Story Ever Written. Lennard Bickel tells the tale of Sir Douglas Mawson, who survived for weeks trekking alone across the Antarctic in **wind**, **snow**, and cold; he was thirsty, starving, and suffered from snow blindness. This account is seconded by David Roberts's excellent ***Alone on the Ice***, published 36 years after Bickel's rendition. (See also **Books** and the books sidebars.)

McGraths Flat. In this area in New South Wales, **Australia**, not far from the **Strzelecki Desert**, paleontologists have discovered a valuable deposit of 15-million-year-old fossilized **arachnids**, **insects**, **fish**, and flowers (Tamisiea).

Mecca is the holy city of **Islam**. It is located in **Saudi Arabia** and is visited by people on the **hajj**.

Medicinal plants. Many hundreds of desert plants have medicinal qualities and have been used for their curative and palliating powers by **indigenous peoples** for millennia. The southwestern **US** is especially rich in these plants (not all of which are **cactuses**) and Michael Moore's **guidebook**, *Medicinal Plants of the Desert and Canyon West*, is an invaluable **resource**. It includes overviews, maps, and line drawings of the plants, color plates, an extraordinary 20-page glossary, and an extensive "therapeutic and use index." The text is arranged alphabetically beginning with **acacia**, rolling through 68 other plants, and concluding with **yucca**. Each entry offers the same structure; therefore, the data is consistent. So, for pineapple weed, there are other names, appearance, collecting, preparation, and uses (anti-inflammatory, sedative) (86–88). Depending on the plant, the roots, branches, vines, leaves, pods, or flowers may be used to make a powder, a poultice, or especially a tea. Some of these plants are listed individually in this encyclopedia. (See the "Medicinal Plants" sidebar.)

MEDICINAL PLANTS
(partial listing)

• Acacia	• Night-blooming
• Agave	cereus
• Chaparral	• Ocotillo
• Cypress	• Passion flower
• Desert willow	• Prickly pear
• Echinacea	• Prodigiosa
• Manzinita	• Sagebrush
• Mesquite	• Sangre de drago
	• Yucca

(Moore)

Meherengarth Fort is an exquisite fort in Rajasthan, **India**. David Miller claims, and with good reason, that this complex with its seven concentric walls is "one of the finest and most impregnable fortresses in the world" (250). (See also **Jaisalmer**.)

Meheri are a people who inhabit North African **deserts**.

Memoir. Many desert **explorers**, adventurers, scientists, and others write autobiographical accounts, or memoirs. A fine example is *The Camel's Nose*, Knut Schmidt-Nielsen's memoir, in which he wrote about his time in the **Sahara** investigating camel physiology. Others are noted throughout this encyclopedia.

Mesa Verde National Park is located in Colorado and contains cliff houses including the enormous Cliff Palace, the biggest such structure in North America, as well as many archeological sites. (See **Ruins** and the "Southwest US Indian Ruins" sidebar.)

Mesopotamia. The Fertile Crescent is drying up, turning land into **desert** (Rubin).

Mesquite (tornillo) is a very common small **tree** or **shrub** replete with thorns and pods. It is antimicrobial, good against intestinal problems, and a cleansing wash; the pods are edible (Moore 73–76). Its roots can descend into the **soil** to an extraordinary depth, as much as ca. 180 feet.

Meteor Crater. Located near Flagstaff, **Arizona**, this impact crater is the first proven meteorite impact site on earth. The crater is almost a mile in diameter and more than 550 feet deep. There is a visitor center at the lip, but one is not allowed to descend (*Meteor* brochure).

Meteorites are found in **deserts** both because they have collected there for thousands of years and because the land is often open and unencumbered. Hamed Pourkhorsandi and his 10 coauthors report on a site in Iran's **Lut Desert** that yielded some 200 specimens; these chondrites (unmodified stony meteorites) are susceptible to terrestrial **weathering** (Pourkhorsandi abstract). Some 45,000 of these extraterrestrial objects have been found in **Antarctica**.

Mexican hat (coneflower). The yellow or maroon flowers droop from an extended cone. It is found in **Arizona** and **Mexico**, among other locations (*Desert Wildflowers*, 25).

Mexico is a large North American country south of the **US**. The **Chihuahuan Desert**, which partially extends into the US, is found primarily in Mexico.

Microclimate is the climate of a small area; it differs from that of adjacent areas. An example is **rain** below but **snow** at higher elevations.

Microorganisms. One of the most bizarre and incredible current scientific contentions is that a tiny amount of earth contains billions of microorganisms (**bacteria**, viruses, spores), and so it is too for barren, ostensibly lifeless **deserts**, where unknown microbes may also be found. An additional belief holds that the total weight of these entities, found on surfaces, in crevices, within animate bodies, indeed everywhere, is truly enormous. This is because there are 5 million trillion trillion, or 5 by 10 to the 30th power, bacteria on earth. Their biomass is 13 percent of the total.

Midden. In these refuse dumps, often exposed under **dunes**, archeologists discover valuable **artifacts** of the (distant) past: bones, shells, stones, and **flora**. A group of **scholars** headed by Guy Bar-Oz investigated some middens in **Negev** caravanserais along the Nabataean-Roman "Incense Route" and found "material culture attesting to wide, interregional connections, combined with archaeobotanical and zooarchaeological data illuminating the subsistence basis of the caravan trade" (Bar-Oz abstract). Note that a midden is a distant relative of the geneza, into which unusable Hebrew religious texts are slipped through a slot into a sealed room, to lie dormant forever. The famous Cairo Geniza, for example, functioned for 1,000 years and was only opened in 1896 to reveal 400,000 manuscripts.

Migration. Migrants from third-world countries attempt to reach the first world in order to better the lives of their families. In the **US**, illegal migrant crossings occur along the southwestern borders of Texas, **New Mexico**, and **Arizona**. Here, many thousands of people often get caught or lost or die in the **desert**. Travelers in North Africa cross the **Sahara** and attempt to traverse the Mediterranean in order to reach Europe, especially Greece, Italy,

Spain, and eventually countries farther north. A different type of "migration" occurs when peoples from various countries (**India**, Pakistan, **Egypt**) live temporarily in the **United Arab Emirates** and other wealthy desert lands in order to earn a living. Foreign workers comprise about 90 percent of the UAE's **population**. (View *Mojados: Through the Night*, a **documentary** that depicts four Mexicans walking 120 miles in order to cross into Texas and eventually work in the US before returning home. See the trailer at https://www.youtube.com/watch?v=OsS9CpVBgUU.) (See **Trash**.)

Milk. All **mammals** nourish their young on milk. **Indigenous peoples** milk their **camels**, **cattle**, **goats**, and sheep and drink or cook with the milk or make cheese. The Masai mix the milk with blood taken from their cattle.

Mines, abandoned. When **ore** runs out, the mining facility is often abandoned. One such example can be found deep in the **Grand Canyon**, on a narrow, unused side **trail**, where a horizontal shaft replete with tracks, ore cart, and shovel have sat unseen for decades. **Death Valley** is replete with hundreds of abandoned mines, and Lost Horse Mine is located in **Joshua Tree National Park**.

Mining, or digging underground for various things (coal, **borax**, **ore** such as **gold** or **silver**, **copper**, **salt**, marble, uranium), often takes place in desert environments. It differs from quarrying, which is done in an open-pit quarry. Some material, such as marble, is mined and quarried. Mining often results in various forms of **pollution**. (See **Bingham Canyon Mine** and **Indigenous peoples**.)

Mirage (also fata morgana) is an illusion that occurs in the **desert**; one thinks they see something (often quite complex such as a replete city), but in reality there is nothing there. Anna Sherman mentions **singing sands** (which are real) and voices (which are not) in western **China** (36). The dunes' sand particles in this area of the **Gobi** produce various sounds collectively termed singing (Miller, 108). Roaring dunes have been encountered in **Nevada**, Hawaii, **Libya**, and **South Africa**.

Mission San Xavier del Bac is located in **Tucson** within the Papago Reservation and is called the "White Dove of the Desert."

Missions. The early Spanish settlers who came up from **Mexico** to the southwestern **US** beginning in the 16th century built missions, e.g., Mission Basilica San Diego de Alcalá, manned by Jesuits and Franciscans; these were large churches that were dedicated to converting Native Americans to Catholicism. There are 21 of these scattered along the southern California coast, with additional examples in Texas, **Arizona**, and **New Mexico**.

Mistreatment. History is replete with the notoriously horrendous mistreatment of **indigenous peoples** by colonialists and those in power; examples include Australian **Aborigines**; North American Indian tribes such as the Sioux and the five tribes (Cherokee, Choctaw, Chickasaw, Creek, and Seminole) that were sent from the east to Oklahoma in what is known as the Trail of Tears; and the **San**.

Mitla Pass is a long pass in the **Sinai**. It is also the title of a 1988 Leon Uris **novel**.

Mogao Caves (Thousand Buddha Grottoes) are located near **Dunhuang** and contain countless treasures (**sculptures**, paintings) from 1,000 years of China's dynasties. (See also **Theft**.)

Mogollon (culture) are an ancient people who lived in the mountainous **desert** of the southwestern **US**. They built adobe and masonry apartment houses that contained 40 to 50 rooms arranged around a plaza (Parrott-Sheffer). They are famous for their black-on-white **pottery**. A **Three Rivers Petroglyph Site** brochure notes that the Jornado Mogollon lived in the desert and the Mimbres Mogollon resided in the **mountains**.

Mohave Museum of History and Arts is located in Kingman, **Arizona**, and is interested in the **preservation** of the heritage of northwestern Arizona.

Mojave Desert (Mohave)

| 25,000 (54,000, 15,000) square miles | SW US, California | Hot desert | Minimal precipitation | Native Americans |

Mojave and Sonoran Deserts map. PUBLIC DOMAIN

One of the **four major American deserts**, it is found in **California**, **Nevada**, **Utah**, and **Arizona** and is bounded by mountain ranges including the Sierra Nevada; it is here that **Death Valley** is located. **Plants** include **creosote bush** and **Joshua trees** (see **Yucca**); **animals** include **bighorn sheep**. **Gold**, **silver**, and iron are **mined**.

Extent: 25,000 square miles
Environment: Hot
Surfaces: Sand, gravel, and salt
People: Mojave, Cahuilla, Chemehuevi, and Serrano [91-92]
Animals: Bighorn sheep, Gila monster, mountain lion, owl, and roadrunner
Plants: Buckhorn cholla, creosote bush, Mojave yucca, and turpentine broom
Resources: Borax, copper, gold, lead, salt, silver, tungsten, and zinc

(See http://mojavedesert.net, http://digital-desert.com, and "The Four Major American Deserts" sidebar.)

Mojave Mystery: Vanished in the Desert is a two-hour 2021 television **documentary** concerning Kenny Veach. Near **Area 51** in the Nevada desert, Veach discovered a cave (the M Cave) that seemed to emanate frightening vibrations. In 2014 he returned to it and was never seen again. Conspiracy theories, naturally, abound (he staged his disappearance, he committed suicide). Four years after he disappeared, a security camera seemed to record his presence but nothing ever came of this. His relatives have various thoughts on what occurred. On the internet, he has become, what one person called, an urban legend. A number of inconclusive videos can be found on Youtube. (Take a look at https://www.youtube.com/watch?v=x-v4PsEfXL1E or https://www.youtube.com/watch?v=C3HaX74Refo.)

Mohave boy (left) and Mohave mother (right) in photos from 1907. (CREDIT UNKNOWN)

1921 photo of two Mongolian women. Their elaborate hairpieces suggest they are married to well-to-do cattlemen. PUBLIC DOMAIN

Mongolia is an Asian country in which a portion of the **Gobi Desert** can be found. Traditionally (and even today), its inhabitants were nomadic herders of **camel**, horse, sheep, **yak**, and other **animals**. Abundant **copper** and **gold** are found in the Gobi. Michael Martin points out that some men continue to use eagles to hunt for smaller game (102–3). (See **Ger**.)

Mongolian Ping Pong is a fictional 2006 **documentary** that depicts the typical life of a young boy and his family near the **Gobi Desert**.

Mongoose can be found in the **Kalahari Desert**.

Monkey. The patas monkey (wadi or hussar monkey) lives in semiarid areas in **Africa**. **Baboons** are found in the **Kalahari** and **Namib Deserts**.

Nomads Day Festival, Mongolia. GABIDEEN, 2017. CREATIVE COMMONS ATTRIBUTION-SHARE ALIKE

Monod, Théodore André (1902–2000) (**Sahara**), whom Goudie calls "the greatest French desert naturalist of the twentieth century," traveled from Algiers to **Timbuktu** and Dakar and visited the **Eye of the Sahara**. He gathered some 20,000 specimens and wrote almost 2,000 works. He was a vegetarian and cared about **animal rights**. Two of his books are *Méharées—Explorations au vrai Sahara* and *Terre et Ciel* (Goudie, *Great*, 314–16). (See **Explorers** and the "Desert Explorers" sidebar.)

Monsoon. During the summer in **Arizona**, extreme **weather** with thunderstorms, high **wind**, **flash floods**, and **heat** is called a desert monsoon.

Monte Desert (1,250 or 137,000, 177,000 square miles) is a lesser-known rain shadow **desert** in Argentina. It is hot and contains armadillo, **flamingo**, guanaco, **mountain lion**, **reptiles**, **grasses**, and **shrubs**.

Montezuma Castle National Monument is located south of **Sedona**, **Arizona**. This five-story dwelling inside an open **cave** was built 900 years ago by the **Anasazi**, specifically the **Sinagua**, who were related to the **Hohokam**. It is one of the best-preserved prehistoric structures in the Southwest (brochures).

Montezuma Well. Just 5 miles (as the crow flies) northeast of **Montezuma Castle** one finds this sinkhole filled with replenishing springwater, a source for the **Hohokam** and Sinagua peoples (brochure).

Monument Valley, located on the **Colorado Plateau** where **Arizona** and **Utah** meet, is a Navaho Tribal Park. This red-sand **desert** contains enormous and impressive sandstone buttes and mesas.

Morenci Copper Mine, southeast of Phoenix, is North America's leading producer of copper at an open pit where 2,000 people work 24 hours a day. Tours allow visitors to observe "spectacular sights" (brochure). (See https://www.atlasobscura.com/places/morenci-mine.)

Mormonism is a Christian denomination founded in the early 19th century by two Vermonters, Joseph Smith and Brigham Young, who eventually led the Latter Day Saints to **Utah**, where the **desert** played a role in its flourishing.

Morocco is a North African country to the west of **Algeria**. The **Sahara Desert** can be found in its eastern and southern regions.

Moss. Much like **lichen**, moss is a simple **plant** that covers surfaces. It comes in various shades of green and can be found in desert environments.

Mound Springs have been the only reliable permanent water sources in the arid Australian **outback** since humans first arrived in the region. (See **Great Artesian Basin**.)

Mountain lion (catamount, cougar, panther, puma). The plethora of names for this **mammal** attest to its former ubiquity and importance in the **US**. It is a powerful predator comparable to the **jaguar** and only exceeded in ferocity in the New World by the grizzly, brown, and **polar bears** and perhaps the sea lion and walrus, although these creatures are not usually found in the **desert**. It still inhabits southwestern areas and occasionally can be spotted in New England. Once in a while, as its traditional territory shrinks, it appears in urban or suburban areas such as Livermore, **California**.

Mount Sinai. See **Sinai, Mount**.

Mountains. The proximity of mountains and **oceans** influences both **desert formation** and **weather**. A limited number of oceans exist but proximate mountain ranges to deserts are uncountable. Here are noted just a few of these: Guadalupe and Organ (**Chihuahuan**), Sierra (**Mojave**), Rockies (**Sonoran**), Panamint (**Death Valley**), Atlas (**Sahara**), and Tien Shan and Altai (**Gobi**). (See **Rain shadow**.)

Mu Us Desert (Ordos) lies south of the **Gobi** (Nathaniel Harris, 70).

Native Americans selling jewelry at the Museum of New Mexico, Santa Fe. ROBERT HAUPTMAN

Mulga is an Australian acacia **tree** or **shrub** of many species. It is also a habitat and a **snake**.

Mulla mulla is an Australian **plant** with pink or lavender frond-like flowers.

Mummies. From 2003 to 2005, archeologists unearthed about 200 well-preserved, mummified bodies of people who had died some 4,000 years ago. They were buried in Small River Cemetery No. 5, located in the **Taklamakan Desert** in China's Xinjiang province, and, astonishingly, they have European features (Wade). The difference between these mummies and those discovered in Egyptian tombs is that the latter are surgically altered and chemically preserved, whereas the Taklamakan examples are merely buried in the **sand**. (See **Preservation**.)

Murgab Oasis is located in Turkmenistan; it employed early irrigation techniques.

Muscat. See **Oman**.

Museum of Indian Arts and Culture (Santa Fe) displays art and cultural material of Native Americans. It has more than 75,000 objects, including **sculptures**, baskets, **pottery**, **jewelry**, textiles, and an ancient 151-foot-long hunting net made of human hair. (Also see **Museums** and the "Museums" sidebar.)

Museum of Indigenous People (formerly Smoki Museum of American Indian Art and Culture) can be found in Prescott, **Arizona**. It collects Native American **artifacts**.

Museum of New Mexico is a group of museums and sites. Its press publishes books that deal with the **desert**.

Museum of Northern Arizona displays indigenous material and natural history specimens from the **Colorado Plateau**.

Museums, international. Like the **US**, other countries have museums dedicated to their **deserts**. Here follow a culling: **Grand Egyptian Museum (Egypt)**, Hotan Cultural Museum (**China**), Ordos Museum (**Mongolia**), Mbantua Aboriginal Cultural Museum (**Australia**), Museo de Sitio de Paracas (**Peru**), Museum of the Atacama

Desert—Huanchaca Ruins Museum (**Chile**), Museum of Egyptian Antiquities (Egypt), Museum of the Living Desert (**Niger**), National Museum of Australia (Australia), Warradjan Aboriginal Cultural Centre (Australia), Sahara Museum (**Tunisia**).

Museums, US. Scattered around the southwestern US are countless museums that deal with the desert environment including human beings and living **plants** and **animals**. A 1986 publication lists 117 for just **Arizona** and **New Mexico**, although some may be peripheral to the **desert**. (See, for example, **Arizona-Sonora Desert Museum**, **Heard Museum**, **Hi-Desert Nature Museum**, **Indian Pueblo Cultural Center**, **Mohave Museum of History and Arts**, **Museum of Indian Arts and Culture**, **Museum of Indigenous People**, **New Mexico Museum of Natural History and Science**, **Palm Springs Art Museum**, and **Yavapai Museum** as well as the "Museums" sidebar.)

MUSEUMS
(partial listing)

- Arizona-Sonora Desert Museum (Tucson)
- Arizona State Museum (Tucson)
- Desert Museum (Loiyangalani, Kenya)
- Eromanga Natural History Museum (Eromanga, Australia)
- Georgia O'Keeffe Museum (Santa Fe)
- Grand Egyptian Museum (Giza, Egypt)
- Havasupai Museum of Culture (Supai, Arizona)
- Hi-Desert Nature Museum (Yucca Valley, California)
- High Desert Museum (Bend, Oregon)
- Museum of the Atacama Desert (Antofagasta, Chile)
- The Museum of Indian Arts and Culture/ Laboratory of Anthropology (Santa Fe)
- New Mexico Museum of Natural History and Science (Albuquerque)
- Ordos Museum (Ordos, China)
- Pima Air and Space Museum (Tucson)
- Pueblo Grande Museum and Archeological Park (Phoenix)
- Yavapai Museum (Grand Canyon National Park)

Music. Unlike some other natural areas (**mountains**, **oceans**), the **desert** has not inspired very many classical composers. An exception is Verdi's **opera** *Aida*. Popular music is more generous. America's catchy "A Horse With No Name" takes place in the desert:

> You see I've been through the desert
> On a horse with no name.
> It felt good to be out of the rain.
> In the desert, you can remember your name
> 'Cause there ain't no one for to give you no pain.

As does Uncle Acid and the Deadbeats' "Desert Ceremony":

> Out of the dusty desert wind
> The ceremony will begin.
> And when the light comes round the bend
> We know the earth will meet its end.

Here you have three long-haired men jumping around and singing unintelligibly: https://www.youtube.com/watch?v=kJOkR9oQuPk. There are many others including Delta Rae's "Somewhere in the Desert," Pat Benatar's "Painted Desert," Peter, Paul and Mary's "On a Desert Island," Sting's "Desert Rose," and Rufus's "Desert Night." (See ***Déserts*** and https://www.ranker.com/list/the-best-songs -with-desert-in-the-title/ranker-music.)

Music: indigenous. Native Americans hold pow-wows at which they sing, **dance**, and **drum**. Traditionally, these are religious and spiritual ceremonies, but sometimes dancing and singing are performed for tourists. The **Navaho** and **Hopi**, among many other tribes, sing and dance as part of their cultural heritage. The Gathering of Nations pow-wow takes place in **Albuquerque** every April; some 565 American and 220 Canadian tribes take part. (See also **Festival au Desert**.)

Ethnic music is performed in the Mongolian **Gobi** (see https://www.youtube.com/watch?v=Fy-FoMtHrV2M); by the **Tuareg** in the **Sahara** (see https://www.youtube.com/watch?v=oZVs-RHeRIGk); by Arabs along with dance (see https://www.youtube.com/watch?v=d3OyuetNAyM); in

the **Kahlahari**, a tribal war chant (see https://www
.youtube.com/watch?v=fGoAtIfVQiA); and in the
same desert, the San Moon Dance (see https://www
.youtube.com/watch?v=dTL_TdONVBs).

Mustang is a wild horse. Miller indicates that it
can be found in **deserts** in Wyoming and **Namibia**
(166). I have seen herds of mustangs in the canyon
leading to the base of Boundary Peak, the Nevada
high point, and some of them may never have
encountered a human being, but, amazingly, also in
non-desert environments in Ohio and on Virginia's
high point, Mount Rogers.

Mustang Desert is a **cold desert** in the Nepalese
mountains.

The Mysterious Lands is a detailed personal
account of Ann Haymond Zwinger's many forays
into the four southwestern US **deserts**. It contains
her experiences, historical remarks, and many of
the author's excellent line drawings of flowers and
some few **animals**. The annotated **bibliography**,
though dated, is useful. **Edward Abbey** was a fan.
(See "The Four Major American Deserts" sidebar.)

Myth. All peoples, from what was once termed
primitive but now undeveloped or native, to the
most sophisticated religionist or atheist, have
mythic beliefs that they hold dear and that to one
degree or another control their lives. Not every
myth concerns dancing goddesses or vengeful spirits. Because there exist many thousands of disparate
desert peoples, tribes, groups, or kinship clans,
their myths are diverse and often noncontiguous.
The extent of myth generally can be seen in Pierre
Grimal's *Larousse World Mythology*, a ca. 600-page
compendium of mythic thinking. The landscape in
which a people are located, the **plants** and **animals**,
the **weather**, and the mode of being (gathering,
hunting, trading) all influence how myths develop,
and the superimposition of external **religion**
(e.g., **Islam** in **Africa** and **Arabia**, Buddhism and
Shamanism in **Mongolia**) alters original mythic
perception. Consider that the ancient Egyptians
obsessed about death and the afterlife; the Israelites,
like most people, creation and correct action; and

the Australian **Aborigines** about morphological
genealogy. The Bushmen, now **San**, have animal-
spirits (Grimal, 520); most North American Indians
have protective (or evil) spirit animals (449); and
it is here that Grimal notes that myth is the justi-
fication of prohibitions (453). Joseph Campbell is
more complex in his four functions of a mythology:
First, it wakens a sense of wonder; second, it fills
the cosmology with mystical import; third, it val-
idates a moral order; and fourth, it guides people
through the passages of life (8–9). (See **Religion**
and **Songlines**.)

Myth of the desert. This is very different from
the various myths and mythic thinking that can be
found in the cultures of **desert peoples** (noted in
the previous entry). Here, it is the perception of
outsiders that deserts are invariably sandy, barren,
waterless, uninhabited, harsh, and deadly. There
is some truth to all of this (some parts of some
deserts do conform to this stereotype) and that is
why the uninformed may succumb, but it is equally
true that the antitheses of each of these descrip-
tors also obtains. Claude Prelorenzo discusses the
desert myth based on cinematic depictions of the
Sahara, themes of which had first been created by
literary works such as Pierre Loti's *Le Désert*, T. E.
Lawrence's ***Seven Pillars of Wisdom***, and Paul
Bowles's ***Sheltering Sky***, among other **novels**, as
well as tourist **guidebooks** (21, 22). To the above
list of characteristics, Prelorenzo adds spectacle,
beauty, emptiness, solitude, threat, and a mystical
aspect (23, 24, 26). Subsequently, **films** such as *The
Sheik*, ***Lawrence of Arabia***, ***The Little Prince***, *The
Passenger*, ***Zabriskie Point***, *Syriana*, *Greed*, and
Beau Geste, as well as others (26–27), carried the
myth forward. This legend contrasted the purity of a
hostile desert peopled with idealistic human beings
with civilization (passim). Now, a transformed myth
has turned the desert into a tourist haven (32).

Mzab (**M'zab**) is a region in Algeria's **Sahara**.

Naadam is a yearly Mongolian festival in which young boys race horses over a 10- to 17-mile, undifferentiated course in a frenetic fashion (Miller, 193–95).

Nabataea.net is an extraordinary website that contains many hundreds of brief or longer remarks on material related to the **Nabataeans** (https://nabataea.net).

Nabataeans were an important people from the sixth century BC onward who lived in **Petra** and other locations. They were guides and traders with **China**, **India**, **Egypt**, and Rome.

Nachtigal, **Gustav** (1834–1885) (North Africa, **Sahara**), was a German surgeon who practiced in Tunis. He traveled from Tripoli to southern **Libya**, Tibesti, **Lake Chad**, and Darfur, taking readings and measuring using the many scientific instruments he carried with him. His three-volume *Sahara and Sudan* is an excellent account (Goudie, *Great*, 210–12). (See **Explorers** and the "Desert Explorers" sidebar.)

Nafud Dunes. See **Erg**.

Nagaur is a city in Rajasthan, **India**. It is an area with a forest belt of thorn scrubs that circles the **Thar Desert**.

N!ai: The Story of a !Kung Woman (1980) is a John Marshall **film** that portrays the **!Kung** as a people who have been dislocated by the government and therefore have lost their way in the modern world. (See the trailer at https://www.youtube.com/watch?v=1nLWevhitPM and *Nisa*.)

Namaqua sand grouse. David Miller points out that this unusual **bird** gathers **water** on its feathers and flies long distances back to its nest, where its young drink the water (135). Michael A. Mares observed this bird, which he calls a black-bellied sand grouse (*A Desert*, 150).

Namib Desert

ca. 50,000 (97,000, 36,000, 31,000) square miles | Namibia | Hot desert | Minimal precipitation | Himba people

Located on the western side of **Namibia**, the Namib ("vast area") is 800 miles in length and only 30 to 100 miles wide. The sun beats down here on the oldest **desert** in the world, the **Himba**, its **dunes**, cormorants, zebras, **elephants**, and even seals, as well as **spiders**, wasps, **beetles**, **ants**, and **scorpions**. Tungsten, diamonds, uranium, and **salt** are found here. The desert is expanding its reach into the **ocean**.

Extent: 50,000 square miles
Environment: Hot desert
Surfaces: Bedrock, gravel, sand dunes, and mountains
People: Himba
Animals: Antelope, ostrich, and zebra
Plants: Baobab, buffalo thorn, bushwillow, lala palm, marula tree, mopane tree, and *Welwitschia mirabilis*
Resources: Copper, gold, diamonds, lithium, uranium, salt, wildlife, and zinc

Amazing dune in the Namib Desert. PAVEL SPINDLER, 2014. CREATIVE COMMONS
ATTRIBUTION 3.0

Ostriches in the Namib Desert.
ERIC BÉZINE, 2009. CREATIVE COMMONS
ATTRIBUTION-SHARE ALIKE 3.0

Namibia drought, dying oryx.
DUMBASSMAN, 2020. CREATIVE COMMONS
ATTRIBUTION-SHARE ALIKE 4.0

(See **Fog**, **Paleoclimatology**, and **Skeleton Coast**.)

Namib-Naukluft National Park, the largest game park in **Africa**, covers part of the **Namib Desert** and the Naukluft mountain range. Its **animals** include aardwolf, **baboon**, caracal, **fox**, **gemsbok**, kudu, and leopard.

Namibia is a country, with few people, located to the east of **Botswana** and **South Africa**. Its **animals** include **elephant**, **lion**, seal, and zebra. (See **Namib Desert**, **Kalahari Desert**, and **Skeleton Coast** as well as an enticing **film** at https://www.youtube.com/watch?v=YSMNBrzJWGA.)

Nansen, Fridtjof (1861–1930) (**Arctic**). In 1888 Nansen became the first person to cross the ice cap on Greenland; he later failed to reach the North Pole. He won the Nobel Peace Prize and his books include *Farthest North*. (See **Explorers** and the "Desert Explorers" sidebar.)

Namibia rivers map. PECHCHRISTENER, 2017. CREATIVE COMMONS ATTRIBUTION-SHARE ALIKE 3.0

Napoleon I. Between 1798 and 1801, Napoleon attacked **Egypt** and Syria across the **desert**.

Nashi is a **wind** in Iran and the **United Arab Emirates**.

National parks and monuments. In 1872 the **US** created Yellowstone, the first national park in the world, and followed with additional parks. Other countries eventually did the same thing so that there are now many of these extraordinary natural environments, some of which (**Arches**, Big Bend) preserve **deserts**. Monuments are considered less significant than parks, but when attitudes alter, a monument, like **Death Valley**, becomes a park. (See **State parks** and "Guide to US National Parks" sidebar.)

Fridtjof Nansen. STEPHAN VASILIEVITSJ VOSTROTIN, 1913.

GUIDE TO US NATIONAL PARKS: DESERTS AND THE WEST

- Arches (Utah)
- Badland (South Dakota)
- Big Bend (Texas)
- Black Canyon of the Gunnison (Colorado)
- Bryce Canyon (Utah)
- Canyonlands (Utah)
- Capitol Reef (Utah)
- Carlsbad Caverns (New Mexico)
- Death Valley (California)
- Grand Canyon (Arizona)
- Great Basin (Nevada)
- Great Sand Dunes (Colorado)
- Joshua Tree (California)
- Mesa Verde (Colorado)
- Petrified Forest (Arizona)
- Saguaro (Arizona)
- Theodore Roosevelt (North Dakota)
- Wind Cave (South Dakota)
- Zion (Utah)

(https://magazine.northeast.aaa.com/daily/travel/national-park
-vacations/us-national-parks-deserts-west)

National parks and monuments: brochures. The US government produces exceptionally attractive and useful hard-copy and digital brochures for 423 national parks and 129 national monuments scattered across the country, many of which are in desert locations. They contain information, colorful illustrations, and **maps**. For example, there exists an excellent 1993 brochure for **Death Valley**, when it was a mere monument; when it became a national park, a year later, a new brochure was produced and, astonishingly, it is quite different with varying information, data, and images. Even the maps differ in some respects. But they are both magnificent, and like all such brochures are given to visitors free of charge when entering the park (although there is, of course, an entry fee). Private entities also offer free or sometimes very inexpensive brochures and handouts.

National parks and monuments: videos. Many **films** on US desert national parks and monuments are found on Youtube, e.g., **White Sands National Park** (https://www.youtube.com/watch?v=GhmRmhrs808); **Arches National Park**, an hour-long silent drive—except for the sound of the moving car (https://www.youtube.com/watch?v=1-xaEoKGIxY); **Saguaro National Park** (https://www.youtube.com/watch?v=m2bjj31HIYI); and **Great Basin National Park** (https://www.youtube.com/watch?v=cEHHG1aJnuU). (See also **Desert documentaries**.)

National Park Service app. This app has tools to explore more than 400 national parks nationwide, with interactive **maps**, tours of park places, accessibility information, and much more to plan your park adventures (https://www.nps.gov/subjects/digital/nps-apps.htm).

National Trails Systems includes El Camino Real de Tierra Adentro National Historic Trail, which was employed for 300 years as the major path between Mexico City and **New Mexico**; Santa Fe National Historic Trail, the route for American and Mexican traders; and Route 66 Corridor Preservation Program—Route 66 led from Chicago to Los Angeles partially through the **desert** (brochure).

National Training Center. See **War**.

Native Plants of Southern Nevada: An Ethnobotany. David Rhode's **guidebook** is different because it not only presents descriptions and colorful images of ca. 130 different **plants** found in parts of the **Great Basin** and **Mojave**, but also includes their names in Native American languages, their uses by **indigenous peoples**, and even historical **photographs** of the **Paiute** and **Shoshone** at work on the plants. This is an exceptional ethnobotanical volume. (See individual plants and ***Sonoran Desert Life***.)

Natural area is a geographical area (as in a city) having a physical and cultural individuality developed through natural growth rather than design or planning.

The Nature of Desert Nature is a multicultural collection of 26 often personal pieces by a diversity of authors including Homero Aridjis, Paul Dayton, and Ofelia Zepeda. The headings under which they occur tell the story: "Native Ways of Envisioning Deserts," "Desert Contemplatives," and "Desert as Art," among others (Nabhan). (See other **readers**: *the new desert reader*, *The Sierra Club Desert Reader*, and *Voices in the Desert*.)

Navaho (Navajo; Diné is the preferred usage) is the largest Native American tribe with some 250,000 members. They live in the Southwest—17.5 million acres in the **Great Basin** (Miller, 21), and some can be found in the **Grand Canyon**. The Navaho language (Diné Bizaad), of the Athabaskan group, is extremely difficult to master and so it was used during WWII by the Code Talkers to send messages that could not be deciphered by the Axis powers (Germany, Italy, and Japan). (See **Sand paintings**.)

Navaho National Monument is located in the Navaho Nation in **Arizona**. It contains three cliff dwelling sites (Betatakin, Keet Seel, and Inscription House).

Nazca Lines (Peru). There exist hundreds of enormous **geoglyphs** incised in the ground in the Nazca Desert. They form geometric shapes as well as some creatures (**spider**) and **plants** (**cactus**). They are especially visible from above, in an airplane, for example. (See **Aerial observation** and https://www.youtube.com/watch?v=RorrCkxR2tY.)

Nebkha, a coppice **dune** or dune hummock, forms around clumps of **shrubs** and **grass**.

Nebraska Sand Hills (19,000 square miles). Partially in South Dakota, these are the largest sand dune formations in the Western Hemisphere.

Negev Desert (4,700 square miles). Located in **Israel**, the Negev ("dryness") plays an extremely important role in the **Bible**. Some urban areas, such as **Beersheba**, are located here. **Bedouins** continue to inhabit the desert with flocks, while others pursue agriculture. Daniel Hillel discusses various aspects of **water** including runoff, **floods**,

Navaho men at Grand Canyon. LUIGI SELMI, 2015. CREATIVE COMMONS ATTRIBUTION 2.0

groundwater, **plants**, **animals**, and crops and indicates that there are surprising sources of water in this otherwise arid environment (passim). The Negev has an extensive human history, and Nelson Glueck offers a detailed overview stretching all the way back to the Paleolithic period and later emphasizing the **Nabataeans** (the people responsible for **Petra** in **Jordan**) who controlled water and built cisterns to contain it. Other peoples who have lived here include the Romans and Byzantines (243, 247) but also the Kenites, Kenizzites, Calebites, Yerahmeelites, and Rechabites (141).

Steven A. Rosen discusses the demise of **agriculture** and the abandonment of six towns by the 10th or 11th century in the central Negev Desert. That farming had prospered prior to this is shown by the terrace and catchment systems built to control flash flooding and the resulting water. Subsequently, **desertification** impinged: "The desert had reverted to desert" (45–46). Yael Zerubavel discusses the Negev's mythic aspect, its "symbolic landscape that represented the interweaving of time and space and was constructed in relation to the Promised Land" (13); that is, what it has meant to Israelis from biblical times (the Exodus and its many meanings [19–20]) to its early modern era and then to the present, especially in relation to the desert settlements. Recently, the state has marketed **tourism**, especially to **Eilat** and the **Dead Sea**, in order to improve economic matters (179ff.). Note

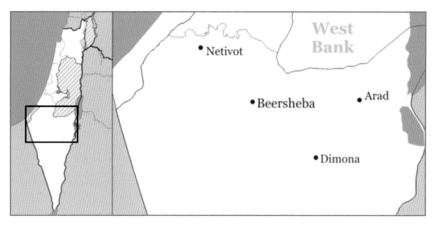

Israel outline, North Negev. YNHOCKEY, 2009. CREATIVE COMMONS ATTRIBUTION-SHARE ALIKE 3.0

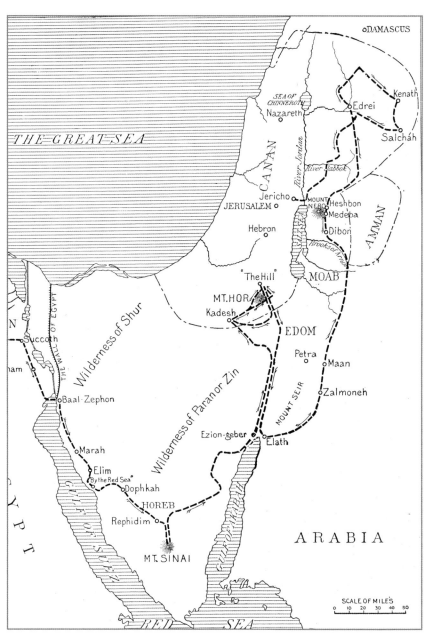

The route of the Exodus. PUBLIC DOMAIN

Horned desert viper, Israel. MINOZIG, 2016. CREATIVE COMMONS ATTRIBUTION-SHARE ALIKE 4.0

Mourning wheatear, Negev Desert, Israel. ARTEMY VOIKHANSKY, 2016. CREATIVE COMMONS ATTRIBUTION-SHARE ALIKE 4.0

that the Negev is crisscrossed by 750 hiking **trails** and contains many hundreds of **plants**. In 2018 an astonishing **flash flood** in the desert killed at least 10 people.

Extent: 4,700 square miles

Environment: Hot desert

Surfaces: Rocky and desert pavement (oldest surface on earth)

People: Bedouin (200,000), settlers

Animals: Caracal, hyena, ibex, leopard, and ostrich

Plants: Acacia, desert iris, *Pistacia atlantica*, *Thymelea ehuman*, and more than 1,000 other trees and shrubs

Resources: Copper, glass sand, granite, gypsum, and phosphate

(See **Bedouins**, **Desert archeology**, **Dunes**, **Kibbutz**, **Midden**, **Precipitation**, **Walking**, and *Where Mountains Roar*.)

Negev Desert panorama. EKSTIJN, 2016. CREATIVE COMMONS ATTRIBUTION-SHARE ALIKE 3.0

Negev Desert Botanical Garden is a planned garden in the **Negev**. (See http://negevdesertbotanical garden.org.uk, **Ein Gedi Botanical Gardens**, and individual **plants**.)

Negev rock art. Located primarily along the southwestern border of the **Negev Desert**, one will find an astonishing 200,000 carvings, some of which are 5,000 years old. There are Thamudic, Nabataean, and Arabian inscriptions, frequently accompanied by engraved elements (https://www.bradshaw foundation.com/negev/index.php).

Nejd (Najd) is an area in **Saudi Arabia**.

Nevada is a state in the western part of the **US**. The **Great Basin Desert** as well as **Death Valley** can be found here. It was an important source of **silver**, and a **museum** celebrating this is located in Carson City.

Nevada State Museum has seven branches in different locations such as Carson City and **Las Vegas**.

It presents materials representing natural and cultural history.

The new desert reader is a second collection of essays following 11 years after the Sierra Club edition. It contains 24 selections from **Edward Abbey**, Horace Greeley, Tony Hillerman, Aldo Leopold, **John Wesley Powell**, and other well-known authors, as well as Indian legends. Each entry is preceded by a substantial introduction by the editor, Peter Wild. Illustrations are scattered throughout. (See other **readers**: ***The Nature of Desert Nature***, ***The Sierra Club Desert Reader***, and ***Voices in the Desert***.)

New Mexico is a state in the southwestern **US** to the east of **Arizona**. Here one finds the Chihuahuan Desert's northern terminus. **White Sands National Park** is located within the desert's boundaries. This state is especially rich in indigenous peoples' sites. (See the "Southwest US Indian Ruins" sidebar, which will lead to individual entries.)

New Mexico landscape. ROBERT HAUPTMAN

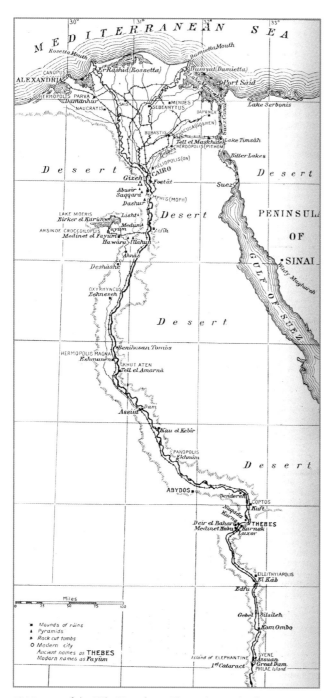

1911 map of the Nile River from Alexandria to the first cataract.
PUBLIC DOMAIN

New Mexico Museum of Natural History and Science (Albuquerque). Its collections include dinosaur **fossils** and an overview of southwestern US **geology** and natural history (brochure).

New Mexico thistle is a tall, thistle-like **plant** with a pink/magenta flower. It is found in the **Sonoran Desert** (Gerald A. Rosenthal, 177).

Ngorongoro National Park as well as **Serengeti National Park** are located in semiarid **desert** in Tanzania. Buffalo, **cheetah**, **elephant**, leopard, and rhinoceros are found here in profusion. (See the "International Wildlife Refuges" sidebar.)

Niger is a large African country in the **Sahara;** it lies to the south of **Algeria**, east of **Mali**, and north of Nigeria and is rich in uranium deposits. In its **Ténéré Desert**, archeologists have found a 5,000-year-old cemetery, formerly located near a lake, replete with human skeletons, tools, **jewelry**, and catfish bones (Fowler).

Night-blooming cereus (queen of the night) is a **cactus** that only flowers a few nights each year. It is extremely fragrant like honeysuckle and the fruit is very tasty. The root and stems produce a heart stimulant for chest pain and shortness of breath (Moore, 79–80).

Night parrot (called ghost bird by the **Aborigines**) is a **bird** previously found in the desolate Australian **outback** but until recently believed to be extinct. Now after 140 years, some 300 have been rediscovered (Ham).

Nile River. See **Nile Valley**.

Nile Valley (6,000 square miles). The mighty Nile, the longest **river** in the world, cuts a swath through Egypt's **Eastern** and **Western Deserts** (and **Sudan**), and every summer (until the **Aswan High**

Statues of Memnon at Thebes in a 1922 photo. PUBLIC DOMAIN

Dam was built) its flooding fertilized a broad plain, allowing agriculture to thrive. The 100-mile-long Nile Valley is home to many important cities and cultural **artifacts** including the **pyramids**, Dandara, Thebes, Valley of the Kings, Karnak, the Aswan High Dam, **Abu Simbel**, and **Cairo**.

Nisa: The Life and Words of a !Kung Woman is Marjorie Shostak's 1981 ethnographic account of Nisa, who lives and suffers in the **Kalahari Desert**. (See *N!ai*.)

Nitrate. See **Sodium nitrate**.

Nomad (G: Nomade; F: nomade; S: nómada; I: hirðingja; Af: nomade; H: נווד [navad]; Ar: بدوي (badawi), M: нүүдэлчин [nüüdelchin]; C: 游牧民族 [Yóumù mínzú]). Many of the peoples who have traditionally inhabited desert environments have been nomads, those who do not have fixed dwelling places. This is the case, for example, for the Saharan **Tuareg** and in the past, the **Apache**, but not for the ancient **Anasazi** or their progeny, the **Hopi**, or for the Puebloans, who built stone structures. Like the **Navaho**, Iranian nomads weave carpets. Lloyd Cabot Briggs observes that Saharan nomads raid each others' tribes and that the result is the ongoing blood feud (120ff.). In the recent past, some nomads, e.g., Australian **Aborigines**, followed the same traditions and ways of life that their distant ancestors did. Over the years, a host of Westerners have become obsessed with nomads and visited their environments: Lady Hester Stanhope, Jane Digby, **Sir Richard Burton**, **T. E. Lawrence**, **Wilfred Thesiger**, Lady Anne Blunt, Bertram Thomas, and H. Saint John Philby all traveled to or with nomadic **Bedouins** in the **Arabian Desert**, and John Ure recounts their experiences in some detail (77ff.). He does the same for those who visited with nomadic Mongols in the **Gobi** (138ff.) and the Tuareg in the **Sahara** (177ff.). Victor Englebert's *Wind, Sand & Silence* is a beautifully illustrated, oversize volume that recounts the author's **travels** with Africa's last nomadic peoples: the Tuareg, Bororo, Danakil, and Turkana. (See also **Crafts**, **Ethnobotany**, **Pastoralist**, and *Tracks* as well as *Nomadic Tribes*

of the Sahara at https://www.youtube.com/watch?v=CUHEc1fQhwo.)

Nomadic Peoples (1997–). This is a **journal** that takes as its domain the current circumstances of the world's nomadic peoples and how they will fare in the future.

North America. Parts of the western third of the US, **Mexico**, and parts of Canada are covered by **desert** or **tundra**. (See the "The World's Most Important Deserts" sidebar.)

Northern desert nightsnake can be found in the southwestern **US**. It is a smaller, nonpoisonous **reptile**.

Novels. Many works of fiction revolve around or occur in the **desert**. Examples include *King Solomon's Mines* (H. Rider Haggard), ***Dune*** (Frank Herbert), ***The Sheltering Sky*** (Paul Bowles), ***The Stranger*** (Albert Camus), *The Survivor* (Thomas Keneally), and ***Voss*** (Patrick White). (See the "Desert Novels" sidebar, which leads to individual entries.)

Nubian Desert (154,000 [163,000] square miles) is a sub-desert of the **Sahara** in the **Sudan** and Eritrea. Stoppato indicates rocky surfaces, **dunes**, and **wadis** predominate. **Wind**, called the *haboob*, rolls through here. Some people live in wadis and grow dates and wheat (152–53).

Nullah (nala) is a dry riverbed, **gully**, ravine, or **wadi** in **India**.

Nullarbor Plain is a limestone karst plain south of Australia's **Great Victoria Desert**.

Nunatak (glacial island) is the protruding top of a **mountain** surrounded by an ice field or **glacier** in the **Arctic** or **Antarctica**.

Oak Creek Canyon, south of Flagstaff, near **Sedona**, is a scenic deep gorge (similar to the **Grand Canyon**) with lush foliage. The **desert** is nearby.

Oasification is the antithesis of **desertification**. Here the eroded areas that have been turned into **desert** are resuscitated. Similarly, **Saudi Arabia** is turning its desert into arable land and is now able to export substantial amounts of wheat, fruits, and vegetables (https://www.youtube.com/watch?v=_8MimufcV64). **Israel** began a similar successful project 60 or more years ago, although some disagree that it made "the desert bloom." And **China** is building a great wall of **trees** (reforestation) to halt the Gobi's encroachment (https://www.youtube.com/watch?v=KtpaJn22w4I). (See also **Iceberg** and **Desalination**.)

Oasis is a small or occasionally larger environment in the **desert** where **water** and **plant** (and **animal**) life can be found. In the **Sinai**, many oases and underground water as well as **rain** irrigate the palms and other planted cereals, vegetables, and fruit **trees** the **Bedouins** maintain (Siliotti, 22–23). Multiple oases can be found in various deserts including the **Sahara**. The water that comes up from underground springs derives from distant rain and melted **snow** that flow downhill underground and is trapped in **aquifers**. (See also **Wadi**.)

Oasis Sanctuary. Located in Cascabel, **Arizona**, this sanctuary protects hundreds of parrots and other psittacine tropical **birds** (https://the-oasis .org). (See a short video at https://www.youtube .com/watch?v=il1v51yh1D4.)

Oasis de Huacachina, Ica, Peru; 115 people live here. DIEGO DELSO, 2015. CREATIVE COMMONS ATTRIBUTION-SHARE ALIKE 4.0

Obelisk is a monumental, man-made stone column that tapers toward its top. Examples range in size and can reach 100 feet in height. Obelisks are indigenous to **Egypt** but were removed (stolen), and of the 21 extant columns, 4 famous examples adorn New York City, Paris, Istanbul, and the Vatican. (This author has probably seen all four of them but sadly paid little attention.)

Ocean. The colder ocean, surprisingly, provides a nearby environment in which **deserts** develop. **Fog** replaces inland **precipitation**, and the interior **Skeleton Coast** of the **Namib** is a result. (See **Rain shadow**.)

Ocotillo (couchwhip) is a large **shrub** with 6- to 20-foot "viciously spined stems"; the flowers are red and abundant. It is useful against hemorrhoids, enlarged prostate, and other ailments (Moore, 81–83).

Ocucaje. In this small Peruvian **desert**, there is an extraordinary trove of marine **fossils**. The megalodon teeth and large penguin bones are quite valuable and so they are collected, but it is illegal for them to be exported and so smuggling occurs (Romero and Zarate). (See also **Gebel Kamil Crater**, **Cactus trade**, and **Wildlife trade**.)

Off-road adventure. It is possible to use one's own four-wheel-drive car or truck or rent an all-terrain vehicle (ATV) in order to penetrate into difficult or hard-to-reach desert areas. The Off-Highway Motor Vehicle Recreation Division of California State Parks offers a guide, and the Bureau of Land Management and the US Forest Service sell guides and **maps**. (See **Dumont Sand Dunes** and **Sports**.)

Oil is found in desert locations. When a well runs dry, it is often abandoned and owners are difficult to find. Sadly, these old wells continue to pollute, and the US government is attempting to seal some of them. (See **Petroleum**.)

Okavango Delta is an area in Botswana's **Kalahari** that is rich in wildlife. It is considered one of the "Seven Natural Wonders of Africa." The Moremi Game Reserve, a **national park**, is on the eastern side of the delta. (See "International Wildlife Refuges" sidebar.)

O'Keeffe, Georgia (1887–1986), is a well-known painter whose work includes large flowers but also **deserts**. She was married to the photographer Alfred Stieglitz and lived in northern **New Mexico**. A **museum** in **Santa Fe** is devoted to her work. (See also *A Brush with Georgia O'Keeffe*.)

Olgas are a group of rock domes visible from **Uluru**.

Olive trees can be found in the **Sahara** near the **Nile**.

Oman, a small country, with its capital Muscat, lies south and east of **Saudi Arabia** on the Indian Ocean. It is desertlike and the **Empty Quarter** impinges in its northern border area. Along the **ocean** are beaches where sea turtles lay eggs and **flamingos**, herons, and egrets flock in large numbers. The ancient tomb of Job is located here (Vanhoenacker).

The Opal Desert: Explorations of Fantasy and Reality in the American Southwest is a series of 13 sometimes lengthy essays by Peter Wild (editor of *the new desert reader*) on desert authors such as **Edward Abbey**, **Mary Austin**, **Joseph Wood Krutch**, and **John Wesley Powell** followed by a comprehensive eight-page bibliography.

Opera. Examples of operas that take place in a **desert** are *Aida*, *Manon Lescaut*, *Moses and Aron*, and *Kais: Or, Love in the Deserts*.

Ordos Desert lies south of the **Gobi** and east of the Yellow River.

Ore is any valuable solid material mined. **Gold**, **silver**, and lead are excellent examples. **Mining**, naturally, often occurs in the **desert**, where old, abandoned **mines** can still be found.

Oregon Dunes National Recreation Area runs along the Oregon coast near Coos Bay. It is one of the largest expanses of temperate coastal sand dunes in the world.

Organ Mountains–Desert Peaks National Monument is located near Las Cruces, **New Mexico**, and protects prehistoric, historic, geologic, and biologic **resources** in a volcanic landscape of cinder cones, lava flows, and craters.

Organ pipe cactus is a straight-line, tall **cactus** that proliferates in extraordinary abundance, as in its **national monument** (see next entry).

Organ Pipe Cactus National Monument. Located in **Arizona**, along the Mexican border in the Sonoran Desert's grasslands, this monument is replete with organ pipe cactus as well as 27 other species, including **saguaro**, plus many **birds**, **reptiles**, and **mammals**. Nearby one will find Ruby, a real ghost town (Mulvihill).

Ornithologist is a biologist who specializes in **birds**. Many people enjoy spotting and watching birds in the **desert** but also in any environment; those who become addicted to this hobby will travel thousands of miles to see an erratic, e.g., an arctic tern, far from its natural environment. There are about 1,100 bird species the **US** and almost 11,000 worldwide. This author once knew an Australian who lived in America who had some 7,000 birds on his life list.

Oryx. See **Gemsbok**.

Osmoregulation is the regulation of **salt** and **water** balance in the body, which is especially crucial in a torrid desert environment. (See also **Thermoregulation**.)

Ostrich is the largest **bird**; it is found in the **Kalahari Desert**. The local San people use its eggs to carry **water** (Miller, 139).

Oued. See **Wadi**.

Organ pipe cactus, El Pinacate y Gran Desierto de Altar, Sonora.
KYLE MAGNUSON, 2014. CREATIVE COMMONS ATTRIBUTION 2.0

Outback (ca. 2 million square miles) is the popular term that indicates Australia's large central desert areas (the Center), at the heart of which one will find **Uluru**. It is arid and barren but, nevertheless, plant and animal life abound, and for many thousands of years **Aborigines** have lived and thrived here. (See "The "10 Australian Deserts" and "The World's Most Important Deserts" sidebars.)

Owl is a **bird** of about 250 species. It has special features (large eyes, ability to turn its head 270 degrees) and hunts at night. It is often difficult to spot in the wild. Some, such as the great horned, desert, and ferruginous pygmy, inhabit desert environments.

Owl's clover. An abundance of red/purple flowers spotted with yellow surround a short stalk. It is found in **Arizona** and **California** (*Desert Wildflowers*, 22). In appearance, though shorter, it is similar to **lupine**.

Owyhee Desert is in **Nevada**, Oregon, and Idaho, and is considered the most undeveloped location in the contiguous **US**.

"Ozymandias" is the famous Shelley poem that depicts the fall of a powerful monarch and concludes, "The lone and level **sands** stretch far away."

Pacific Crest Trail. Seven hundred miles of this 2,650-mile trail roll through the **desert**. Very sadly, much of it is now devastated by **global warming**, torrid days, lack of **water**, and **fire**, which has resulted in a charred landscape.

Pack rat. These rats live in burrows, **caves**, and stick and cactus houses that they construct. They are found in the Arizona **desert**.

Painted Desert, located in **Petrified Forest National Park**, southeast of Flagstaff, contains red, orange, pink, and other colorful badlands and buttes. Some **scholars** claim that this is a fifth major American **desert**. (See "The Four Major American Deserts" sidebar.)

Paiute is a Native American tribe that lives in the **Great Basin Desert**. Others include the Washoe and Western Shoshone.

Pakistan has five **deserts**: the **Thar** plus Cholistan, Thal, Kharan, and Katpana.

Paleoclimatology. Desert climate, naturally, has altered over time; for example, a 50,000-year pollen record from the **Namib Desert** presents data that "are used to reconstruct vegetation change and quantitative estimates of **temperature** and **aridity**. Results indicate that the last glacial period was humanized by increased water availability [with more vegetation] at the site relative to the Holocene" with its high temperature and thus "increased aridity and an expansion of the Desert Biome" (Lim abstract).

Paleomagnetism is the study of magnetic fields in the earth's distant past. This is but one of innumerable *paleo*-technical terms devised by **scholars** that are tangentially relevant to desert studies. Here are a few others: paleocene, paleoclimate, paleoecology, paleolith, paleology, paleosol, paleosome, paleotype, and paleozoology.

Palm Desert is a city in **California** that can be extremely torrid: July, for example, has an average high **temperature** of 107 degrees Fahrenheit. (See **San Andreas Fault**.)

Palm Springs is an artificial California town that movie stars favor.

Palm Springs Art Museum contains Native American **artifacts** and material from the natural sciences.

Palm trees, of which there are multiple species, prefer watery areas and grow in many desert environments including **California**, the **Sahara**, and the **Negev**. (See **Date palm** and **Oasis**.)

Israel palm plantation in the southern Negev Desert. ZAIRON, 2014.

Temple of Bel, Palmyra, Syria. JAMES GORDON, 2009. CREATIVE COMMONS ATTRIBUTION 2.0

Palmyra (Tadmor) (The Bride of the Desert) is a beautiful ancient city in the **Syrian Desert**. The site is replete with baths, mausoleums, and temples. Especially noteworthy is the Sanctuary of Bel, a temple that has served various peoples (Sahner) and like many Greek and Roman structures is in disrepair, although, naturally, its majestic grandeur is still evident.

Palo verde (green stick) is an unusual desert tree. It has leaves only in the spring, and photosynthesis occurs through its green bark.

Pan is a small geological **basin**. (See **Sebkha** and **Kalahari Desert**.)

Pan de Azúcar National Park (Chile) is a nature reserve in the **Atacama Desert**. Here one will find **cactuses** and **animals** such as guanaco and **fox**, and because the **ocean** is nearby, white sand beaches, pelican, and Humboldt penguin, which is an endangered species.

Papago ("The Desert People") is an Indian tribe that inhabits the **Sonoran Desert**. In 1970, they still made use of saguaro fruit and limbs (Larson, *Deserts*, 233, 238.)

Parakeelya is an Australian succulent herb with dense, bright green, fleshy foliage. Flowers are hot pink/purple with five petals.

Paris–Dakar Rally. See **Dakar Rally**.

Pasolini, Pier Paolo (1922–1975). The **films** of this director as well as those of Michelangelo Antonioni, Claire Denis, and Wim Wenders can be oriented around the **desert**. Antonioni's *Il Deserto Rosso* (*The Red Desert*) is a good example.

Passion flower (ehumaniza, maypops) is a vine with flowers that represent Christian iconography. Helpful as a sedative and against morning sickness and hypertension (Moore, 83–85).

Passion in the Desert is a 1997 **film** that depicts an artist in 18th/19th-century **Egypt** who, lost and suicidal in the desert, befriends a leopard.

The Pastoral Tuareg: Ecology, Culture, and Society. In this exquisite oversize, two-volume set consisting of almost 900 pages, replete with hundreds of black-and-white and color images, Johannes and Ida Nicolaisen lovingly describe the Tuaregs' life, including stock breeding, culture, politics, kinship, and marriage. A ca. 1,600-item vocabulary list defines Tuareg terms, and the volumes conclude with an 800-item **bibliography**.

Pastoralist. Pastoralists are people who raise **camels**, cattle, sheep, and **goats**, sometimes as desert **nomads** and more recently as people settled in a community (Salzman, passim). Some of these nomadic herders, in the Sahara area, for example, are turning to farming. In the Ikh Nart Nature Reserve located in Mongolia's **Gobi**, there is a conflict between herders and wildlife: "For argali [wild sheep], pastoral activities decreased food availability, increased mortality from dog predation, and potentially increased disease risk." Humans and wolves also compete with each other (Ekernas abstract).

Patagonian Desert

| 300,000 (260,000) square miles | Argentina | Cold desert | Little precipitation | Tehuelche |

The Patagonian Desert is an arid, windswept, treeless plateau often covered with **lava** or **grass** and **shrubs**. **Mountain lion**, armadillo, guanaco, **rodents**, and skunk roam here (Nathaniel Harris, 128–29 et al.).

    Extent: 300,000 square miles
    Environment: Cold
    Surfaces: Gravel and lava
    People: Tehuelche and others
    Animals: Mountain lion, armadillo, guanaco, rodents, and skunk
    Plants: Cushion plant, hierba ehum, mate negre, pallo ehumani, and peppertree
    Resources: Petroleum and some minerals in Patagonia but perhaps not in the actual desert

Pavement. See **Deflation**.

Peary, Robert Edwin (1856–1920), was an important Arctic **explorer**. (See the "Desert Explorers" sidebar.)

Pecos National Historical Park. Located in the high **desert** northeast of **Albuquerque**, this monument is devoted to archeological sites, Indian pueblos, and Pecos culture, as well as an 1862 Civil War battleground.

Pelton, Agnes (1881–1961), was a visionary painter of rather abstract desert images. She had a major show at New York's Whitney Museum (*Agnes Pelton: Desert Transcendentalist*) in the spring of 2020 (Roberta Smith).

Penguin is a flightless **bird** that enjoys the frigid **waters** and bitingly cold environment of the **Antarctic**; other species prefer the warmth of a Hawaiian seashore. Tui De Roy's *Penguins: The Ultimate Guide* presents 400 photos of all 18 species. (Also see the 2005 **documentary** *March of the Penguins*, https://www.youtube.com/watch?v=L7tWNwhSocE.)

Penitentes. These tall ice formations that look like praying monks and which one finds on **mountains** and on **glaciers** also occur, amazingly, high in the **Atacama** when it **snows** (Klein).

Peoples. See **Desert peoples**.

Perennial snakeweed is a **shrub** with innumerable small yellow flowers. It is so toxic that it can kill cattle. It is found in the **Sonoran Desert** (Gerald A. Rosenthal, 125).

Perpetual ice cave. There are various locations where perpetual ice is found in **caves** or caverns, even in warm **climates**. One is near the Bandera **volcano** in New Mexico's **El Malpais National Monument**. (See https://www.ice-caves.com and https://www.youtube.com/watch?v=QJ4SDBmfHN0.)

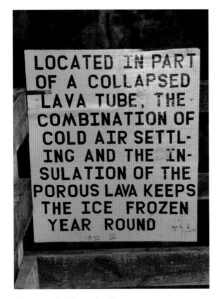

Sign explaining the ice cave phenomenon in a New Mexico cave. ROBERT HAUPTMAN

Frozen ice in a desert cave. ROBERT HAUPTMAN

Personal Narrative of a Pilgrimage to Al-Medinah and Meccah is Richard Burton's famous 1855–1856 account of his visit to these two cities. Since non-Muslims are not allowed here, he was disguised as a Pashtun on a **hajj**. Had his ruse been discovered, he would have been killed.

Peru is a country in **South America** where ca. 7,000-year-old irrigation canals were discovered in the **desert**. (See **Ica**, **Sechura**, **Peruvian Coastal Desert**, and **Nazca Lines**.)

Peruvian Coastal Desert (Tacama, Paracas, or Nazca Desert) may merely be an extension of the **Atacama Desert**. Here are history, culture, and **dunes**.

Petra, an ancient city in **Jordan** built by the **Nabataeans** more than 2,000 years ago, contains the archetypes of a "temple" ("the Monastery," "the Treasury") carved out of the side of a cliff. It is different from erected structures in **caves** because the "cave" in this case did not exist naturally. Here, there are many temples, tombs, and altars (Rodrigues). (See also **Charles Montagu Doughty** and **Mada'in Saleh**.)

Petrification. See **Dunes, petrified**.

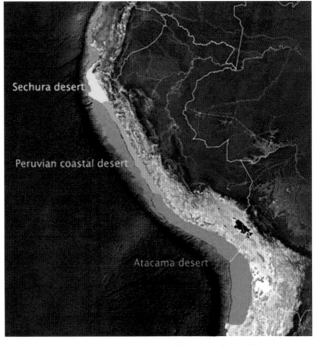

Deserts of Peru, based on ecological and geological characteristics: Sechura Desert, Peruvian Coastal Desert, Atacama Desert. HOOKERY, 2014. CREATIVE COMMONS ATTRIBUTION-SHARE ALIKE 4.0

Petrified Forest National Park, southeast of Flagstaff, contains petrified logs, **petroglyphs**, and both the **Painted Desert** and the Rainbow Forest.

Petroglyph (G: Petroglyphe; F: pétroglyphe; S: petroglifo; I: steinist; Af: rotstekening; H: פטרוגליף [petrogliph]; Ar: صخري [sakhri]; M: хадны сүг зураг [khadny süg zurag]; C: 岩畫 [Yánhuà]).

Geometric petroglyph image. ROBERT HAUPTMAN

Human petroglyph image. ROBERT HAUPTMAN

Petroglyphs are ancient incised images on rock walls. They sometimes are located in distant locations deep in **deserts** or other environments. In the Mojave Desert's Naval Air Weapons Station China Lake (**California**), not far from **Death Valley**, there are 100,000 Coso petroglyphs (human figures, dogs, bear paws, **desert bighorn sheep**, geometric patterns), some of which are 16,000 years old (David Page), and 21,000 at **Three Rivers**. Petroglyphs can also be found in the **Atacama**, **Sahara**, **Egypt**, **Jordan**, **Namibia**, **South Africa**, **Australia**, and the **Negev**. (See **Pictographs**, **Geoglyphs**, **Negev rock art**, **Rock art**, **Rock art symbols**, and the "Petroglyphs and Rock Art" sidebar.)

PETROGLYPHS AND ROCK ART
(partial listing)
- Dinosaur National Monument (Colorado, Utah)
- Hohokam petroglyphs (Arizona)
- Naval Air Weapons Station China Lake (California)
- Negev Rock Art (Israel)
- Petroglyph National Monument (New Mexico)
- Three Rivers Petroglyph Site (New Mexico)

Petroglyph National Monument (northwest of **Albuquerque**). Volcanic cones, archeological sites, **animals**, **insects**, and geometric designs, among 25,000 images, are what one will find at Petroglyph National Monument on West Mesa. Some of these carvings may be as old as 3,000 years (*Petroglyph* and brochure). (See the "Petroglyphs and Rock Art" sidebar.)

Petroleum is found in abundance in desert environments such as **Saudi Arabia** and Iran. Middle Eastern countries, which otherwise would be dirt poor, are extremely wealthy because of the **oil** that lies buried under the desert **sands**. (See **ARAMCO**.)

Philip L. Boyd Deep Canyon Desert Research Center is one of 39 sites in the University of California Natural Reserve System. Various projects dealing with **cactus**, nematodes, and wasps are undertaken here

Photographs play an inordinately important role in desert books, articles, and online sources. But keep in mind that many **indigenous peoples** do not allow themselves to be photographed because they believe it dehumanizes them or steals their souls. (Some dissenting voices claim that this is not entirely accurate.) Nevertheless, the internet is replete with, for example, Australian Aborigines, Navaho, and Hopi individuals. (See the note immediately preceding the entries.)

Phreatophyte is a **plant**, such as a **cottonwood** or **willow tree**, that has longer roots in order to draw up deeper groundwater.

Physical aspects. Excluding **plants**, **animals**, and humans, there are seven physicalities that require scrupulous consideration: **climate**, **geography**, **geology**, **geomorphology**, and **weather** including **precipitation** and **temperature**.

Namib Desert. CREATIVE COMMONS ATTRIBUTION-SHARE ALIKE 3.0

Pictographs are ancient artworks painted on cliff faces, in Ennedi, **Chad**, or **Tassili n'Ajjer**, for example. (See **Petroglyphs**.)

Piedmont is an expanse at the base of a **mountain**.

Pima are a people who live in the **Sonoran Desert**. They are descended from the ancient **Hohokam**.

Pima Air and Space Museum (Tucson) is an enormous, privately owned aerospace museum with some 400 aircraft. Their website features a series of interesting videos (https://pimaair.org/about-us/news-and-media-center).

Pincushion cactus is a small ball- or barrel-shaped **plant** with pink flowers. It is found in the **Sonoran Desert**.

Pinnacles Desert is located in Nambung National Park, **Australia**. It features limestone monoliths (up to 20 feet high) that are actually fossilized dead tree roots (Miller, 113). (See also **Dunes, petrified**.)

Plants. Many thousands of plants (**wildflowers, shrubs, cactuses, trees**) grow in desert environments. Note that not all of them are cactuses or **succulents**. They have adapted to the often-harsh environment in many ways, e.g., by having ribs to increase surfaces or by storing **water**. There exist many **books** on various countries' desert

plants. (See **Climate change, Egypt, Huntington Botanical Gardens, Medicinal plants, Survival, Wildflowers**, and individual species—only some of which are listed separately—via the "Desert Wildflowers (US)," "Most Abundant Wildflower Locations," and "Cactuses" sidebars as well as https://www.youtube.com/watch?v=oSSUA2r-cXk.)

Photo of a Hopi chief from a 1911 issue of *National Geographic*. PUBIC DOMAIN

Playa is a flat section of **desert**, once a lake. **Water** does not last long here because it evaporates. (See **Sebkha**.)

Poems concerning the **desert** are not as numerous as **novels** or **films**. Nevertheless, here follows a short list of desert poetry:
- "Crossing the Desert" by E. Richard Shipp
- "Desert" by Arthur Crew Inman
- "The Desert Flock" by Grace C. Howes
- "A Desert Memory" By Bertrand N. O. Walker
- "Desert of Arabia" by Robert Southey
- "Desert Places" by Robert Frost
- "The Lure of the Desert Land" by Madge Morris Wagner
- **"Ozymandias of Egypt"** by Percy Bysshe Shelley
- "Ozymandias of Egypt" by Horace Smith
- **"The Rime of the Ancient Mariner"** by Samuel Taylor Coleridge
- "Sand of the Desert in an Hour-Glass" by Henry Wadsworth Longfellow

(In part from https://discoverpoetry.com/poems/desert-poems.)

Poison. See **Toxins**.

Polar bear. See **Bear**.

Polar depression sounds like a weather event but it is rather a mental ailment that comes on from being trapped in the polar regions for many months or even years.

Pollution sullies the natural and human worlds and results from human activities such as industrial production, vehicle usage, and in **deserts**, petroleum extraction and **mining**, which may produce sulfuric acid, harm waterways, and last for thousands of years. Kirk Johnson describes the situation in **Nevada** (the driest state), where gold mining has wasted **water**, some 10 million gallons a day: The water is pumped from the **aquifer**, so that the mine does not flood, and resides in pools but it is contaminated. **Toxins** such as mercury, arsenic, and selenium have been found here. Indeed, Nevada releases more toxins than any other state (Harding).

Valerie L. Kuletz has written an entire monograph (*The Tainted Desert*) on nuclear pollution in Nevada. Note that the Yucca Mountain Nuclear Waste Repository will not be used. Researchers in Iran have discovered microplastics in the **Lut** and **Kavir Deserts**; these polymers arrive on the **wind** and in water runoff (Abbasi abstract). Additionally, at times smog may reach a desert environment but the desert also produces pollution in the form of **dust** that can contaminate distant cities. (See also **Dust**, **Sandstorm**, and **Wind**.)

Polo, Marco (1254–1324), was an **explorer** and trader who initiated the **Silk Road** in his quest to travel from Italy to **India** and **China** in order to discover and trade for spices. He remained in China for 17 years, at the court of Kublai Khan. His *Travels* recounts his adventures. (See **Explorers**, *Marco Polo*, and the "Desert Explorers" sidebar.)

Poor-will. Jake Page describes in great detail instant **hibernation**, an astonishing **adaption** of this amazing **bird**, which inhabits North American **cold deserts** (and neither bears the brunt of winter, as does the cardinal in the northeastern **US**, nor migrates, as the Canada goose does). It hibernates so deeply that neither heartbeat nor breath are in evidence, but when heated, it flies off; when returned to a cold environment, it immediately reverts to its hibernating state (118–19).

Population of the world's **deserts** varies dramatically depending on many factors including type, **aridity**, **oases**, population of the country, and so on. Egypt's population is increasing at such a rate that **scholars** suggest that the **Western Desert** should be reclaimed for settlement, since the Nile Valley's population is 300 times the population density of the desert. Ecological settlement will reclaim the desert and make the Bahariya oasis fully habitable (Khalifa abstract).

Pottery. Because of its nature (hardened [fired] clay, adobe), pottery and its shards have a long shelf life and so it is extremely important to archeologists, who can deduce a great deal about the **ancient desert peoples** who created the (bean, seed) pots, bowls, ollas, pipes, pitchers, statues, masks, plates, and so on. The pottery of some coeval non-desert peoples, such as the ancient Greeks and Romans, is so extraordinary that it may be almost impossible for current potters to reproduce the amphora, hydria, kraters, and so on that line the shelves in the Kunsthistorisches Museum in Vienna. (I was emotionally overcome when I saw the proliferation of these pots here more than half a century ago.) (See also **Dwellings**.)

Powell, John Wesley (1834–1902). As important as **John Charles Frémont** in the **exploration** of the western **US** (and its **geology**), Powell, with only one arm, went down the Green and **Colorado Rivers**, through the **Grand Canyon**, and explored **Utah** and **Arizona**. His books include *The Lands of the Arid Region of the United States* (Goudie, *Great*, 279–83). (See **Explorers** and the "Desert Explorers" sidebar.)

The Power of the Dog is a 2021 **film** that partially takes place in the **desert**.

Precipitation. One of the defining characteristics of a **desert** is the maximum amount of received precipitation (**rain**, **snow**), which is noted at 10 inches per

year, although in some cases more may occur but the land remains arid. The driest US cities (by average annual precipitation) are, naturally, located in the **desert**: **Yuma** (3 inches), **Las Vegas** (4.5), Bishop, **California** (5), **Palm Springs** (5.2), and so on. The driest world locations are **Africa**, **Chile** (**Atacama**) (.03 inch), South Pole (.08), Wadi Halfa, **Sudan** (**Sahara**) (.10), and so on (Burt, 105, 113, 114). It is said that it has not rained in **Antarctica** or parts of the **Atacama** in a million years. **Afforestation** (in desert plantations) is used to sequester carbon dioxide, and weather modification methods, like cloud seeding, are applied in order to counter water shortages (Branch abstract). (See also **Quillagua**, **Drought**, **Locust**, and **Negev Desert**.)

Preservation. The desert environment is arid and this dryness helps to preserve things that otherwise would fully desiccate, rot, decay, decompose, disintegrate, or crumble. This is why ancient papyri, manuscripts, and **mummies** are as well preserved as the bodies and scrolls buried by and encased in **lava** at Pompeii and Herculaneum.

Prickly pear (many varieties, including **beavertail** and silver dollar) is a well-known **cactus** whose parts (nopalitos) resemble pancakes with short

Prickly pear cactus, Tularosa Valley (Basin), Chihuahuan Desert. ROBERT HAUPTMAN

Opuntia (prickly pear) cactus heart. FACTSOFPHOTOS, 2008.
PUBLIC DOMAIN

spines. The flowers are quite colorful, and the nopalitos are edible. Parts are used as a poultice, and a diuretic, to reduce inflammation, and so on (Moore, 89–91). A jelly is made from the fruit. Doves, thrashers, finches, and woodpeckers eat its seeds (*Desert Bird Gardening*, 26).

Primrose. There exist many species of this **plant**, and even evening primrose comes in many varieties. With four white, rose, or yellow petals, they open in the evening and can carpet the **Mojave Desert**. **Indigenous peoples** ate Hooker's evening primrose seeds (Rhode, 120ff.).

Prodigiosa (bricklebush) is an "herb-bush" that may reach 3 feet. It is useful against diabetes and prevents gallstones (Moore, 94–95).

Pronghorn (antelope). The antelope designation is a misnomer. Jaeger observes that it can fly along at 30 to 40 miles per hour (50), but its numbers have been reduced by farming and **hunting** (55).

Przewalski's horse was found in the **Gobi**. It is 4 feet high and has a black mane. People are attempting to reintroduce it in the wild (https://www.deser-tusa.com/du_gobi_life.html).

Pseudo-deserts receive more than the allowable 10 inches of **rain** per year, but in all other respects they appear to be **deserts**—arid to some degree, hot, and often barren with **sand** and **dunes**. (See **South America**.)

Publishers. Naturally, many general publishing houses offer **books** on the **desert**. So too do those that specialize in coffee table (**art**) books such as Abrams. But a few publishers that emphasize the desert are worth noting: Desert Publisher; the Universities of Arizona, New Mexico, and Oklahoma Presses; Stanford University Press; and **Western National Parks Association**.

Pueblo. This is a complex and probably misunderstood word today. Long ago, the **indigenous peoples** of the southwestern **US** lived in large communities called pueblos, some of which were houses carved into cliffsides. Its older clan system was matrilineal. Today, groups of people live in 100 pueblos, e.g., **Taos**, San Ildefonso, **Acoma**, **Zuni**,

Chief of the Laguna Pueblo, 1921. PUBLIC DOMAIN

and **Hopi**, primarily in **New Mexico**. These are in a sense closed communities because if an outside woman marries a pueblo dweller and lives there, after his death she must leave. Tribes include Hopi, Zuni, Keres, and Jemez. There are currently 19 major pueblos in New Mexico and like many Indian **tribes**, they are **sovereign nations**. The Puebloans' creations are gathered together in Nancy Wood's sumptuous *Serpent's Tongue*, which contains prose, poetry, and art by a host of well-known people including Paula Gunn Allen, Franz Boas, Edward S. Curtis, D. H. Lawrence, Barry Lopez, N. Scott Momaday, and Leslie Marmon Silko. The photos and paintings are exquisite. (See **Pueblo Bonito** and "The 19 Pueblos" and "The 19 Pueblos' Languages" sidebars. Also see http://www.santafenm.info/pueblos.htm for an overview of eight northern New Mexico pueblos.)

THE 19 PUEBLOS

- Acoma
- Cochiti
- Isleta
- Jemez
- Laguna
- Nambe
- Ohkay Owingeh
- Picuris
- Pojoaque
- Sandia
- San Felipe
- San Ildefonso
- Santa Ana
- Santa Clara
- Santo Domingo
- Taos
- Tesuque
- Zuni
- Zia

(https://indianpueblo.org/new-mexicos-19-pueblos)

Pueblo Bonito is in **Chaco Canyon** and was the center of the extensive area inhabited by the Chacoan people. It contained 600 rooms and 40 **kivas**. (See https://www.youtube.com/watch?v=Y6XyuWd1foo.)

Pueblo Grande Museum and Archeological Park is a Phoenix museum that features the **Hohokam** at pre-Columbian archaeological sites and **ruins**. (See also **Arizona–Sonora Desert Museum**.)

Pueblo Indian Cultural Center, located in **Albuquerque**, is a **museum** devoted to the 19 Pueblos and their culture, history, and art as well as performances. (See "The 19 Pueblos" and "The 19 Pueblos' Languages" sidebars.)

THE 19 PUEBLOS' LANGUAGES (1979)

Tanoan Language
TIWI dialect:
- Isleta (Tuei)
- Taos (Teotho)
- Sandia (Na-fiat)
- Picuris (We-lai)

TEWA dialect:
- San Juan (O'Kang)
- Santa Clara (ka-'p-geh)
- San Ildefonso (Po-'sogeh)
- Nambe
- Tesuque (Tet-sugeh)
- Pojoaque (Po-joageh)

TOWA dialect:
- Jemez (Wala-towa)

Keresan Language
- Laguna (Ka-waik)
- Acoma (Ako-me)
- Santo Domingo (Khe-wa)
- San Felipe (Koots-cha)
- Cochiti (Ko-'chits)
- Zia (Tsia)
- Santa Ana (Tamaya)

Zuni Language
- Zuni (She-we-na)

(Wood [ascribed to Joe S. Sandos—Jemez], 16)

North wall of the Pueblo Bonito in a 1921 photo. PUBLIC DOMAIN

Puma. See **Mountain lion**.

Puna. See **Altiplano**.

Pupfish is a rare, small fish that inhabits pools in **Death Valley** and other desert locations. Devil's Hole pupfish are extremely endangered: In late 2022 there were fewer than 300 of them left.

Puquio. Built by the Nazca Indians in **Peru**, this is a trench (or tunnel) that runs slightly downhill for a long distance; it carries **water** (from **aquifers** or underground **rivers**) via gravity to a storage unit (a kocha) or directly to the irrigation canals in the fields. The puquios, whose lengths vary from 246 to 1,480 meters, are analyzed and often illustrated with **maps** and **photographs**. They enabled the ancient inhabitants of Nazca to transform the **desert** into a verdant paradise. There are still 36 functioning puquios (Schreiber 35–36, 50, 158). (See also **Qanat**.)

Purnululu National Park (Bungle Bungle), with chasms and gorges, is, according to Nathaniel Harris, a great Australian wonder (158).

Purple aster. Three feet tall, this southwestern US plant's many flowers are blue in color (*Desert Wildflowers*, 8).

Purple sage is a medium-size **shrub** with powerful purple flowers. It is used medicinally for stomach problems, colds, and sore throats, as well as in a smoke mixture for prayers (Rhode 94–95). *Riders of the Purple Sage* is Zane Grey's well-known Western **novel**.

Pushkar Mela is an annual festival and auction of camels in the **Thar Desert** (Miller, 204).

Pustynya is the Russian word for **desert**.

Puye Cliff Dwellings, west of Española, **New Mexico**, are the **ruins** of an abandoned **pueblo**, located in Santa Clara Canyon on Santa Clara Pueblo land. (See the "Southwest US Indian Ruins" sidebar.)

Pygmy owl (cactus ferruginous pygmy owl) is an endangered owl species that nests in **saguaros**.

Pyramid is a triangular-shaped building favored by the early Egyptians as a tomb for their elite, in **Giza** and **Luxor**, for example, but also by the Nubians (in **Sudan**) and Mesoamericans (Olmec, Maya, Aztec). Examples include the earliest, Sneferu's unsophisticated step pyramid, the Great Pyramid of Cholula, near Puebla, **Mexico**, and various buildings at Teotihuacan in central Mexico. These step pyramids are different, especially as exemplified in the Pyramid of the Niches. When this author visited here, the site was empty and we were the only observers as the **Voladores** flew. Pyramids lure fortune seekers and have been plundered for their immense treasures wherever they exist. The most famous of these depredations, and one which does not seem to elicit any criticism, is Howard Carter's 1922 discovery of King Tutankhamun's tomb. Some of its treasures were owned by New York's Metropolitan Museum of Art, but they were returned. Exhibitions are occasionally presented at various museums.

Qaidam is a small **desert** north of **Tibet** with salt lakes and natural gas (Michael Martin, 98).

Qanat. In the Middle East, this is an irrigation tunnel dug by hand. It carries **water** from **mountains** to distant locations. (See also **Karez** and **Puquio**.)

Qashqai are a nomadic group in Iran.

Qatar is a diminutive, oil-rich Persian Gulf country (half the size of Vermont) where part of the **Empty Quarter** is located. In 2022 it sponsored the World Cup, soccer's most important competition.

Qattara Depression is located in Egypt's **Western Desert**. Here one finds **salt pans**, sand dunes, and salt marshes.

Quarantine is Jim Crace's 1997 **novel** concerning Christ's time spent in the **desert**.

Queen of the Desert is Werner Herzog's 2015 **film** concerning the British adventurer **Gertrude Bell** starring Nicole Kidman and James Franco. (See the trailer at https://www.youtube.com/watch?v=jdbT2aU7J8Y.) Not to be confused with *The Adventures of Priscilla, Queen of the Desert*.

Quicksand. Sometimes what appears to be hard-packed sand or slightly softer material turns out to be quicksand, which can suck one in, and escape (despite instructions) can become impossible. This substance is meaningful to the author of this encyclopedia, because his wife was grasped by its claws and he was very lucky to be able to extract her.

Quillagua (Chile), in the **Atacama Desert**, is considered the driest location in the world. It never rains here. (See **Precipitation**.)

Quiver tree (kokerboom). Found in the **Namib** and **Kalahari**, this tree's branches are covered with a white substance that reflects the sun's rays and its leaves resemble daggers (Miller, 125).

Quiver trees, Namib Desert. OLGA ERNST, 2014. CREATIVE COMMONS ATTRIBUTION-SHARE ALIKE

Rabbit is a small hopping creature that lives in a burrow. A hare is larger and lives aboveground.

Races. The "4 Deserts Ultramarathon Series" of footraces take place in the **Gobi**, **Sahara**, **Atacama**, and **Antarctica**. In each race participants run 155 miles over a seven-day period and the runners are unsupported, even though **temperatures** can reach 120 degrees Fahrenheit. See two truly extraordinary videos: first, https://www.youtube.com/watch?v=BE01kvbaBcs and then of the 2018 Gobi version at https://www.youtube.com/watch?v=4c-94geeJuZA. (Also see **Sports** and **Dakar Rally**.)

The Racetrack, in **Death Valley**, is an area in which medium-size rocks (up to 600 pounds) roll or slide over the flat surface seemingly of their own accord. Apparently the **wind** is responsible.

Railway. Bahrain, Kuwait, **Oman**, Qatar, **Saudi Arabia**, and the **United Arab Emirates** formed the Gulf Cooperation Council, which decided to build a railway to link all six countries. The estimated cost is $100 to 250 billion. The project is underway (https://www.youtube.com/watch?v=oNsn3ds2op8). (See also **China-Tibet Railway**.)

Rain. See **Precipitation**.

Rain shadow is the area on the inland side of a **mountain** or range that does not receive much **precipitation**. The Sierras in **California** provide such a barrier so that on the western side there is rain and arable land and in the east there is a relief **desert**. (See **Ocean**.)

Rainbow Basin Natural Area. Located in the **Mojave**, north of Barstow, **California**, this area contains beautiful multicolored rock formations and wildlife such as the **desert tortoise** as well as **fossils** of **camels**, **flamingos**, and mastodons.

Rainbow Bridge National Monument. At 290 feet in height, this is an enormous natural bridge near **Lake Powell** in **Utah**. It is revered by the **Navaho**.

Rajasthan (Indira Gandhi) Canal system irrigates the **Thar Desert**, which now produces wheat, mustard, and cotton (Haynes, *Desert*, 19).

Rally Jameel is an international all-woman road race through the Saudi Arabian **desert**.

Ramadan is the monthlong fasting period in **Islam**. (See **Eid**.)

Ramsey Canyon Preserve, southeast of **Tucson**, is a beautiful preserve with **trails** and ca. 150 bird species, including 14 types of hummingbirds, plus reptiles and 45 mammal species. The wonderful Ramsey Canyon Preserve Check List (https://www.inaturalist.org/check_lists/95991-Ramsey-Canyon-Preserve-Check-List) presents some 600 species of **insects**, **arachnids**, **reptiles**, **birds**, and **plants** illustrated with a little **photograph**; each of these leads to additional images.

Rann of Kachchh. See **Thar Desert**.

Ras Mohammed National Park. Located at the southern tip of the **Sinai Peninsula**, this extraordinary protected preserve attracts tourists because of its 170 types of coral and 1,000 species of **fish** (Siliotti, 36–37).

Rattlesnake (rattler) is a poisonous **snake** (a pit viper), such as the diamondback, found in the southwestern part of the **US**, although its range stretches as far east as New England and as far north as central Vermont. The **Sonoran Desert** is home to many species. Like all snakes, which are **cold-blooded**, it prefers warm **temperatures**. It is infamous for making a rattling sound when disturbed or before it strikes. (See **Reptile** and the "Reptiles" sidebar.)

Raven is a crow-like creature, but larger, that preys on smaller **animals** including other **birds**. It has the reputation of being the smartest bird and is now overpopulating the **Colorado** and **Mojave Deserts**. (On Mount Denali's snowfields and **glaciers**, it will locate a buried cache of **food** by noting a flagging wand, dig it up, and crack it open, even if it

is a wooden crate.) It also may be considered an evil omen, and Edgar Allan Poe's "Nevermore"-spouting raven is rather depressing.

Rayless goldenrod (jimmyweed) is a **shrub** with many orange flowers; it is toxic to **livestock** whose **milk** can harm humans who consume it. It is found in the **Sonoran Desert** (Gerald A. Rosenthal, 128).

Readers are collections of essays, often by literary authors. See *The Nature of Desert Nature*, *the new desert reader*, *The Sierra Club Desert Reader*, and *Voices in the Desert*.

Red Desert (9,300 square miles) is a high-elevation desert in Wyoming. Here one will find colorful badlands, sandstone towers, deep canyons, and shifting sand dunes.

The Red Desert (*Il Deserto Rosso*) is a 1964 Michelangelo Antonioni **film** starring Monica Vitti

Desert wildflowers on Seedskadee National Wildlife Refuge, Wyoming. From top left and going clockwise: cowboy's delight, tufted evening primrose, prickly pear, blue penstemon, prairie pink, and desert paintbrush. TOM KOERNER, 2016. PUBLIC DOMAIN

and Richard Harris. Its critical acclaim is much vitiated by its flimsy plot held together by shots of toxic, urban wastelands and alienated people. (The author of this encyclopedia was tempted to walk out of the theater.) (See the trailer at https://www.you tube.com/watch?v=0uVPQG01JHk.)

Red Rock Canyon National Conservation Area (Nevada) is a popular location that contains colorful rock formations, **birds, tortoise, bighorn sheep**, and wild **burro**. There are also **pictographs**, some of which are 1,000 years old, but since new **roads** have opened the area to people other than serious hikers and climbers, vandals have desecrated the rock walls with graffiti. So, in 2011, parts of the area were closed (Medina). (See also https:/ /www.blm.gov/programs/national-conservation -lands/nevada/red-rock-canyon.)

Red Sea is a body of **water** that begins at the southernmost tip of **Israel**, west of **Jordan**; a second branch, between **Sinai** and **Egypt**, leads to Suez. It is biblically important because this is where the Israelites crossed the parting waters when escaping from Egypt. It is so named because **bacteria** alter its color, but it may also be because the surrounding red Eilat Mountains are reflected in the sea.

"Red Wind" is a story by Raymond Chandler that emphasizes the Santa Ana desert **wind**.

Reference books. A handful of substantial reference tools warrant mention here: Gordon L. Bender's ***Reference Handbook on the Deserts of North America***, Ramon Folch's ***Encyclopedia of the Biosphere***, and Michael A. Mares's ***Encyclopedia of Deserts***.

Reference Handbook on the Deserts of North America is Gordon L. Bender's mammoth (ca. 600 pages), data-rich overview of the major US **deserts**, the Arctic desert, and other entities. Multiple figures, tables, appendices (e.g., desert research institutes, federal experiment stations), and several indices (e.g., author, scientific, subject) make this a mandatory research tool and this despite its datedness, since it was published more than 40 years

ago. It is especially useful for extensive duplicative checklists of **plants** and **animals** including **mammals**, **birds**, **fish**, **amphibians**, and **reptiles**. It is sparsely illustrated throughout. (See also **Bibliography**, **Reference books**, and "The Four Major American Deserts" sidebar.)

Refugee. Refugees are people who are displaced by alterations in game (**animals**), agricultural practices, **precipitation** and water flow, and conflicts or **war** and are forced to move (migrate) to other locations. An excellent example are the people displaced by the decades-long war in the **Western Sahara**.

Reg is flat desert surface covered with tightly packed pebbles. (See **Desert types**.)

Registan is a sandy desert area in Afghanistan.

Religion. Three of the world's seven major religions began in desert environments: the **Abrahamic religions**, **Judaism** and **Christianity**, in Israel's **Negev** and **Islam** in the **Arabian Desert**. Bethlehem and Jerusalem are of extreme importance for the first two and **Mecca** and Medina for Islam. Moses, Jesus, and John the Baptist all had desert experiences. And even Buddhism and Hinduism have ties to the desert. All indigenous **desert peoples** have religious beliefs, rites, and rituals. Indeed, these may play a more important and influential role in their lives than analogous beliefs do in Westernized countries. David Jasper ties literature, art, and culture to Christian theology in his *Sacred Desert*. After chronicling the history and evolution of religion in ancient **Egypt**, Mesopotamia, and Syria-Palestine, Glenn S. Holland concludes *Gods in the Desert* with the following denouement: In the Roman period, "both Isis worship and Judaism at the turn of the age offered their devotees the fulfillment of the perennial religious yearnings of the peoples of the ancient Near East" (284). The desert is responsible for this need and its gratification. (See **Abstract expressionism** and especially **Myth**.)

Reno is a Nevada desert city, smaller than **Las Vegas** but still very popular with people who enjoy gambling at its many casinos. A welcoming archway proclaims it to be "The Biggest Little City in the World."

Reptile is a class, Reptilia, of **animals** that contains some 1,200 genera and more than 10,000 species of these cold-blooded vertebrates with scaly bodies, e.g., **snake**, **lizard**, crocodile, and **tortoise**, all of which are found in desert environments. There are more reptile species in central **Australia** than anywhere else (Haynes, *Desert*, 52). In mid-2022 a study published in *Nature* found that more than 2,800 species of reptiles risk extinction. Human-produced **agriculture**, development, logging, and **climate change** are responsible (Einhorn, "From Tiny"). (See the "Reptiles" sidebar.)

REPTILES
(partial listing)
- Desert crested lizard
- Fringed-toed sand lizard
- Gila monster
- Gopher snake
- Horned lizard
- Leopard lizard
- Racer (snake)
- Rattlesnake
- Side-blotched lizard
- Sidewinder
- Sonora coral
- Whiptail lizard

Research. Serious research concerning all aspects of the **desert** takes place in situ but also in academic, corporate, and government facilities. The results, often esoteric and pointed studies, are disseminated in peer-reviewed general and specialized **journals** as well as in monographs and collections. (There are more than 8 million hits in Google Scholar, although these do include "just deserts" as well as food. Articles are noted throughout this encyclopedia.) An example of a collection can be found in *Progress in Desert Research* (Berkofsky), a 1987 volume containing 21 papers that run the gamut from the simplistic (porcupine behavior in moonlight) to abstruse mathematical formulations (soil moisture), which begins with a listing of 67

symbols in Latin and Greek characters (e.g., "c volumetric heat capacity of soil," "K hydraulic conductivity of soil," "L latent heat of vaporization") complete with super- and subscripts, used in the following complex equations and without which even Stephen Hawking could not have followed nor comprehended the mathematical arguments. (Also see **Desert Research Institute** and **Jacob Blaustein Institutes for Desert Research**.)

Resources. The **desert** is rich in **gems**, minerals, **petroleum**, **antiquities**, and other beneficial materials but its resources go beyond the immediately useful and include **water**, **animals**, **birds**, **plants**, cultural **artifacts** and practices, and peoples. (See **Coober Pedy**, **Harm**, and the "Resources" sidebar.)

RESOURCES
- Atacama—copper, gold, iron, nitrates, etc.
- Sechura—copper, lead, oil, silver, etc.
- Patagonia—oil
- Great Basin/Mohave—boron, iron, tungsten, etc.
- Arabian—gold, gypsum, oil, phosphates, salts, etc.
- Karakum—aluminum, copper, gas, sulphur, zinc, etc.
- Takalmakan—molybdenum, oil, tungsten
- Gobi—coal, copper, gold, silver
- Thar—coal, gypsum, limestone, salts
- Sahara—iron, mercury, gas, oil, zinc, etc.
- Namib—uranium
- Australian—coal, copper, gold, iron, nickel, zinc, etc.
- Antarctic—oil

Resurrection plants (135 species including rose of Jericho and stone flower) are amazing **flora** that can be completely desiccated (for a century) but come back when rehydrated. They can live for more than 200 years, in the **Chihuahuan Desert**, the **Sahara**, or southern **Africa**, for example. (See https://www.youtube.com/watch?v=qK4tFicSbzc.)

Revillagigedo Archipelago (Mexico) is a **national park** located southeast of **Baja California** created to protect marine creatures such as rays, whales, and turtles (https://www.weforum.org/agenda/2017/11/mexico-has-created-the-worlds-largest-marine-park-to-protect-endangered-species). (See the "International Wildlife Refuges" sidebar.)

"The Rime of the Ancient Mariner" is Coleridge's famous poem concerning a sailor who kills an albatross and then feels guilty. It involves Arctic ice.

Rimth saltbush (*Hammada elegans*) is an undergrowth **plant** in Wadi Baba in the **Sinai** (Zaharan, 284). It is an anti-inflammatory and anti-diabetic.

Rio Grande (Rio Bravo). This enormous (1,900-mile) **river** divides **Mexico** from the **US**.

River (G: Fluss; F: fleuve; S: río; I: ánni; Af: rivier; H: נהר [nawhawr/nahar]; Ar: نهر [nahr]; M: гол [gol]; C: 河 [hé]) (and other bodies of water). קום, חצה את הירדן הזה (Arise, cross this Jordan). This biblical text indicates that the Jordan River is located in the proximity of the **Negev**. And likewise, many rivers and other bodies of **water** can be found in or near **deserts**: the Drâa, Ziz, and Guir south of Morocco's **Atlas Mountains** in the **Sahara**, where the **Nile** also flows; the Namib's **Skeleton Coast** to the east of the Atlantic Ocean; dry rivers and the **Okavango** in the **Kalahari**; **Wadi Hajr** and the Tigris in the **Arabian Desert**, which lies to the north of the Persian Gulf; the Amu Dar'ya (and the Aral Sea) in the **Kyzylkum**; Lake Tashk in the **Iranian Desert**; the Ob' in the **Arctic**; the Indus in the **Thar**; the Piura and Lake La Niña in the **Sechura**; the Loa in the **Atacama**; the Negro in Patagonia; the Fitzroy in the **Great Sandy**; **Lake Eyre** in the **Simpson**; the **Colorado**, Columbia, and **Rio Grande** in the **Mojave**, **Great Basin**, and **Sonoran**, respectively; the Amargosa River and **Badwater Basin** in **Death Valley**; the Salton Sea and the Salt and Gila Rivers in the **Sonoran**; and the Virgin and Mojave in the **Mojave**. (See https://journeyz.co/are-there-rivers-in-a-desert for a detailed explanatory discussion.)

Riyadh is the capital city of **Saudi Arabia**.

Road train (also truck train). In the Australian **outback**, enormous tandem trucks transport material. These truck trains are twice, three times, or four times as long as the typical 53-foot trailer (plus cab) that one finds in the **US** or Europe. They do not stop for wandering **Aborigines** (who sometimes lie down in the road), **automobiles**, or **kangaroos**; they do not stop at all. Sometimes their roadways are flooded, when a severe storm drenches the **desert**. This, naturally, causes a great deal of trouble.

Roadrunner (great roadrunner) is a strange-looking **bird** that thrives in the **desert** because of specific biological **adaptions**. It runs extremely fast (thus its name) and can kill a **rattlesnake**.

Roads now cross **deserts** so that sedans (rather than Land Rovers) can travel easily, though it is dangerous to do so without adequate provisions, as one can see from the abandoned vehicles that litter the desert. (See **Automobiles/trucks** and **Survival**.)

Rock art consists of **petroglyphs** and **pictographs** created by ancient peoples in most parts of the world. Polly Schaafsma discusses various ethical issues in relation to rock art in *Images and Power.* Among these are vandalism, **theft**, commercialization, obfuscation of data (v–vi), and cultural bias, reinvention, ownership, and secrecy in interpretation (2). Additionally, interpretations may differ because Western concepts of time and space clash with the mythic perceptions of **indigenous peoples**, the heirs of those who created the rock art long ago (13), and data collected by researchers tends to desanctify the spiritual aspect of the images and so sometimes secrecy is requested (20, 34). Cultural appropriation and reinvention are real problems (55), as is overvisitation of sites (70). The Bradshaw Foundation website leads to an extraordinary wealth of information on the world's petroglyphs (https://www.bradshawfoundation.com/sitemap.php). (See **Tassili n'Ajjer National Park** and **Tsodilo Hills**.)

Rock art symbols. Some rock art consists of arbitrary images such as handprints or geometric scribblings, but many **petroglyphs** are distinguishable as **animals**, humans, and images that signify or

Kokopelli, the flute player. ROBERT HAUPTMAN

symbolize. In an excellent, diminutive **guidebook**, Rick Harris presents almost 200 southwestern US symbols under rubrics such as **migration**, clans, people, spirituality, **maps**, heavens, and animals. Perhaps the most famous image is Kokopelli, the flute player. The 30 clan symbols (**bear**, hawk, black **spider**, etc.) can be abstract or look precisely like a **snake** or eagle; some squiggles may be maps of, for example, **mountains** or streams.

Rodent is a small **animal** that can often be found in desert settings. Species differ from each other as can seen by viewing a mouse, gopher, porcupine, and beaver, among many other possibilities. Rodents gnaw on things and so their teeth continue to grow. If a beaver, for example, stops cutting down **trees**, its teeth grow into the opposite gum.

Rommel, Erwin (1891–1944), was a German general, known as the Desert Fox, who operated in North Africa. He is perhaps the only Nazi military or political representative respected by the Allies.

(See the **film** *Rommel* with a trailer at https://www .youtube.com/watch?v=mzdSJT4Upjk.)

Roswell (**New Mexico**) is the purported location of a UFO crash. Conspiracists claim that Area 51 in **Nevada** is a secret government facility at which flying saucers and aliens are secreted.

Rub' al Khali. See **Empty Quarter**.

Rugs. Southwestern US Indians as well as **indigenous peoples** in North Africa, Iran, and Turkey wove by hand (and continue to weave) exceptionally beautiful rugs. At times, these can be found at very reasonable prices in Near Eastern bazaars, in Istanbul, for example. Sadly, some of these now may be mechanically produced. Rugs play an important cultural and religious role for Muslims and Mongolian **nomads**. (See **Chimayo**.)

Ruins. In various parts of the desert world, **indigenous peoples** built houses and temples and then often abandoned them. The southwestern **US** is

Ancient ruins, New Mexico. ROBERT HAUPTMAN

especially rich in these structures, some of which were built partially within **caves**. In Nepal, Bhutan, and **China** (Gaochang, Turpan), some of these ancient structures are still in use. (See https://americanswobsessed.com/indian-ruins-in-arizona, https://www.youtube.com/watch?v=Epcl6I9neLw, and the "Southwest US Indian Ruins" sidebar.)

SOUTHWEST US INDIAN RUINS
(partial listing)

- Anasazi
- Antelope House
- Aztec Ruins National Monument
- Bandelier National Monument
- Canyon de Chelly
- Casa Grande
- Chaco Canyon
- Chaco Culture National Historic Park (near Nageezi)
- Chimney Rock
- Edge of the Cedars State Park
- Gila Cliff Dwellings National Monument (near Silver City)
- Jemez National Historic Landmark

(continued)

- Mesa Verde
- Montezuma's Castle
- Moonhouse Ruin
- Navaho National Monument
- Pueblo Bonito
- Wupatki

(See also https://www.onlyinyourstate.com/utah/unbelievable-ruins-in-ut.)

INTERNATIONAL RUINS

- Casas Grandes (Paquimé) (Mexico)
- El Tajin (Mexico)
- Luxor (Egypt)
- Masada (Israel)
- Palmyra (Syria)
- Petra (Jordan)

Ruins of Desert Cathay is Marc Aurel Stein's 1912 two-volume overview of his trip to and discoveries in Turkestan's **desert**.

Russian thistle (tumbleweed) is a **plant** that eventually disassociates from its roots and that the **wind** causes to roll around in the **desert**.

Sabkha. See **Sebkha**.

Sacred datura (thorn-apple) is a southwestern US, night-blooming, white **plant** with 6-inch trumpet-like (lavender) flowers. It is toxic and can cause a rash or death. **Indigenous people** used it to produce visions (*Desert Wildflowers*, 11).

Safari is a trip that people take in order to view or hunt **animals** in **Africa** and its **deserts**. It has expanded to the Arabian Desert area and, indeed, any other global location.

Sagebrush (chamiso), a well-known **plant**, grows 3 to 4 feet high, and has a rank odor. It is effective against diaper rash and chafing and as a disinfectant. It can be burned to purify the air (Moore 103–4). It puts down roots as deep as 75 feet and is ubiquitous in the **Great Basin Desert**. Big sagebrush, according to David Rhode, was an exceptionally important medicinal plant drunk as tea or chewed for stomach pain, headache, sore eyes, etc., and used to make textiles, rope, and **clothing** (61ff.).

Saguaro (suh *waa* ro) (candelabra) is an extraordinary large and long-lived (ca. 200 years) **succulent** (**cactus**) often with many arms, whose height ranges from 10 to 52 feet. Saguaros are found in the **Sonoran Desert** among other locations. White flowers that bloom at night result in red fruit that along with seeds, nectar, and pollen attract doves, mockingbirds, oriels, woodpeckers, and **owls**, some of which nest in the trunk (*Desert Bird Gardening*, 10). It is possible that saguaro mortality can be linked to browning (a change in the color of the plant's visible tissues) and this, in turn, can be ascribed to one or more of the 40 Sonoran species of **termites** that have now, for the first time, been found on the visible members of the saguaro (Castellanos, 172, 177). A small moth carries a bacterium that causes necrosis in the **plants** (Larson, *Desert*, 128–29). The excessive **heat** during the summer of 2023 harmed and killed some of these iconic cacti. (See **Cactus trade** and **Papago**.)

Saguaro-Juniper Covenant: A Bill of Rights for Human Occupancy and the Private Governance of Wildlands is an agreement among corporations to protect the lands under their control.

Saguaro National Park is located on either side of **Tucson** and contains many examples of the mighty **saguaro** as well as other **plants**, ca. 50 cactus species, **birds**, **animals**, Hohokam **petroglyphs**, and 128 miles of **trails**. Many videos are available on Youtube; here is one: https://www.youtube.com/watch?v=DevwfwfjRoE. (See **Tohono O'odham** and the "Petroglyphs and Rock Art" sidebar.)

Sahara. Based on a Clive Cussler **novel**, this 2005 adventure **film** concerning treasure hunters and environmentalists stars Matthew McConaughey and Penélope Cruz.

Sahara and Sudan is Gustav Nacthigal's 1881 three-volume account of his six-year journey in these areas. Originally published in German, it has been translated into a number of languages.

The sandy Sahara Desert on the border of Morocco and Algeria. DR. ONDREJ HAVELKA, 2021. CREATIVE COMMONS ATTRIBUTION-SHARE ALIKE 4.0

Sahara Desert

ca. 3,500,000 square miles | North Africa | Hot desert | Very little precipitation | Berber, Tuareg

Note: Some countries within the Sahara are also in its southern neighbor, the **Sahel**.

Sahara means "deserts" in Arabic, and it is the largest and perhaps the single most famous—that is, well-known—**desert** in the world. It also happens to be partially a **white-sand desert** with **dunes**, unlike, say, the **Negev** or **Mojave**, which have **scrub** growing in places, but it also can be rocky, mountainous, rather sandless, and with salt flats despite its many **oases** and **wadis**. Very long ago and then again around 14,000 BC, and for many centuries, it was lush and green. Now, it is often devoid of **clouds** and very hot and windy, and therefore little vegetation is in evidence. The variation in **temperature** from extreme daytime **heat** to colder nights is typical of desert environments. Judith Scheele and James McDougall insist that the Sahara is much more than an empty and ecologically and politically devastated expanse. Instead it has a variety of **languages**, cultures, societies, and livelihoods plus sophisticated **agriculture** and complex transport systems; rather than a barrier, these authors see it as a connector, a commonality (4).

In addition to **animal** and human bones, whale bones have recently been discovered here, which indicates that an **ocean** once inundated much of the area. It covers a vast swath of **Algeria** and other North African countries and is encroaching on Sahel lands such as **Chad**, despite the **Nile River**, which runs through it. It is sometimes divided into western, central, and eastern sections. **Acacia, antelope**, and **gazelle** are some of its life forms. The nomadic **Tuareg**, among millions of people, live (in **oases**) and travel here. *The Sahara*, Jeremy Swift's general overview, is an excellent place to begin the study of this enormous desert; the book is very generously illustrated, and the many images of rock surfaces (27ff.) and **rock art** (53ff.) are especially invaluable.

Extent: The Sahara is the world's largest desert; it covers an enormous tract of land that measures about 1,000 by 3,000 miles, so it is at least as large as the continental **US**. Within its boundaries are eight North African countries and many designated sub-deserts including the **Great Eastern Erg** and the **Libyan**.

Environment: Thousands of years ago, the Sahara was fertile with some arable land (Briggs, 34), but this altered and the landscape became sandy, gravely, or rocky.

Surfaces: Gravel, sand, and rock

People: Many indigenous peoples including Bedouins (Taureg, Teda) inhabit the Sahara.

Animals: Gazelle, fox, rodent, lizard, and ostrich (now exterminated) and domestic camel, cattle, horse, sheep, and goat

Plants: Acacia, grasses, date and doum palm, shrubs, trees (olive, cypress), and thyme

Resources: Iron ore, petroleum, phosphate, and uranium

(See **Climate change**, **Desertification**, **Gobi Desert**, **Salt**, *Sahara Overland*, **Slavery**, its many sub-deserts via the "Sahara Sub-Deserts" sidebar, and the "Countries in the Sahara" sidebar.)

SAHARA SUB-DESERTS
(Some are questionable.)

- Adrar of the Iforas
- Aïr Massif
- Chalbi
- Danikil
- Darfur
- Eastern—Egyptian
- Ennedi
- Erg ehum
- Fezzan
- Great Eastern Erg
- Great Western Erg
- Hoggar
- Kalahari
- Libyan
- Mauritania
- Namib
- Nubian
- Sahel
- Sinai
- Somalia
- Tanezrouft
- Tassili n'Ajjer
- Ténéré
- Tibesti
- Western—Egyptian
- Western Sahara

(Stoppato, 102ff.)

COUNTRIES IN THE SAHARA
(Some are also in the Sahel.)

- Algeria
- Egypt
- Libya
- Mauritania
- Morocco
- Sudan
- Tunisia
- Western Sahara

Sahara Desert Project is an amateur but nevertheless useful website that contains information on **animals**, **plants**, diseases, and other topics (https://saharadesertproject1.weebly.com).

Sahara: The Forbidden Sands is a large-format volume of exquisite **photographs**. Interspersed among them are short essays by eight authors including **Théodore Monod** (Durou, passim).

Sahara mustard (and other mustards) is a bush-like **plant** with diminutive yellow flowers. It is found in the **Sonoran Desert** (Gerald A. Rosenthal, 147).

Sahara Overland: A Route and Planning Guide is an astonishingly replete guide to traveling through and across the **Sahara Desert**, a venture, even with a tour and guide, not to be taken lightly. A third of this book's 544 pages are allocated, in excruciating detail, to cars and motorcycles and legitimately so: Your life depends on a dependable vehicle and your knowledge of what to do when it breaks down 1,000 miles from anywhere because you forgot to take along extra oil or avoid land mines and bandits. Precisely explicit itineraries for nine countries provide essential information for the independent traveler (Chris Scott). Had I had a book such as this when I crossed **Death Valley**, the **Negev**, and the **Australian Desert**, things would have gone more smoothly and we would have been considerably safer; e.g., our second Death Valley hitchhiker might not have gotten lost and been forced to spend the night alone in the dark, cold desert and with only an emergency aluminum blanket to comfort him. (See **Guidebooks** and **Walking**.)

Saharan art is different than that of the **Sahel**. *Caravans of Gold, Fragments in Time: Art, Culture, and Exchange across Medieval Saharan Africa*, a 2020 show organized by Northwestern's Block Museum of Art, takes into account the influence of **Islam**, something that a simultaneous exhibit in New York does not (Cotter). (Also see **Sahel sculpture**.)

Saharan tribes. Of the many tribal peoples who inhabit the **Sahara**, some are nomadic (e.g., **Tuareg**) while others are sedentary (e.g., Dauada). Many agricultural enclaves are also inhabited by smiths (*maalmin* in Arabic, *duti* to the **Teda**, and *enaden* to the Tuareg). They are iron, **copper**, and brass workers who live outside the tribal society physically and spiritually (Briggs, 70–71). Surprisingly, some tribes had **slaves**. (See Lloyd Cabot Briggs's detailed and generously illustrated *Tribes of the Sahara* and the "Saharan Tribes" sidebar.)

SAHARAN TRIBES

- Chaamba
- Chouchan
- Dauada
- Fezzanese
- Haratin
- Imraguen (Hawata)
- Jew
- Mozabite
- Nemadi
- Teda
- Senoussi
- Tidjani
- Tuareg

Sahel. (*Note:* Some countries in the Sahel are also in the **Sahara**.) The Sahel ("coast," "shore") (1,900,000 square miles) is the semiarid, transitional area south of the Sahara stretching from Senegal through Burkina Faso, Cameroon, **Chad**, the Gambia, Guinea, Mauritania, **Mali**, **Niger**, and Nigeria to the **Sudan**. All of these lands have **deserts** and are sorely affected by **desertification** through which the Sahara is moving southward. For centuries, the Sahel has suffered from severe **droughts**: During the 1970s, 1980s, and again in 2012, there were horrendous droughts here that resulted in animal loss and death. (See Salopek in the bibliography and the "Countries in the Sahel" sidebar.)

COUNTRIES IN THE SAHEL

(Some are also in the Sahara.)

- Burkina Faso
- Chad
- Eritrea
- Ethiopia
- Mali
- Mauritania
- Niger
- Senegal
- Sudan

Sahel sculpture. The sculpture of the Sahel region (**Mali**, **Niger**) that has not been lost consists of small human figures (figurines), sometimes distorted (with an elongated neck, for example) and often astride an **animal**. Large human forms also exist. In early 2020, New York's Metropolitan Museum of Art mounted *Sahel: Art and Empires on the Shores of the Sahara.* There are, naturally, differences between the Sahel works and the sculpture of Benin or Nigeria, but there are also some

similarities—similarities that do not exist between this sculpture and that found in more distant locations such as Greece, Rome, or **India**. (See Cotter in the bibliography and **Saharan art**.)

Salada. In the US **deserts**, this is a former lakebed now covered with **salt**. (Also see **Salar** and **Sebkha**.)

Salar is a salt-encrusted depression (in **Chile**). (Also see **Salada** and **Sebkha**.)

Salar de Uyuni is an enormous salty plain in Bolivia.

Salina is a **salt pan**, salt lake, or salt marsh. (Also see **Sebkha**.)

Salinas Pueblo Missions National Monument (New Mexico). The ancient Anasazi and Mogollon cultures were located in the Salinas Valley. The inhabitants wove, made baskets and **jewelry**, and traded. The monument contains Las Humanas (Gran Quivira) Pueblo and the **ruins** of four mission churches (brochures).

Salmon Ruins and Heritage Park, southeast of Farmington, **New Mexico**, is an ancient Chacoan and Pueblo site with some 300 rooms and **kivas**.

Salt (sodium chloride) (G: Salz; F: sel; S: sal; I: salt; Af: sout; H: מלח [malakh]; Ar: ملح [milh]; M: давс [davs]; C: 鹽 [Yán]). A necessary concomitant for life, salt helps to maintain the electrical system in mammalian bodies. When **animals** crave it, they do strange things; e.g., **goats** seem to eat tin cans, but they are simply licking the salt on the exterior, and cattle lick salt blocks that farmers put out. For millennia, salt has been transported across **deserts** in camel caravans. It was long thought that there was a time when salt was as valuable as gold, but this is untrue, though it certainly was extremely valuable. In 2006 Michael Benanav published *Men of Salt* (its 23 color photos are edifying), an account of his unusual journey accompanying a traditional camel caravan (of white gold) that transported large blocks of salt from Taoudenni (where the mines are located) to **Timbuktu** in **Mali**. This is only a 400-mile venture, or 800 miles round-trip, but it is a very hard 800 miles in torrid, devastating **heat**, with few sources of **water**, encounters with other **caravans**, confusing shifts from one group to another, and various people who seem to think it is acceptable etiquette to request (demand) watches and money from strangers. In the past the **Tuaregs** simply killed travelers.

Salt bush in the **Arabian Desert** grows to between 3 and 6 feet in height and tolerates salty **soil** well (https://arabiandesert1.weebly.com/plants.html).

Salt desert (salt pan). Here, salt covers the desert floor. Examples include **Salar de Uyuni** (Bolivia), Bonneville Salt Flats, **Great Salt Lake**, and Chile's Salar de Atacama. (See https://10mosttoday.com/10 -most-amazing-salt-flats.)

Salt flat, Death Valley. ROBERT HAUPTMAN

Plant growth in salt flat rubble.
ROBERT HAUPTMAN

Camel caravan circa 1907. PUBLIC DOMAI

Salt Lake City is a city in **Utah** founded by the Mormons.

Saltation occurs when particles (**sand**) travel in flowing air or **water**.

Salton Sea, located in the **Colorado Desert**, is a toxic lake that is drying up.

Saltwort (batis) is a **plant** that thrives in the Gobi's salty areas; its leaves have a hairy covering that prevents evaporation (https://www.discovermongolia.mn/blogs/plants-of-the-gobi-desert).

Samarkand is a trading city on the **Silk Road**.

San (formerly Bushmen, now derogatory) are a people who inhabit **deserts** in **Botswana**, **Namibia**, and **South Africa**. In the past the San were mistreated, murdered, and put on display (alive and dead) in England (Isaacson, 85ff.). *The San*, Jiro Tanaka's illustrated 1980 study, describes in great detail these people's **ethnographic** and ecological lives: how they live and interact socially, how they forage and hunt, kinship, marriage, and so on. Implements (bow, poison, spear snare, and many other items), **plants (acacia**, aloe, etc., for a total of 94) (173ff.), and **animals** (mongoose, leopard, green mamba, etc., for a total of 86) (177ff.) are listed and discussed. In 2007, so 27 years later, Alan Barnard published *Anthropology and the Bushman*, a historical overview of anthropologists' changing perspectives on these peoples of southern Africa (1). (This is a small volume with a ca. 400-item bibliographic listing.) The language and people designations are quite complex:

Khoe-speaking peoples:
Khoekhoe or Hottentot (herders): Namibia
Khoekhoe-speaking Bushmen or San: Namibia
Central Bushmen or San: Botswana

Non-Khoe-speaking Bushmen or San:
Northern Bushmen or San (**!Kung**): Botswana, Namibia
Southern Bushmen or San: originally South Africa, now Botswana (6–7)

Note that though controversial, many hold that "Bushman" and especially "Hottentot" are derogatory. Tanaka does not use these terms. (See also **Kalahari**, **Khoikhoi**, **Language**, *The Lost World of the Kalahari*, and the "Indigenous Desert Peoples (International)" sidebar.)

San Andreas Fault is a geologic (tectonic) fault that runs through **California** and touches Desert Hot Springs and **Palm Desert**.

San Xavier del Bac, the White Dove of the Desert, is a mission near **Tucson**.

Sand (G: Sand; F: sable; I: sandur; Af: sand; H: חוֹל [khol]; Ar: رمل [ramil]; M: элс [els]; C: 沙 [shā]) consists of silicate particles that accumulate in large numbers and create a beach (along a shore) or a **desert**. Most sand is white or tan, but black sand beaches exist, in Hawaii, for example. **Wind** moves the particles around and creates **dunes**. It would, at first, appear impossible but Michael Welland devotes a 343-page disquisition to sand (*Sand: The Never-Ending Story*). It is here that one learns that

San, Kalahari Desert, Namibia. FRANK VASSEN, 2009. CREATIVE COMMONS ATTRIBUTION 2.0

Al-Dahna, eastern Saudi Arabia, parallel dunes. RICHARD MORTEL, 2020. CREATIVE COMMONS ATTRIBUTION 2.0

sand derives from abraded rock, quartz (silica), and biogenic material (shells, coral); comes in various coarsenesses; and 22 animal phyla can be found in "the spaces in between sand grains" (3, 4, 18ff., 9, 60). (See **Abrasion**, **Ralph Alger Bagnold**, **Dust**, and **White-sand Desert**.)

Sand paintings. Navaho medicine men create very complex figurative paintings in the **sand** that are both part of religious rites and also esthetically pleasing, although not to the participants. Gladys A. Reichard's *Navaho Medicine Man Sandpaintings* explicates the Bead and Shooting Chants used in healing **ceremonies** and presents 25 full-color plates. After the ceremony the paintings are purposely destroyed. Nevertheless, some not used in a religious context do exist in **museums** and collections. Indeed, the author of this encyclopedia owns two miniature examples. Authentic paintings sell for many hundreds of dollars. Though different, these works are reminiscent of the **dot paintings** of the Australian **Aborigines**.

Sandboarding is a little-known sport, growing in popularity, in which people fly down steep sand dunes on waxed boards. Locations include **White Sands National Park (US)**, Concón **(Chile)**, Monte Kaolino (Germany), **Namib Desert (Namibia)**, **Great Sand Sea (Egypt)**, **Tottori Sand Dunes** (Japan), and Kangaroo Island **(Australia)**. (See

https://sandboard.com and https://www.surfertoday.com/surfing/the-best-sandboarding-dunes-in-the-world.)

Sands of the Kalahari is a 1965 Hollywood **film** concerning some plane crash survivors. **Baboons** attack and one person wants to get rid of the other men. Theodore Bikel is one of the actors. A short trailer in Spanish is available at https://www.youtube.com/watch?v=Ke_ntqHcFZc. The complete film may be viewable online without paying a fee.

Sandstorm. Wind whips up particles of **sand** and blasts them around so that they sting exposed skin and the eyes and it may become impossible to see anything. This is why white-sand desert dwellers often wear face coverings. Sometimes a sandstorm can inundate a distant city such as Beijing or **Dunhuang**. Early in 2022, **Iraq** was hit by severe sandstorms that caused pulmonary problems for residents who live outside desert areas. In August of 2022, Iraq, Syria, and Iran were inundated by yet another storm that resulted in hospitalizations and deaths. These storms also occur in the **Sahara**. (See **Black blizzard**, **Dust**, **Dust Bowl**, **Gobi Desert**, **Karaburan**, and **Clothing**.)

Sangre de drago (dragon's blood). There exist three varieties of this **plant**—two **shrubs** and an herb—and they grow as high as 8 feet. Depending on the type, it is effective as a laxative or anti-inflammatory (Moore, 104, 107–8).

Santa Ana is a hot, dusty **wind** that comes off the **desert** in western **California**. It fans **fires** and stimulates irritability in people.

Santa Fe is a well-known, arid New Mexico city.

Santa Fe Opera House is an architecturally famous building in **Santa Fe**.

Saqqara (Sakkara). Thirty miles from **Cairo**, in **Giza**, one will find this extraordinary necropolis with many tombs and burial sites in tunnels. (See https://www.youtube.com/watch?v=cU3rO-9blpRI, https://www.youtube.com/watch?v=1WmN-RJAKlE, and https://www.youtube.com/watch?v=CxgHeh9Mlrg.)

Saudi Arabia is a large, tribal Middle Eastern country whose extensive desert land is rich in **petroleum**. As with most countries, Saudi Arabia is replete with modern cities such as Jidda, **Mecca**, and Medina but there still exists a distant, rural, and poor life led by **Bedouins**. (See **Arabian Desert**, **Islam**, **Rally Jameel**, and **Religion**.)

Saxaul (black saksaul, saksaul, saxaoul) is a **shrub** or **tree** in Asian **deserts**, especially the **Gobi**. It offers shelter and **food** for various creatures and human beings and stores **water** in its bark. It is used to build windbreaks against **desertification** (https://www.discovermongolia.mn/blogs/plants-of-the-gobi-desert).

Scholarly disciplines. Anthropology, archeology, architecture, biology, botany, ecology, exploration, geography, geology, geomorphology, and history as well as agronomy, astronomy, chemistry, climatology, demography, engineering, entomology, ethnography, ethology, horticulture, meteorology, paleontology, political science, sociology, and zoology all contribute to our understanding of the **desert** and its peoples. It is good to keep in mind William Blake's poetic observation,

To see a World in a grain of sand,
And a Heaven in a wild flower,
Hold Infinity in the palm of your hand,
And Eternity in an hour.

This may be merely metaphysical musing, but it contains the truth that if one could know all there is to know about a grain of **sand**, one would thereby know everything.

Scholars are those people who concern themselves with various aspects of the **desert**. (See **Scholarly disciplines**.)

Scorpion. Many species of this **arachnid** are harmless, but a few such as the Arizona bark and the deathstalker are especially toxic.

Scott, **Robert Falcon** (1868–1912), is the great Antarctic **explorer**; he died on his return after reaching the South Pole. (See ***The Last Place on Earth: Scott and Amundsen's Race to the South Pole***, *Scott's Last Expedition*, **Explorers**, and the "Desert Explorers" sidebar.)

Robert Falcon Scott. J. THOMPSON, CA. 1900. PUBLIC DOMAIN

Scotty's Castle, Death Valley. ROBERT HAUPTMAN

Scotty's Castle (Death Valley Ranch) is a bizarre, complex villa built in Grapevine Canyon, an **oasis** in **Death Valley**. It was closed for many years due to flooding and a **fire**.

Scrub (chaparral) is a low bush or **shrub**.

Sculpture, naturally, is an enormous category. Because sculpted works are made of wood, metal, stone (including alabaster), and ceramic, they tend to last, whether ensconced in an Egyptian tomb or buried in the Sahel sand. Many sculpted works are small replicas of human beings, others are forms on sarcophagi or stelae (at times painted), and still others are monumental stone figures, at **Luxor**, for example. One negative aspect associated with African (**desert**) sculpture is that it is looted from sites and sold illegally around the world; some of these are now being returned. Another is that replicas are produced en masse and sold both as replicas and

Columns of the Temple of Luxor in a photo from 1922. PUBLIC DOMAIN.

also unethically as authentic pieces. (See **Obelisk, Egyptian sculpture**, and **Sahel sculpture**.)

Searles Lake (Slate Range, Borax) is dry lake in the **Mojave**. A crystal hunt occurs every October.

Sebkha (shebka, sabkha) (Arabic), in the **Sahara**, is a salt flat (**pan**) or marsh. In the **Maghreb**, schott; in **Asia**, **kavir** or **takyr**; in southern **Africa**, **vleis** or pan; in the **US**, **playa** or sink; in **South America**, salar or **salina** (Michael Martin, 312).

Sechura Desert (1,900 square miles) runs along the northwest coast of **Peru** and contains many sand dunes. It is extremely rich in phosphate. Earlier inhabitants include the Vicus and Mochica (Nathaniel Harris, 122–23). Farther south, the Pan-American Highway is bordered by enormous dunes.

Sedona is an Arizona town known for its surrounding red rocks and **mountains**. It is considered "The Most Beautiful Place on Earth" and for good reason. The architecturally modern Chapel of the Holy Cross is built within natural rock formations and is quite awe-inspiring. (See **Oak Creek Canyon**.)

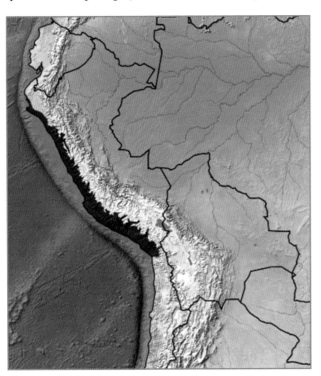

Sechura Desert ecoregion. TERPSICHORES, 2012. CREATIVE COMMONS ATTRIBUTION-SHARE ALIKE 3.0

Sediment is silt, sludge, or dregs that settle at the bottom of a liquid.

Seguia is a water channel.

Seif is an isolated **dune** with a knife-edged ridge (Michael Martin, 315).

Selima Sand Sheet lies less than 100 miles west of the **Nile** in southern **Egypt** and northern **Sudan**. It is composed of **sand**, granules, and **dunes** and contains some military camps.

Serengeti National Park (Tanzania). See **Ngorongoro National Park** and the "International Wildlife Refuges" sidebar.

Seri are an indigenous group of ca. 1,000 people who inhabit a corner of the **Sonoran Desert** in Punta Chueca and other communities in northwestern **Mexico** (López Torres).

Serir is sandy gravel larger than on a **reg**. (See **Desert types**.)

Seven Cities of Gold is a Spanish **myth** or legend that Cibola, El Dorado, and other cities were located in New Mexican **pueblos**.

Seven Pillars of Wisdom is T. E. Lawrence's exceptional autobiography; it covers his time as a British military officer in **Jordan**.

Sex. See **Tourism**.

Shackleton, Sir Ernest (1874–1922), was the leader of three Antarctic expeditions. The *Endurance*, one of his ships, was stuck in the ice and crushed (and rediscovered in 2022, 107 years later). He and two men left, and after a harrowing journey, he returned to Elephant Island and saved all 27 of his men. This is the single most astonishing rescue in all of recorded history. (See the extraordinary historical footage at https://www.youtube.com/watch?v=gqJDqjS8RLE.)

Ernest Henry Shackleton. Г. HURLEY, 1920. PUBLIC DOMAIN

Shamal (shemaal, shimal, shumal, barih) is a hot, dry northerly **wind** in **Iraq**, Iran, and the Arabian Peninsula.

Shamos. See **Gobi**.

Shar Lut. See **Yardang**.

Sheikh is a strong leader in **Saudi Arabia**.

The Sheltering Sky is Paul Bowles's superb **novel** (and **film**) concerning some young travelers lost in an alienating world. The Tunisian Sahara plays a seminal role.

Sherlock Holmes. Some of the Sherlock Holmes tales take place in a desert setting: "The Great Alkali Plain" in the second part of *A Study in Scarlet*, for example.

Ship Rock is a **volcano** located on the **Colorado Plateau**.

Shoshone. The Timbisha Shoshone people continue to inhabit lands within **Death Valley National Park**. (See the "American Desert Indian Tribes" sidebar.)

Showy milkweed, like its sister namesakes, exudes a white sap and its seeds are attached to silky strands. According to David Rhode, it is found in the **Great Basin** and Native Americans used it to make string, textiles, and **clothing** and as an antiseptic and against respiratory diseases (128ff.).

Shrub is a small **plant**. There is a bizarrely shaped, bulbous shrub in the **Atacama Desert** that is 2,000 years old (Wolfe).

Side-blotched lizard is the most common lizard in the southwestern **US**.

Sidewinder. This viperous **rattlesnake** moves sideways by lifting and twisting its body, an **adaption** that allows these types of **snakes** to keep part of their body off the scorching ground.

The Sierra Club Desert Reader, edited by Gregory McNamee, is a wonderful collection of 62 excerpts from literary works by a superb group of authors including **Edward Abbey**, **Gertrude Bell**, Jorge Luis Borges, **Richard Byrd**, Bruce Chatwin, Charles Darwin, **T. E. Lawrence**, John Muir, Pablo Neruda, and Slavomir Rawicz. (See other **readers**: *The Nature of Desert Nature*, *the new desert reader*, and *Voices in the Desert*.)

Silk Road is a network of ancient but diverse routes from Eastern Europe to **China**, which **Marco Polo** followed in order to explore and trade. It runs through **deserts** in what are now many lands. Locations along the way include Samarkand, Bukhara, and Tashkent as well as ancient lost or buried cities. It is littered with the bones of **camels** and **donkeys** that have died; sometimes these carcasses are the only markers to indicate the route. Geordie Torr, in *The Silk Roads*, a comprehensive history with a replete bibliography, points out that there were many routes (including a maritime one) and they have existed for thousands of years. The desert portions are bounded by mountain ranges such as the Altai. Crops that moved along the Silk Road include wheat, millet, hemp, cotton, alfalfa, tea, grapes, and so on. Diseases include the Black Death, measles, leprosy, smallpox, diphtheria, and Behçet's disease (Silk Road disease) (passim and 212ff.). In *The Silk Road*, Tim Winter attributes the term to Carl Ritter in 1838 (1) and declares that his study traces the term's evolution and critically interrogates its merits as a historical narrative (2). (See **Marc Aurel Stein**.)

Silk Road exhibitions. Tim Winter lists 88 exhibitions starting in 1910 and concluding in 2021. Examples include *The Grand Exhibition of Silk Road Civilizations* in Nara in 1988; *Travelling the Silk Road: Ancient Pathway to the Modern World*, at the National Museum of Australia in Canberra in 2012; and *Cave Temples of Dunhuang: Buddhist Art on China's Silk Road* at the Getty Center in Los Angeles in 2016 (185–88).

Silk Road organizations include Silk Road Cities Alliance, Silk Road Forensic Consortium, and Oxford University Silk Road Society, among many others (Winter, 189).

Silver is a precious metal whose ore is mined in **deserts**. A **museum** featuring silver is located in Carson City, **Nevada**.

Silver cholla is a **cactus** whose structural portion remains long after the **plant** dies. It can be found in Joshua Tree National Park's Cholla Cactus Garden (brochure). Also called golden or staghorn cholla, it has long, sharp spines—used for needles and awls—and lime green flowers. It also produces medicinals for cuts and wounds as well as **food** (Rhode, 106ff.).

Silverleaf nightshade is a large, five-lobed purple flower whose seeds and foliage are toxic. It is found in the **Sonoran Desert** (Gerald A. Rosenthal, 230).

Simpson Desert

56,000 (116,000) square miles	Australia	Hot desert	Little precipitation	Aborigines

One of Australia's major **deserts**, the Simpson lies to the northeast of the **Great Victoria Desert**. It is filled with ancient dunes and a few **Aboriginal** people (Nathaniel Harris, 161).

  Extent: 56,00 square miles
  Environment: Hot and dry
  Surfaces: Red, sandy plain and dunes
  People: Aborigines
  Animals: Camel, dingo, fox, frog, Gila monster, mouse, sand goana, and thorny devil
  Plants: Acacia, grasslands, river red gum, and ehumani
  Resources: Few
  (See **Australian Desert**.)

Sinagua is a southwestern US farming people who lived in the Verde Valley of **Arizona** (brochure). (See **Montezuma Castle**.)

Sinai (Peninsula) (23,000 square miles) is the Egyptian desert area located to the west of **Israel** and the Gulf of Aqaba, up to the Red Sea's western arm that leads north to **Suez** and the Mediterranean. Alberto Siliotti, in his superb **guidebook**, *Sinai*, notes that in the north one finds **dunes** and **wadis** and farther south rock outcrops and **mountains**

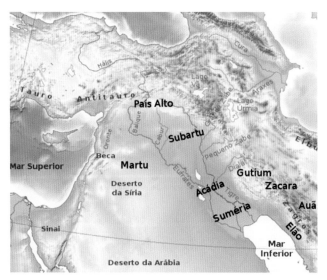

Near East topographic map, Syrian Desert. YIYI, 2013. CREATIVE COMMONS ATTRIBUTION 3.0

Monastery of St. Catherine. PUBLIC DOMAIN

(6–7). The peninsula is very rich in **animals** and **birds** including **gazelle**, the incredibly horned **ibex**, **lizard**, gull, falcon, heron, **fish**, and coral as well as innumerable migratory **birds** and mangrove and acacia trees (8–13). Nomadic and now sedentary Bedouin peoples have made Sinai their home. They have partially altered their traditional way of life and now earn their livelihoods through **tourism** (32–33); the Sinai Trail now makes a 350-mile loop and brings in hikers, whom the Bedouins guide (Patrick Scott). St. Catherine's Monastery, with its extraordinary collection of 4,570 biblical manuscripts including a small part of the *Codex Sinaiticus* as well as 13,000 **books**, is located here. These priceless materials are being digitized. (See **Oasis**, **Ras Mohammed National Park**, **Trails**, *Where Mountains Roar*, and https://www .egypttoday.com/Article/4/112845/Digitization -of-St-Catherine-s-Monastery-Library-protects -4000-rare.)

Sinai, Mount (Mount Horeb and Jebel Musa, the Mountain of Moses), is the **mountain** in **Egypt** from which Moses received the Ten Commandments. (See **Karkom**, **Mount**.)

Singing sands. See **Mirage**.

Jebel Musa. PUBLIC DOMAIN

Sinkiang is a desert region near the Tien Shan and Pamirs, not far from the **Gobi**, in which the **Taklamakan** is located (Michael Martin, 95, 97).

Sipapu is a small hole at the bottom of a **kiva**. In Puebloan mythology, it is thought that humans first came into the world through this entryway.

Sirocco is a Saharan wind that travels to Europe.

Siwa is an extraordinary **oasis** located 350 miles southwest of **Cairo**.

Skeleton Coast is the 300-mile coastline of **Namibia**, with the **ocean** to the west and the foggy **Namib Desert** to the east. It is a treacherous area replete with many shipwrecks. The desert contains a reserve with 200,000 seals and great **salt pans** (Genna Martin). (See also **Atacama Desert** and **Fog**.)

Skeleton Coast National Park is located in **Namibia**. It contains desert creatures such as **elephant**, giraffe, **jackal**, **lion**, rhino, seal, and zebra. (Also see **Skeleton Coast**.)

Skeletons on the Sahara is a 2006 History Channel **documentary** that reenacts an 1815 West African shipwreck. The Americans become **slaves** but are eventually freed. The commentary, in part, derives from descendants of the survivors. This is also covered in Dean King's *Skeletons on the Zahara*.

Sky City. See **Acoma**.

Sky Islands are some 55 mountain ranges in the Arizona, New Mexico, and Mexico **deserts**.

Slavery has been widespread, and so it is no surprise that it occurred in the Sahara, the Arabian, and US **deserts**. What is astonishing is that it still exists in the 21st century in its original form or in a less stringent manner where migrants, prostitutes, and other poverty-stricken people are treated as chattel. As an example, "Mauritania's endless sea of sand dunes hides an open secret: An estimated 10% to 20% of the **population** lives in slavery. But as one woman's journey shows, the first step toward freedom is realizing you're enslaved" (Sutter, and see the video at https://www.youtube.com/watch?v=5yQlOPD8mNo). Slaves were also transported across the **Sahara**. There exist many encyclopedias that deal exclusively with slavery but neither Paul Finkelman's edited *Macmillan Encyclopedia of World Slavery* nor Junius P. Rodriguez's edited *Historical Encyclopedia of World Slavery* emphasize the desert or particular examples such as the Sahara or **Mojave** or tribal slave holders such as the Cherokee.

Sleepy catchfly has small, pink/purple/lavender flowers emerging from an unusual tubular calyx. It is found in the **Sonoran Desert** (Gerald A. Rosenthal, 182).

Snake. Snakes are **cold-blooded** creatures and therefore **deserts**, which are usually extremely hot during the daytime, are an excellent habitat for them. Ten of the most virulently poisonous snakes can be found in **Australia**, much of which consists of nonarable desert. Many snakes, such as the garter, are harmless, but others are not. A strike by a hemotoxic creature, such as a **rattlesnake**, can be harmful or fatal, but a young person with a strong immune system can fight off the **toxin** even without **antivenin**. A strike by a neurotoxic snake, such as America's coral or India's cobra, may be fatal within 30 minutes unless appropriate antivenin is administered. (See Steve Irwin's snake hunt on which he locates a fierce snake [the small-scaled snake or the inland taipan, the most venomous in the world] at https://www.youtube.com/watch?v=0cV3iFxzIVc.)

Snapdragon penstemon (many penstemon species) is a very tall, yellow flower that looks like the open mouth of a dragon; it is found in the **Sonoran Desert** (Gerald A. Rosenthal, 170).

Snow occasionally falls in desert environments, even in the torrid **Sahara**. (See https://www.youtube.com/watch?v=4iLO3xo0EDI.)

Rare snowfall in the Mojave Desert. JESSIE EASTLAND, 2012. CREATIVE COMMONS ATTRIBUTION-SHARE ALIKE 3.0

Snow leopard. Some of Mongolia's ca. 700 to 1,700 of these elusive cats cross the **Gobi**. Tom McCarthy managed to capture and tag five of them (Man, 49). They can leap as much as 30 feet and eat **goats** and sheep (https://www.desertusa.com/du_gobi_life.html). However, they are so hard to spot that Peter Matthiessen, who wrote *The Snow Leopard* (1978), never saw one.

Sodium nitrate is mined in desert environments.

Soil. There exist many types of soil, some rich in natural nutrients, some not. In certain areas the topsoil is negligible, as in the Vermont mountainous woods where it is riddled with rock and conifer roots when not directly overlying bedrock. However, in Topeka, along the Kansas River, at one time there was 30 feet of rich topsoil. Soil develops from **weathering** of rock and decomposition of organic matter such as leaves. Where there is soil in **hot deserts**, it consists of aridisols (arid soils) (Allaby, *Deserts*, 33). But we are running out of good topsoil.

Solar installation. In order to install solar panels (at a cost of $2 billion) in the California **desert**, a company must first relocate **animals** such as the **desert tortoise** nearby (Woody).

Solar salt pond is a solar energy collector, generally fairly large in size, that looks like a pond.

Soleri, Paolo. See **Arcosanti**.

Songlines trace the journeys of ancestral spirits as they created the land, **animals**, and lore of the Aboriginal people of **Australia**. The mythic/genealogical recitation of one's songline is critical, because an error can result in death. Bruce Chatwin's *Songlines* presents an excellent overview of this complex subject. (Also see **Dreamtime**.)

Sonoran Desert (Yuma)

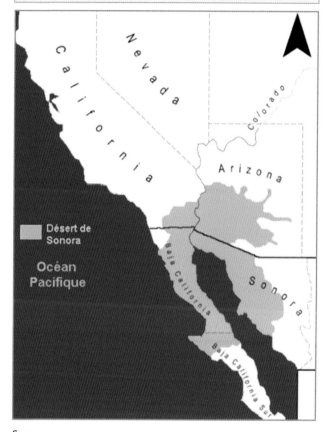

106,000 (120,000) square miles | SW US, Mexico | Hot desert | Minimal precipitation | Native Americans

Sonora map. URBAN AT FRENCH WIKIPEDIA (UPLOADER), 2005. CREATIVE COMMONS ATTRIBUTION-SHARE ALIKE 2.5

One of the **four major American deserts**, it is located in **Arizona**, **California**, **Baja California**, and **Mexico**. With some 2,700 species of **plants** (Wall, 133), its **trees** include **mesquite**, **ocotillo**, **willow**, ash, and walnut as well as **saguaro cactus**. **Animals** such as **desert tortoise**, **chuckwalla**, **antelope**, **coyote**, bobcat, **mountain lion**, **desert bighorn sheep**, quail, and elf owl make this desert their home. The Sonoran's vegetation leads to seven subdivisions: Lower Colorado Valley, Arizona Upland, Plains of Sonora, Foothills of Sonora, Central Gulf Coast, Vizcaíno Region, and Magdelena Region (Larson, *Deserts*, 23).

Extent: 106,000 square miles
Environment: Arid, hot
Surfaces: Sand and dunes
People: Hohokam, Seri, and Tohono O'odham

Saguaro cactus. ROBERT HAUPTMAN

Mojave and Sonoran Desert map. MAXIMILIAN DÖRRBECKER. CREATIVE
COMMONS ATTRIBUTION-SHARE ALIKE 2.0

Animals: Antelope, chuckwalla, coyote, desert
bighorn sheep, desert tortoise, and mountain lion
Plants: Mesquite, ocotillo, and willow as well as
saguaro cactus
Resources: Copper, lead, and zinc
(See **Colorado Desert** and "The Four Major
American Deserts" sidebar.)

SONORAN DESERT BIRDS
- Hawks, eagles, kites, and allies; order
  Accipitriformes
- Nightjars, swifts, hummingbirds, and allies;
  order Caprimulgiformes
- New World vultures; order Cathartiformes
- Pigeons and doves; order Columbiformes
- Cuckoos; order Cuculiformes
- Falcons and caracaras; order Falconiformes
- Landfowl; order Galliformes
- Perching birds; order Passeriformes
- Woodpeckers, barbets, and allies; order
  Piciformes
- Owls; order Strigiformes

(https://www.inaturalist.org/guides/669)

Sonoran Desert Life. Disguised as a text, Gerald
A. Rosenthal's volume is in reality a magnificent
guidebook to the **plants** of the **Sonoran Desert**.
Its ca. 400 large photographic images of the plants'
flowers help to make identification much easier
than is afforded by the typical guide. The book is
arranged by flower color, which helpfully extends
to the book's fore-edge. There are sections on
cactuses and **yucca**, followed by the variously
hued flowers, and finally **shrubs** and **trees**. (See
individual flowers and the "Cactuses," "Death
Valley Wildflowers," "Desert Wildflowers," and
"Medicinal Plants" sidebars.)

Sonoran Desert National Monument, located
southeast of Phoenix, is a protected area of the
Sonoran Desert. It is currently closed.

Sonoran Quarterly (1947–) is a **journal** that presents material on the **Desert Botanical Garden**,
plants, and the **Sonoran Desert**.

Sophora flavescens. The roots of this Gobi ever-green **plant** are used in traditional Chinese medicine (https://www.discovermongolia.mn/blogs/plants-of-the-gobi-desert).

Sossusvlei in the **Namib** is a beautiful lake bed in Namib-Naukluft National Park.

Souk (souq, soq, sooq, suk, suq) is an Arab outdoor market or bazaar.

Sources for this encyclopedia. See comment in the preliminary notes to the entries.

South Africa is a large country in southern **Africa** where three **deserts**, the **Namib**, **Karoo**, and **Kalahari**, can be found.

South Africa San Institute is a nongovernmental organization (NGO) that supports the revival of the cultural identity and heritage of the San people of **South Africa** and promotes the socioeconomic development of the **San**.

South America is a continent with three major **deserts**: the **Atacama**, **Patagonian**, and **Sechura**. The anonymous author of an excellent article, "11 Incredible Deserts in South America," divides the 11 into 6 authentic deserts (Brazil's Jalapao, Argentina's **Monte**, Peru's Sechura, Columbia and Venezuela's La Guajira, Chile's Atacama, and Argentina and Chile's Patagonian) and 5 **pseudo-deserts** (Venezuela's Médanos de Coro, Brazil's Lençóis Maranhenses, Bolivia's Salvador Dalí, Brazil's Rosado Dunes, and Columbia's La Tatacoa). The deserts in the latter group receive more than the allowable (by definition) 10 inches of **rain** per year. Even some of the real deserts here are quite small. The pseudo-deserts are often tiny; e.g., Médanos de Coro covers only 35 square miles ("11 Incredible"). (See the "The World's Most Important Deserts" sidebar.)

The South Pole: An Account of the Norwegian Antarctic Expedition in the "fram," 1910–1912. Here **Roald Amundsen** recounts his journey to the South Pole between 1910 and 1912 in the *Fram*, the same ship that **Fridtjof Nansen** had sailed. (See also **Books** and the books sidebars.)

Southwest Classics is Lawrence Clark Powell's 1974 collection of essays on 26 literary authors and their works concerning southwestern US desert areas. Here one will find remarks on D. H. Lawrence and his *Plumed Serpent*, Willa Cather and *Death Comes for the Archbishop*, Zane Grey's *Riders of the Purple Sage*, John C. Van Dyke's *Desert*, and Joseph Wood Krutch's *Desert Year*. Each piece is preceded by a photo and concludes with an excellent and diverse reading list.

Southwestern wild carrot (rattlesnake weed) is a tall, edible **plant** with white flowers found in the **Sonoran Desert** (Gerald A. Rosenthal, 64).

Sovereign nations. There are almost 600 Native American tribes in the **US** and they all are sovereign nations. What this means is that they do not fall under local jurisdictions for matters such as licensing **automobiles** or criminal prosecution and they have certain hunting and fishing rights that nonmembers do not possess. Federal crimes, however, are prosecuted in federal courts. A typical license plate might read:

Sovereign Nation
3249
Mille Lacs Band of the Ojibwe

This band resides in Minnesota but no state is listed on the plate.

Spain does contain some desertlike environments, the Tabernas Desert, for example. But as early as 2008, it was observed that encroaching **desertification** in the southeastern corner (which is where the Tabernas is located) was a real problem caused by **global warming** and poorly planned development that gobble up the little available **water** (Elizabeth Rosenthal, A1). Another **desert** is called the Bardenas Reales, which can be found in Navarra.

The Sphinx in a 1913 photo. PUBLIC DOMAIN

Sphinx is a giant **sculpture** of a **lion** with a human head. Someone vandalized the nose so that it is now missing. The sculpture resides in **Giza**.

Spider is a diverse species of **arachnid** that thrives in most **climates** and locations.

Spinifex is a **grass** that grows on **dunes**.

Spiny mimosa is a showy **shrub** with purplish-pink flowers that blooms in **Arizona**. Its seeds feed **birds** such as **Gambel's quail** (*Desert Bird Gardening*, 17–18).

Spirit Mountain in southern Nevada's **Mojave Desert** is a holy site for the Fort Mojave tribe. President Joe Biden vowed to protect a larger area in addition to Avi Kwa Ame, which is already under protection (Davenport).

Sports. Many desert areas provide flat surfaces for auto and motorcycle racing, land speed racing (go-karts with sails), and sand skiing. The **Dakar Rally** is an example of an organized race. (See **Off-road adventure** and **Races**.)

Spreading fleabane is medium-size, purple/lavender/white, multi-rayed flower that spreads over an area. It resembles a typical fleabane and is found in the **Sonoran Desert** (Gerald A. Rosenthal, 178).

Squirrel is a small ubiquitous **rodent** of more than 200 species. It has a bushy tail and collects seeds and overwinters. Some species live in burrows, others in **trees**.

St. Catherine's Monastery. See **Sinai**.

St. Jerome in the Desert is Giovanni Bellini's well-known ca. 1480 painting.

Star Wars. Many of the planets and moons (Tatooine, Jakku, Jedha, Byss, Hok, and lots of others) that exist in the *Star Wars* saga contain **deserts**. (See https://starwars.fandom.com/wiki/Category:Desert_planets and https://www.youtube.com/watch?v=VQGuPgSYGXM.)

Stark, **Freya** (1893–1993) (**Arabian**), was an **explorer**, **geographer**, **historian**, **archaeologist**, and traveler who worked for the British Ministry of Information in Aden, Baghdad, and **Cairo** during World War II. Among her many books are *East Is West*, *The Southern Gates of Arabia*, and *A Winter in Arabia*. (See **Gertrude Bell**, **Explorers**, and the "Desert Explorers" sidebar.)

State park (SP). In most countries, the **national parks** and wildlife areas are administered at the federal level, which is also the case in the **US**. But individual US states also manage public lands and parks. So, for example, **Arizona** has 31 state parks, some of which, such as the Sonoran Desert's Lost Dutchman SP, are in the **desert** (brochure). (See https://azstateparks.com and https://www.youtube.com/watch?v=VD-hFe4N0hA.)

Station is the term used in **Australia** (and New Zealand) for a livestock ranch (or farm), many of which are located in **deserts**. The largest is Anna Creek, which at 9,000 square miles is bigger than the country of **Israel**.

Steatopygous is as an adjective denoting steatopygia, the quality of having extra tissue (fat) on the thighs and buttocks. It is common among Khoi (formerly Hottentot) women.

Steele Burnand Anza-Borrego Desert Research Center is located in southern **California**. It promulgates research on local desert **animals** and **plants**.

Stein, Sir Marc Aurel (1862–1943) (**Taklamakan, Thar**, and others), is a renowned archeologist, ethnographer, photographer, and **explorer**. He discovered an AD 868 copy of *The Diamond Sutra*, the oldest dated printed book in the world, and removed it and thousands of other scrolls from the sealed "Cave of a Thousand Buddhas" in **Dunhuang, China**. His books include ***Ruins of Desert Cathay*** and *Sand Buried Ruins of Khotan*. (See **Silk Road, Theft**, the "Desert Explorers" sidebar, https://www.youtube.com/watch?v=FUWvCSUpBdM, and https://www.youtube.com/watch?v=QlExF6jxPFk.)

Steppe is a semiarid area with **grass** or **shrubs**. (Also see **Tundra**.)

Stillwater National Wildlife Refuge (east of **Reno, Nevada**). In the midst of this desert environment one finds marshland with 161 species of **birds**—hundreds of thousands of duck, goose, swan, pelican, egret, heron, and waterfowl—as well as **mountain lion**, **bat**, muskrat, **lizard**, toad, **frog**, garter snake, and **rattlesnake**. **Hunting** is permitted (Wall, 58–64). The refuge is the US's largest water-rights holder. (See https://www.fws.gov/refuge/stillwater.)

Stolen Generation are the Aboriginal children taken from their families in order to become part of the dominant society, which is white.

Stone pavement. See **Deflation**.

Stony Desert. See **Sturt Stony Desert**.

Sir Marc Aurel Stein. THOMPSON, 1909. CREATIVE COMMONS ATTRIBUTION 4.0

Storms are usually associated with coastal areas where hurricanes or cyclones rage or mountainous environments where **wind** and snow blizzards engulf and destroy. But windstorms in the **desert** can be just as devastating, if humans or **animals** happen to be present. (See **Sandstorm**.)

The Story of the Weeping Camel is documentary fiction and concerns camel herders in the **Gobi**. The emphasis is on a baby camel rejected by its mother (Ramsey).

Stovepipe Wells is a location on the **road** that crosses **Death Valley**. In the early days, it was the only source of **water**, and when **sand** covered it, a stovepipe was emplaced to mark the location.

The Stranger. This and *The Plague* are Albert Camus's masterpieces. *The Stranger* concerns Meursault, who flouts convention, wantonly kills an Arab, and is sentenced to death.

Strzelecki Desert is a **Central Australian Desert** with **dunes**. Its wildlife includes **crow**, **gecko**, martin, mouse, **raven**, robin, and skink and its **plants**, yellow button and **grasses**.

Sturt, Charles (1795–1869) (Australian, **Simpson**), according to Goudie, "one of the greatest explorers of the interior of **Australia**," discovered southeastern Australia's twin rivers and almost reached Australia's center. And he treated the indigenous **Aborigines** with respect (which was most unusual in the past). He is responsible for *Narrative of an Expedition into Central Australia* (Goudie, *Great*, 103–7).

Sturt Stony Desert is a small **desert** to the southeast of the **Simpson**, between it and the **Strzelecki** in the Australian **outback**.

Succulent is a **plant** that can store **water** in its thick leaves and roots. **Cactuses** are a primary example.

Sudan is a large country just south of **Egypt**. **Khartoum** is its capital. In 2011 it was divided in half so that South Sudan is now an autonomous country with its capital located in Juba.

Suez Canal is a 120-mile, lockless waterway carved from the Egyptian **desert** between Port Said on the Mediterranean and the **Red Sea**. It was completed on August 18, 1869. The political and financial necessities and construction are covered in Zachary Karabell's *Parting the Desert: The Creation of the Suez Canal.* The following overly long **film** contains scenes of the original construction: https://www.youtube.com/watch?v=Abc-AepM0JA.

Sumer is an ancient land in southern **Iraq**. It is said that "history begins at Sumer." In early 2023 a 4,500-year-old palace was uncovered at Girsu, a Sumerian city (Zahid).

Sunset Crater Volcano National Monument, northeast of Flagstaff, contains a volcano, lava flows, and archeological **ruins** (https://www.arizona-leisure.com/sunset-crater-volcano.html).

Sunshot: Peril and Wonder in the Gran Desierto is an oversize volume that contains Bill Broyles's personal essays with Michael P. Berman's 101 intercalated black-and-white **photographs**. The activity takes place in the western section of the **Sonoran Desert** near and along the excruciatingly hot **El Camino del Diablo**, which the author plans to walk. Another adventure seduces him into walking many dangerous miles when his truck dies. (See **El Gran Desierto del Altar**.)

Survival. Certain **plants** and **animals** thrive in the **desert** because they have acquired various survival techniques. These **adaptions** are beneficial here but would be of little use in other environments. They include ephemerality and miniaturization of and water storage in plants, which also have smaller leaves or spikes and extensive root systems, and high tolerance to **heat** and the ability to estivate for all types of animal life; additional adaptions are transparent eyelids and webbed feet. For humans, survival depends on protecting against the sometimes extreme heat (the record is 136 degrees

WATER REQUIREMENTS FOR WALKING

Temperature	DISTANCE			
	25 miles	50 miles	75 miles	100 miles
	LITERS			
80° F	2	4	8	13
90° F	2	8	15	21
100° F	6	17	28	39
110° F	8	21	34	48

(After Piantadosi, 80 reduced.)

Fahrenheit); maintaining **body temperature** and avoiding **hyperthermia** by drinking adequate amounts of **water** (what is now colloquially termed *hydrating*); and avoiding harmful **mammals (mountain lion, kangaroo**, wild dog), **birds (ostrich), reptiles (rattlesnake**, desert **death adder), arachnids** (death stalker scorpion), and **insects** (killer bee, tarantula hawk wasp). Correct equipment including warm **clothing** (to be used when the **temperature** drops dramatically at night), a compass and **map** (GPS), sunblock (sunburn can send one to the ER), and flashlight will increase one's chances of surviving in this harsh environment.

The key is to always keep in mind that **distances** are deceptive: What appears to be a mile to the distant **mountain** may actually be 10. Professional survivalists have various tricks that they call upon when they find themselves in precarious desert situations. These may include cutting into a **cactus** for its stored water or building a **fire** (with wet wood and without matches) to boil contaminated water. Sometimes these tricks fail. It would be wiser to carry a lighter or iodine pills. In a very different setting, described in Jack London's "To Build a Fire," when the temperature plummets in the Yukon and the protagonist steps in cold water, this knowledge is crucial. But the adventurer fails to build the fire and perishes.

David A. Bainbridge's desert restoration volume concludes with a 14-page guide on how to avoid and treat desert maladies such as heatstroke, water poisoning, rattlesnake bite, **desert rheumatism**, tick bite, and viral encephalitis (333–46). David Alloway, in *Desert Survival Skills*, prioritizes

necessities for survival: water, **food**, fire, shelter, helping searchers, deciding to stay or not, creating tools and weapons, and debriefing (11ff.). He presents a very long list of things one should bring along; even more extensive is the automobile kit that includes tools, foxhole shovel, extra fan belts, water pump, and so on (17ff.). (When I considered traveling by car in the **Australian Deserts**, I knew that I would have to take all of this stuff along including extra water, gasoline, and two spare tires mounted on the roof—of a Land Rover or Cruiser. I went by bus.) Next come the many plants (134ff.) and animals (155ff.) one might eat and how to acquire them—without poisoning oneself or accidentally wounding a companion.

Claude A. Piantadosi, a medical doctor, analyzes in great detail the biological aspects of survival in harsh environments including the desert, where the physiological limits (1) for human existence is sorely tested by water and salt loss (41ff.), heatstroke (76–77f.), and hypothermia in the **Arctic** and **Antarctic** (89ff., 99ff.).

Some hints on how to avoid the necessity of search and rescue can be found at https://www.you tube.com/watch?v=aC_uK7BbJwk. (See **Climate change, Pablo Valencia, Poor-will, Temperature variation**, and Adolph as well as Howard in the bibliography.)

Sweetbush (chuckwalla's delight) is an unusual flower that is often yellow but other colors as well; it is cylindrically shaped and bunched. The **chuckwalla** likes to eat it, and it is found in the **Sonoran Desert** (Gerald A. Rosenthal, 118).

Syrian Desert. YIYI, 2013. CREATIVE COMMONS ATTRIBUTION 3.0

The Swiss Family Robinson is Johann David Wyss's well-known **novel** (and a 1940 and 1960 **film**) of a family marooned on something of a desert island after a pirate attack and shipwreck. This is the 1960 trailer: https://www.youtube.com/watch?v=O3qehNT3KjY.

Syrian Desert (201,000 square miles) extends into Syria, **Jordan**, **Iraq**, and **Saudi Arabia**. Stoppato informs readers that plateaus, rock, **sand**, and few **wadis** are found here. The **khamsin** blows and brings **dust**. **Oases** including **Palmyra** allow **nomads** to raise **camels** and horses (182–83). (See **Bedouin**.)

Taboo is a social or religious prohibition against doing or saying something. Some desert examples follow: Many Native American tribes (**Navaho**, Cherokee) restrict contact with bodies of the dead. The Cherokee are not allowed to kill a wolf, eagle, or snake. Navaho taboos include never standing in a doorway, nor whistling at night, nor touching a sister. Mongolians do not bring weapons into a **ger** (yurt) or lean against its posts. The **Tuareg** do not eat blood. **Berbers** avoid **fish**, animal heads, eggs, and cocks. **Aborigines** do not speak of the recently dead. The **Himba** do not go back to a grave after burial or own hornless cattle. However, the **Aymara** have few taboos. (See https://www.youtube.com/watch?v=ZRJ4Ylwa8JY.)

Tafoni are hollowed-out rocks, in Antarctic valleys, for example.

Tagelmoust (tagelmust, teguelmoust, cheich, litham) is the long, variously colored (but usually blue) veil that Tuareg men use to cover their faces. Its position indicates social rank. Tuareg women are unveiled but do wear a headcloth.

Taklamakan Desert (Sea of Death)

125,000 (104,000, 115,000, 600,000) square miles | China | Cold desert | Little precipitation | Uyghur

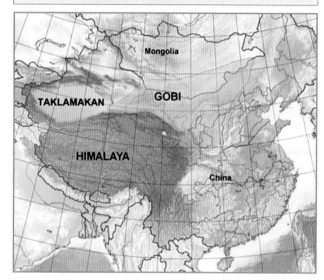

Map showing location of Taklamakan Desert. THE DRIVE, 2016. CREATIVE COMMONS ATTRIBUTION-SHARE ALIKE 4.0

This is a substantial **desert** between the Kunlun and Pamirs and the Tian Shan **mountains** and the **Gobi** in **China**. It contains enormous (1,000-foot) **dunes** of different varieties. The Silk Road routes ran through it (Nathaniel Harris, 68–69). In order to halt its advancing **sand**, billions of poplar and jujube **trees** have been planted along its periphery. **Mummies** have been found here.

Extent: 125,000 square miles
Environment: Cold desert
Surfaces: Barren, dunes, salt, and sand
People: About 55,000 Uyghur
Animals: Field mouse, gerbil, jerboa, and rabbit
Plants: Camel thorn, poplar, tamarisk, and scrub

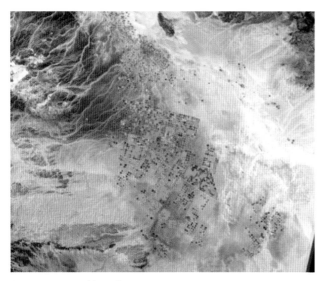

Dust storm in Taklamakan. JEFF SCHMALTZ, NASA

Resources: Gas, groundwater, and oil
(See **Desertification**, **Hotan**, **Kashgar**, and
Great Green Wall.)

Takyr (takir), in **Asia**, is a smooth plain of clay
rubble, like a salt flat. (See **Sebkha**.)

Tamanrasset (Tamanghasset) is an oasis city in
southern **Algeria** that was a caravan trading center.
The **Tuareg** grow peaches, apricots, dates, almonds,
figs, cereals, and corn here.

Tamarisk (salt cedar). In the southwestern **US**,
tamarisk is an invasive **shrub** or small **tree** that eats
water, incredibly as much as 200 gallons per day.
Therefore, as long ago as 2004, in order to combat
drought, efforts were undertaken to eliminate it
(Sink). It is also found in the **Sahara**, where, with
its white, pink, or red flowers, it can reach more
than 50 feet in height; here, it is used for building
and fuel (https://saharadesertproject1.weebly.com/
plants.html).

Tamazight (**Amazigh**) are the Berber **languages**.

Tanami Desert. This small, gravelly desert can be
found northeast of the **Great Sandy**. Its primary
feature is a gold mine (Michael Martin, 137). (See
Australian Desert.)

Tanezrouft Desert

58,000 square miles | Algeria, Mali | Hot desert | Little precipitation | Very few Tuaregs

A subsection of the **Sahara** in the southern part of
Algeria and northern **Mali**, the Tanezrouft ("Land
of Thirst," "Land of Terror") is a **reg**, a stone-
covered plateau with sandstone, canyons, and **wind**.
Some **acacia**, **jojoba**, **jackal**, **gazelle**, and **reptiles**
are in evidence (Stoppato, 110–11). It is desolate
but an auto route crosses it.

Taos is a high desert town in northern **New Mexico**.
It is famous for its art community and galleries. The
Taos Pueblo is nearby.

Taos Pueblo (**New Mexico**) is the largest existing
multistory Pueblo structure in America. Its people
create **pottery**, silver and turquoise **jewelry**, and
moccasins (brochure). (See **Pueblo** and "The 19
Pueblos" sidebar.)

Tarantula is a scary-looking and hairy **arachnid**
found in **deserts** but basically a harmless **spider**.
Some people keep them as pets. Compare **scorpion**.

Tarantula hawk is a wasp that feeds on
tarantulas.

Tashelhit (Tashelhiyt, Tashlhiyt, Shilha) is a Berber
language spoken in **Morocco**.

Tassili means "plateau" in **Berber**.

Tassili n'Ajjer Desert

31,000 square miles | Algeria | Hot desert | Minimal precipitation | Tuareg

Tassili n'Ajjer ("Plateau of Rivers") is located on
a plateau in southern **Algeria** bordering on **Libya**,
Niger, and **Morocco**. This region of the **Sahara** is
extremely arid, mountainous, and festooned with
wadis. Lynx, eagle, and **reptiles** along with cypress
trees are found here (Stoppato, 112–15). (See
Tassili n'Ajjer National Park.)

Tassili n'Ajjer National Park (Algeria). About 15,000 rock paintings are located here, some 10,000 years old. (See **Rock art** and the "International Wildlife Refuges" sidebar.)

Tataouine is a desert city in southern **Tunisia** known for the unique cave **architecture** of its Berber **population** (https://www.loveexploring .com/galleries/85625/secret-wonders-hidden-in -the-worlds-largest-deserts?page=17). (See also **Tatooine**.)

Tatooine is a desert world in the *Star Wars* saga.

Teda of the Tibesti (Toda, Todaga, Todga, Tuda, Tudaga) is a **Saharan tribe** of clans analogous to the **Tuareg**. They are nomadic but also sedentary and raid other peoples. They also have **slaves**. In 1960, there were about 10,000 of these tribal people.

Teddy bear cholla (jumping cholla). Touching this **cactus** causes spines to stick to the skin, but they do not jump from the **plant** to the victim. It can be found in Joshua Tree National Park's Cholla Cactus Garden (brochure).

Tehama. In **Saudi Arabia**, this is a 30-mile-wide heat-racked wasteland between the **mountains** and the **Red Sea** (Dale Walker, 47).

Teddy bear cholla. ALICE WONDRAK-BIEL, 2011. PUBLIC DOMAIN

Tehuelche (Aónikenk) are a nearly extinct group of **indigenous people** in Patagonia and the **Atacama Desert**.

Telescope Peak. The difference in altitude between **Badwater Basin**, the lowest point in the **US** (−282 feet), and nearby Telescope Peak (11,049 feet) is one of the greatest in the US. (See **Death Valley**.)

Tellem Burial Caves. In the Dogon country of **Mali** there exist 1,000-year-old burial **caves** first used by the Tellem people who, 700 years ago, were replaced by the Dogons. The caves contain skulls, bones, and amulets (Hammer, "Hiking").

Temperature in **deserts** varies dramatically depending on the time of day. At 2:00 p.m., when the **heat** is at its most intense, it can rise to more than 130 degrees Fahrenheit, but after the sun goes down it may drop well below freezing. **Antarctica** allows for the anomalous situation that the lowest recorded desert temperature is −128.6 degrees, which occurred at Vostok Station. The world's hottest temperature (134 degrees) set in 1913 in **Death Valley** was superseded in 1922 in the Libyan **Sahara**, when it reached 136. Christopher Burt notes that the hottest US cities in July on average are all in the southwestern and California deserts— **Palm Springs**, 108.3; **Yuma**, 107; Phoenix, 106; **Las Vegas**, 104.1 (15)—and the hottest world locations are also in deserts: **Algeria**, 135; **US**, 134; **Tunisia**, 131; **Mali**, 130; **Israel**, 129; **Australia**, 128; and so on (24–26). Alissa J. Rubin and her many coauthors insist that the world is getting hotter and so lives are altering in urban areas because of the extreme heat; in deserts, in the US, **Africa**, **Saudi Arabia**, **India**, and Australia, the number of days during which dangerous heat levels (above 103) occur are increasing (passim).

Warm-blooded **animals** such as **mammals** and **birds** maintain a fairly constant **body temperature**. When it drops or rises precipitously, very bad things occur. Cold-blooded animals such as **reptiles** (**snakes**, **lizards**) vary their temperature according to how cold (or hot) it gets. On a freezing morning, a **gecko** must warm up as the sun arrives in order to function effectively. (See **Weather**.)

Temperature variation. Jake Page observes that by altering location, **animals** can dramatically diminish the ambient temperature: When the surface is 167 degrees Fahrenheit, a jerboa's underground burrow, 5 feet below, is 86; at almost 4 feet above the surface, a camel's long legs allow its lower torso to be at 122 and its head, above 7 feet, less than 100; and a **bird** flying at 1,000 feet is at a comfortable 80 degrees (107). I discovered that this is helpfully the case: When I crossed **Death Valley** on foot, the ground temperature was almost 160 degrees, but the air temperature hovered at a mere 100. (See **Survival**.)

Temple of Tulán-54 is an early site in the **Atacama Desert**. "Evidence of **ceremonies** and ritual activities, such as feasts and offerings, demonstrates that Tulán-54 was the scene of important cultural and economic transformation, from hunter-gatherers to early pastoralist communities" (Núñez abstract).

Ten Tall Men is a 1951 **film** concerning the **French Foreign Legion**; it stars Burt Lancaster.

Ténéré Desert (Great Emptiness) (154,000 square miles) is part of the **Sahara** within **Niger**, north of the Lake Chad basin. It contains **ergs** and **rock art**. **Desertification** results in a vegetationless environment in which the **Tuareg** reside (Stoppato, 138–39).

Termite is a small, social **insect** that eats **plants** and can digest wood, even if part of a **dwelling**. (Also see **Saguaro** and **Termite mound**.)

Termite mound. Termites are tiny, wood-consuming **insects** that live in mounds (in the **Australian Deserts**, for example). These architectural wonders can be fairly small (4 feet high) or quite enormous (20 feet). Termites are able to digest cellulose because they have specialized **bacteria** and protozoans living in their guts. (When termites penetrate and inhabit a **dwelling**, they eat the wooden structure, which obviously is extremely detrimental.) Miller observes that kingfishers

Namibia Kalahari Desert termite mound. ALEXANDER KLINK, 2018.
CREATIVE COMMONS ATTRIBUTION 4.0

frequently nest in termite mounds (147). (See also **Circles**, **mysterious**.)

Terra Incognita: Travels in Antarctica is Sara Wheeler's comprehensive account of her personal experiences and detailed historical overview of Antarctic **exploration** in this most arid of continents. "As Scott noted in his diary, the bloated body of Arctic literature contrasted sharply with the skeletal material on its southern counterpart" (28). This was long ago, but additional publications did not deter Wheeler from composing this substantial tome. And while in the Antarctic **desert** she read Wilfred Thesiger's *Arabian Sands* (59), thereby connecting **cold** and **hot deserts** at least metaphorically. The working bibliography is useful in a disciplinary area so replete with excellent material (343–46).

Terrain. In **white-sand deserts** one finds fairly continuous if undulating terrain, although large **dunes** can obscure **distances**. But in other desert environments, there are undulations, formations, valleys, hills, and additional topographical features.

Thaj. The **ruins** of Thaj are about 50 miles west of Al-Jubail in **Saudi Arabia**. It is a small village situated beside a dry lakebed generally known as a **sabkha** (https://nabataea.net/explore/cities_and _sites/thaj).

Thar Desert

77,000 (83,000) square miles │ India, Pakistan │ Hot desert │ Little precipitation │ Banjaras nomads

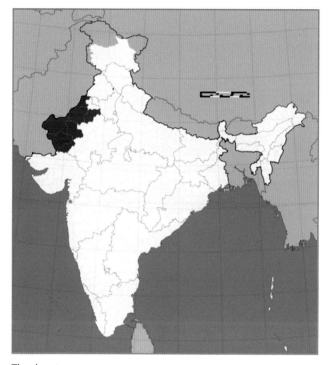

Thar locator map. PLANEMAD, 2006. CREATIVE COMMONS ATTRIBUTION-SHARE ALIKE 3.0

A traditional building in the Thar Desert. JI-ELLE, 2019. CREATIVE COMMONS ATTRIBUTION-SHARE ALIKE 4.0

The Thar (Great Indian Desert, "The Land of Death") is located in northwestern **India** and southeastern Pakistan. It contains 500-foot-high sand dunes and many inhabitants. Sandy plains (bhakars) and salt lakes (dhands) are in evidence. The Marusthali Desert is located in the eastern Thar, and the **Rann of Kachchh** is a large salt mudflat adjacent to or part of the Thar (Nathaniel Harris, 66–67). Calotropis bushes, **antelope**, peacocks, and wild dogs are part of the environment (Juskalian). According to Michael Martin, despite deforestation, overgrazing, and the resultant **desertification**, the Thar is the most densely populated and most cultivated desert in the world (79, 81).

Extent: 77,000 (96,000) square miles
Environment: Hot desert
Surfaces: Sand and dunes
People: Banjaras nomads
Animals: Antelope rat, Bengal tiger, fox, great Indian bustard, leopard, shrike, wolf, and antelope

Plants: Amaranthaceae, herbs, shrubs, thorny bushes, and trees
Resources: Feldspar, grasses, gypsum, kaolin, and phosphorite
(See **Desert dermatoses**, **Desertification**, **Desertification: restoration**, **Jaisalmer**, and "The World's Most Important Deserts" sidebar.)

Thebian Desert Road Survey investigates ancient caravan routes and oasis settlements that have been ignored in the past. In 2010, a 3,500-year-old settlement (now called Umm Mawagir) at **Kharga Oasis** in the western Egyptian **desert** was discovered. Large quantities of bread were baked there (Wilford, "Desert").

Theft. Archeologists, collectors, smugglers, and thieves ("tomb robbers") have removed treasures from **pyramids**, tombs, **caves**, and other locations (**museums**, Acropolis, **obelisks**), especially from third-world countries, usually by merely taking but perhaps sometimes purchasing, and gave or sold them to the wealthy or museums in Europe and America. A time has finally arrived when some of these treasures are being repatriated. It is worth at least noting (and not as an excuse) that it is not impossible that some material may be saved when a country goes on a rampage and destroys ancient treasures, which occurred during China's Cultural Revolution (something that could have harmed the **Mogao Cave** treasures, for example) and in

Wilfred Thesiger. FLAVIALONER, 2018. CREATIVE COMMONS ATTRIBUTION-SHARE ALIKE 4.0

Sunrise over the sands of Rub' al Khali. SRK60, 2021. CREATIVE COMMONS ATTRIBUTION-SHARE ALIKE

Afghanistan when the Taliban blew up the large Bamiyan Buddhas.

Thermoregulation is the term that denotes how a body regulates its **temperature**. Special **adaptions** for **animals** that live in extreme **heat** are necessary for **survival**. These include estivating, sweating, vocal panting, and heat-calling in some **birds**. (See also **Osmoregulation**.)

Thesiger, Sir Wilfred P. (1910–2003) (**Sahara, Arabian**), was a military officer, adventurer, photographer (38,000 images), and **scholar** who preferred **deserts** to other environments. He is probably the greatest of all desert **explorers** because he spent much of a hard life living in this harsh environment. His many **books** include *Arabian Sands*, *The Last Nomad*, and *Across the Empty Quarter*. Goudie indicates that Thesiger spent five years in the unexplored **Empty Quarter** enjoying various privations (hunger, thirst, **heat**, cold). He later traveled in **Iraq**, **Ladakh**, **Morocco**, Pakistan, and other places (Goudie, *Great*, 88–89). Paul Lewis, in his Thesiger obituary for the *New York Times*, repeats that his adventures were attempts at self-realization. (See **Explorers** and the "Desert Explorers" sidebar.)

Thistle. With large lavender flowers, thistle can reach 4 feet and can be found in the southwestern **US** and contiguous areas (*Desert Wildflowers*, 31).

Lesser goldfinches enjoy the seeds. The down lines some bird nests (*Desert Bird Gardening*, 13–14).

Thobe (thawb) is a long white or black garment worn by men in **Africa**. It is similar to an Egyptian galabia. (Also see **Ghutra**.)

Thomas, Bertram Sidney (1892–1950), was an English diplomat and explorer and the first documented Westerner to cross the Rub' al Khali (**Empty Quarter**). His **books** include *Arabia Felix: The First Crossing, from 1930, of the Rub Al Khali Desert by a Non-Arab*.

Thorn tree (honey locust, thorny locust) is a tree with spiky thorns found in **Africa** and **Arabia**. Despite the thorns, it provides **food** for desert creatures.

Threatened species. Christopher Norment discusses some threatened species that live in the **Death Valley** area. A salamander, four **pupfish**, and a toad are dependent on **water** that humans also require; these creatures inhabit small watered environments and are isolated as human necessity encroaches (Norment, "In.")

Three Rivers Petroglyph Site (north of **Alamogordo**, **New Mexico**). Here one will find 21,000 Jornada Mogollon **petroglyphs** (geometric

Three Rivers landscape. ROBERT HAUPTMAN

shapes, masks, **desert bighorn**, **birds**, and tracks) along a winding, half-mile **trail** (*Guide*). (See "Petroglyphs and Rock Art" sidebar and color **photographs**.) [157, 158, and 159]

Thyme grows in the **Sahara**, where it is eaten by **animals**. It is used medicinally for treating indigestion, respiratory infections, and spasms (https://saharadesertproject1.weebly.com/plants.html).

Tibesti is a mountain range in the central **Sahara**, primarily located in the extreme north of **Chad**. Stoppato notes that it is also a small **desert** (39,000 square miles) with **volcanoes**, **lava**, **wadis**, and **rock art** (144–45).

Tibet is a country southwest of (or in) **China**, which assimilated it. The Tibetan Plateau (Qinghai–Tibet Plateau, Qing–Zang Plateau, or Himalayan Plateau) is high and arid.

Tien Shan (Tian Shan, Tengri Tagh, Tengir-Too) ("Mountains of God") is a mountain range in **China**, Kyrgyzstan, and Kazakhstan.

Tile. Northeastern Brazil suffers from extreme **desertification** due to **climate change** and the production of tile, which requires clay from **soil** (Nicas, A1).

Three Rivers petroglyphs. ROBERT HAUPTMAN

Djenne, Timbuktu's sister city, 1907. PUBLIC DOMAIN

Timbuktu is a famous desert city in **Mali**, where traders met to exchange **salt** for gold ore and **slaves**. It is sometimes cited metaphorically as a very distant and esoteric location. (See **Libraries** and **Manuscripts, rescued**.)

Tinaja is a large porous water jar for cooling **water** by evaporation, and by extension, in the southwestern **US**, a pothole with water.

Tirari Desert (6,000 square miles) is a small Australian **desert**.

Tobacco tree is native to **South America** but found in the **Sahara** as well. Its many colored flowers often open at night. The leaves are **toxic**, and smoking them leads to dizziness and hallucinations (https://saharadesertproject1.weebly.com/plants.html).

Tohono O'odham (Papago: rejected term) ("desert people"). This Native American tribe lives on a reservation near both **Organ Pipe Cactus National Monument** and **Saguaro National Park**. (See **Borders** and **Kitt Peak**.)

Topography. See **Geomorphology**.

Tor (kopje) is a rock extrusion that lies on flatter land.

Tortoise. See **Desert tortoise**.

Tottori Sand Dunes (Tottori Sakyu) are surprising **dunes** near Tottori, Japan. They run for 10 miles along the Sea of Japan and are up to about 1 mile wide and 165 feet high. They are part of Sanin Kaigan National Park.

Tourism has grown from the trips individuals such as Herodotus took, the treks of adventurers such as **Marco Polo**, and the Grand Tours of the 17th and 18th centuries that well-off Europeans made to a cascade—hordes of clamoring crowds from countries such as Japan and the **US** invading, by foot, automobile, bus, train, plane, and cruise ship, every corner of the earth and in this case the **desert**: the **Silk Road**, ancient cities, the heart of the **Sahara**, **Gobi**, or **Sonoran Desert**. This business is far more popular than one might imagine: Tourism is a major economic factor for desert countries. In 2001, for example, **Egypt** earned $2.8 billion; **Israel**, almost as much; **Mexico**, nearly $6.2 billion; and even **Libya** managed $7 million (Allaby, *Deserts*, 158). The author of this encyclopedia knows three normal people who have purposely traveled to the **Namib**, the **Gobi**, and the **Negev**, and he has visited the **Australian**, the **Negev**, and the **four major American deserts**.

But there is another less-discussed aspect to this, one that usually occurs in desert-less countries such as Thailand, and it is the sex trade. Men, but amazingly especially women, **travel** purposely to desert areas in order to engage with **indigenous peoples**. And this is the case that Jessica Jacobs describes in her study of ethnosexual encounters with Egyptians and **Bedouins** in the tourist resorts of **Sinai**. Her conclusions are based on 57 interviews taken in 2002 that implied that the women wished to get closer to wild nature through sex (xiv) but also manifested a "nostalgia for colonial encounters," or, in postmodernist terms, desired a "consumption of the 'other'" (18), and here the imagination plays a significant role, and not just for the women (59–60.). (See also **Travel**.)

Toxins. Toxic substances (and pollutants) have three sources in the **desert**: First are the venoms that **snakes**, **lizards**, **scorpions**, and **insects** produce and administer in order to protect themselves or to paralyze prey; second are naturally occurring elements including arsenic, mercury, and uranium; and third are the poisons that are inadvertently released through **mining** (potassium cyanide), industrial production, and farming (herbicides and pesticides), as well as purposely stored radioactive waste in the future.

Tracks is Robyn Davidson's extraordinary account of her 1977 trek across 1,700 miles of **Australian Desert** accompanied by four **camels**. When they reached the coast, the camels were fascinated by the **water** and one plunged into the **ocean**! Doris Lessing said that it is among the best books about **exploration** and **travel**. But this was not enough adventure for Davidson: More than a decade later, she went to **India** in order to join the annual **migration** of the Rabari **nomads** in the **Thar Desert**, although that way of life was dying out, and when some women did migrate, they traveled by bus, while the men accompanied the camels. (See also **Books** and the books sidebars.)

Trail (track, trace) is at times a partially visible route through areas such as a **desert**. Trails are followed by **animals**, **indigenous peoples**, and hikers. (See **Negev Desert**, **Hayduke Trail**, **Pacific Crest Trail**, **National Trails Systems**, and **Sinai**.)

Trash. Deserts are littered with the remnants of human activity: **antiquities** such as potsherds, coins, and precious metal objects (which people treasure and so remove); abandoned **automobiles;** rubbish; and bodies or skeletons of **animals** and human beings. Vicki Squire, in a bizarrely articulated article (bizarre because it is couched in sometimes incomprehensible postmodern terms), remarks on "the discarded belongings" that migrants in the **Sonoran Desert** leave behind. She wonders, "What is the political significance of humanitarian activist engagements with the discarded belongings of migrants?" Her article explores how "bordering practices between states resonate with bordering practices between the human and non-human. It argues that attempts to transform 'desert/ed trash' into objects of value are nothing less than struggles over the very category of 'the human' itself." She concludes that "humanitarian activists contest processes of ehumanization through the re-configuration of 'desert/ed trash'" (Squire abstract). (See *When the Rains Come*.)

Travel. For millennia, **nomads**, **explorers**, adventurers, traders, researchers, and more recently tourists have engaged with the world's **deserts** by walking, **hiking**, trekking, and traveling through them on and

Tracks recounts Robyn Davidson's 1977 traverse of the Australian Desert. VIAGGIO ROUTARD, 2017. CREATIVE COMMONS ATTRIBUTION 2.0

in various conveyances from **camels** to jeeps. But there is a small breed of scholarly folks who venture forth to participate in the desert's gifts including **plants**, **animals**, food sources, **indigenous peoples** and their ancient habitations, **petroglyphs**, and so on. And this is the case with Diana Kappel-Smith, who recorded her experiences in the southwestern US deserts through the four seasons in *Desert Time*. She cherishes **sand** and **wind**, mesas and reefs, **tortoises** and **bighorn sheep**, and borderlands and **rivers** (passim). In his generously illustrated book, P. T. Etherton describes his journeys across the **Sahara**, **Kalahari**, and **Gobi**. (See **Explorers**, **Food gathering**, **Marco Polo**, **Silk Road**, and **Tourism** as well as *Alone*, *Alone on the Ice*, *Desert Solitaire*, *Libyan Sands*, *Terra Incognita*, and *Tracks*, among many other literary possibilities.)

Travel guiding companies (agencies). Many companies offer tours of desert areas. In my extremely broad **travels** in most parts of the world, I have only used such a company four times, in all cases when it was basically impossible to get to the location I desired, and they were all excellent: Denali National Park (not allowed), **China** (no visa), **Uluru** (no car in the **desert**), and **Masada** in the **Judean Desert** (no car in the desert). Here I include just three such examples: The Sahara Desert Travel Guide (https://www.responsiblevacation.com/vacations/sahara-desert/travel-guide), White Desert Travel (**Egypt**) (https://whitedeserttravel.com), and Wild Bunch Desert Guides Adventure Tour Company (southwestern **US**) (https://www.wildbunchdesertguides.com/wild-bunch-desert-guides-adventure-tours.cfm).

Traveling through Egypt: From 450 B.C. to the Twentieth Century. Deborah Manley and a colleague edited this most unusual compilation of brief excerpts from hundreds of travelers' accounts. Naturally, some areas (**Cairo**, Nile delta) are outside the precise purview of

the **desert**, but even **Aswan** and **Luxor** may obtain and **Sinai** and the deserts are certainly on target: E. H. Palmer enters the St. Catherine convent in 1871 (203–4), Ahmad Hassanein spends a night in the **Western Desert** in 1923 (217–18), Robert Curzon visits a monastery in 1833 (220), the Nobelist William Golding discusses **Egypt** in 1977 (he describes how three men used a series of mirrors to illuminate a tomb's wall paintings) (228–29), and so on. The superb illustrations are from W. H. Bartlett's 1849 *Nile Boat; or, Glimpses of the Land of Egypt*.

Travels in Arabia Deserta is Charles Montagu Doughty's classic 1888 account of his **exploration** of the **Arabian Desert**. This two-volume, ca. 1,400-page work, reminiscent of Richard Burton's *Personal Narrative*, is considered a masterful study, one of the greatest travel books ever written. (A first edition sells for 7,500 pounds!)

Trees. Some trees do grow in the **desert**, e.g., velvet ash, sandbar **willow**, and **Gambel's oak**. (See individual species.)

Trinity Site (northwest of **Alamogordo**, **New Mexico**). Here, on the White Sands Missile Range, is where the first detonation of an atom bomb occurred. It is only open to the public two days each year.

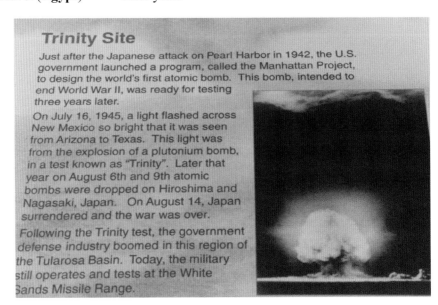

Trinity Site

Just after the Japanese attack on Pearl Harbor in 1942, the U.S. government launched a program, called the Manhattan Project, to design the world's first atomic bomb. This bomb, intended to end World War II, was ready for testing three years later.

On July 16, 1945, a light flashed across New Mexico so bright that it was seen from Arizona to Texas. This light was from the explosion of a plutonium bomb, in a test known as "Trinity". Later that year on August 6th and 9th atomic bombs were dropped on Hiroshima and Nagasaki, Japan. On August 14, Japan surrendered and the war was over.

Following the Trinity test, the government defense industry boomed in this region of the Tularosa Basin. Today, the military still operates and tests at the White Sands Missile Range.

Trinity Site sign. ROBERT HAUPTMAN

Truffles. Surprisingly, truffles can be found in **deserts** from **Morocco** to **Saudi Arabia** but especially in Syria, though they are not as delectable or as expensive as those that pigs root out in France.

Tsamma melon (citron melon) is a Kalahari Desert **plant** that presents one large yellow flower and is cooked and eaten by **indigenous peoples** (https://kalaharidesert12.weebly.com/plants.html).

Tsodilo Hills (Kalahari). In **Botswana** there are more than 4,500 !Kung rock art paintings, some 24,000 years old (Haynes, *Desert*, 91).

Tuareg (European spelling) (Touareg—US spelling, Twareg) ("Abandoned by God") are a nomadic Berber people who live in the **Sahara Desert**. These "Blue Men" call themselves Imouharen or Kel Tamashek (Tamashek speakers). There may be as many as 700,000 of these wandering **nomads**, whose trading **caravans** continue to cross the desert in **Algeria**, **Mali**, and **Niger**. *The Tuareg*, Henrietta Butler's gorgeous, oversize study (with more than 150 illustrations, many color **photographs**), presents an excellent picture of these rather enigmatic people. (See also **Ceremonies** and *The Pastoral Tuareg*.)

Tucson is an important city in **Arizona**.

Tucson Botanical Gardens is a collection of 16 gardens including one that specializes in **cactuses** and **succulents**. (See also **The Arboretum at Flagstaff** and **Desert Botanical Garden**.)

Tularosa Basin (**New Mexico**) is a basin in the **Chihuahuan Desert**.

Tumacácori National Historic Park. Three Spanish mission communities are located in this Arizona park.

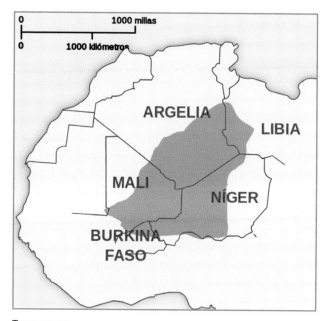

Tuareg map

Tularosa Valley (Basin) in the Chihuahuan Desert. ROBERT HAUPTMAN

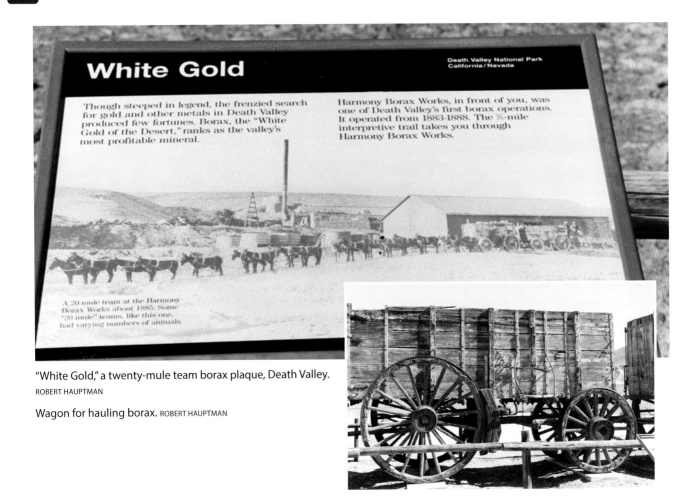

White Gold

Death Valley National Park
California/Nevada

Though steeped in legend, the frenzied search for gold and other metals in Death Valley produced few fortunes. Borax, the "White Gold of the Desert," ranks as the valley's most profitable mineral.

Harmony Borax Works, in front of you, was one of Death Valley's first borax operations. It operated from 1883-1888. The ¼-mile interpretive trail takes you through Harmony Borax Works.

A 20-mule team at the Harmony Borax Works about 1885. Some "20 mule" teams, like this one, had varying numbers of animals.

"White Gold," a twenty-mule team borax plaque, Death Valley.
ROBERT HAUPTMAN

Wagon for hauling borax. ROBERT HAUPTMAN

Tumbleweed. See **Russian thistle**.

Tundra is an Arctic plain without **trees** whose subsoil remains frozen throughout the year. Tundra covers some of Alaska, Canada, Greenland, Iceland, and Siberia.

Tunisia is a small country east of **Algeria** and north of **Libya**. The **Sahara Desert** encroaches on its southern half. In Chenini, a village in the south, 500 Berber farmers live in **caves** carved out of the rocky mountainside but **drought** and the desire for progress are diminishing the **population**. **Tataouine** is nearby (Yee, "A Cave").

Turkestan Desert. See **Karakum Desert**.

Turpan is a desert city near the lowest point in **China**, 500 feet below sea level, where ground **temperatures** higher than 150 degrees Fahrenheit occur (Sherman, 56).

Turpan Desert (19,000 square miles) is a small arid area in **China**, covered in small stones, some **sand**, and **dunes** and prone to **sandstorms** (Stoppato, 192–93).

Turpentine bush is a **shrub** with small yellow flowers; the leaves smell of turpentine. It is found in the **Sonoran Desert** (Gerald A. Rosenthal, 123).

Tuzigoot National Monument. Tuzigoot ("crooked water"), southwest of Flagstaff, is a ca. 700-year-old Sinagua building that had 77 rooms. The ancient people farmed and made rough **pottery** (brochure).

Twenty-mule team. During the late 19th century, **borax** was **mined** in **Death Valley** and drawn out by teams of 18 mules and 2 horses pulling wagons.

U

Uluru (also called **Ayers Rock**) is the Australian Aborigine name for this enormous extrusion, a ca. 1,200-foot-high rock (the largest in the world) with a circumference of almost 6 miles. It consists of steep, striated rock and is 2,831 feet above sea level. On its surface one finds both small **trees** and little pools of **water**, which do not always dry up in the intense **heat**. It takes about 45 minutes to climb to the top, although the record is held by an Australian soldier who ran up in 12 minutes. Because this Aborigine holy site has been desecrated by graffiti, people are no longer allowed to climb it. One of the most amazing things about this formation is that it changes color as the sun slowly sets, and hundreds of tourists photograph it all at the same time. The buzzing flies at the base are ubiquitously abundant but surprisingly benign.

Uluru-Kata Tjuṯa National Park is located 270 miles from **Alice Springs**, **Australia**, and contains both **Uluru** and Kata Tjuta (also called the Olgas), which is made up of 36 giant domes spread over more than 12 miles.

Umm al Samim is a quicksand region in the **Empty Quarter** in **Oman** and **Saudi Arabia**.

Uluru (Ayers Rock).

Hikers next to standing water, Shawka Wadi, UAE. ANGELA MANTHORPE, 2020. CREATIVE COMMONS ATTRIBUTION-SHARE ALIKE 4.0

United Arab Emirates (UAE) is a country composed of a group of emirates; its cities include Abu Dhabi and Dubai.

United Nations Convention to Combat Desertification. See **Desertification**.

United States Geological Survey (USGS) provides science about the natural hazards that threaten lives and livelihoods, and the **water**, energy, minerals, and other natural **resources** we rely on. It maintains at least 54 primary internet home pages that deal with, for example, **birds**, butterflies, earthquakes, water, and volcano hazards. A search for the term *desert* on USGS.gov brings up hundreds of entries, and the USGS Store's "Map Locator" allows one to search and download free digital **maps** in GeoPDF format or order paper maps. (See https://www.usgs.gov and https://store.usgs.gov/map-locator.)

University of Arizona Museum of Natural History. There are more than 25,000 specimens in its mammal collection, many from **Arizona** and **Sonora**, plus herpetological, ornithological (18,000), and ichthyological materials. It is open by appointment.

Ur of the Chaldees is an ancient biblical city in what is now **Iraq**. Its **ziggurat** is a large, attractive edifice that could have been constructed in the 20th century's brutalist style, although it is 4,000 years old.

US. The United States of America is where one finds the **four major American deserts**.

Utah is a state in the **US** that contains five **national parks**. Its **deserts** include the **Great Salt Lake**, the West, and the Sevier, among others.

Utah juniper is a large bush or **tree** that can reach as high as 40 feet in the **Great Basin** and **Mojave**. It produces brown or blue berries, and Native Americans used it medicinally for colds, headache, and asthma as well as to make bows and **clothing**; the smoke produced by burning is used for purification (Rhode, 31ff.).

Valencia, Pablo, is remembered for his 1905 ordeal in the **desert**. He ran out of **water** near **El Camino del Diablo** in **Arizona** but managed to survive in the extreme **heat** for many days of hard **walking**. In the end, his skin was parched, he could not swallow, and he was blind and deaf. He survived because of his excellent physical shape, strong will, and luck. His recovery is considered medical history (Pond, 296ff.).

Valley fever, desert rheumatism (coccidioidomycosis) is a respiratory illness caused by a fungus found in dry desert dust, especially in **California** and **Arizona**. It is often confused with pneumonia and is spreading to other parts of the country. Twenty thousand cases are reported each year. Under normal circumstances one may be infected and unaware, but in some cases it can be fatal. Symptoms include fatigue, cough, headache, muscle or joint pain, and rash on the legs or body. (See www.mayoclinic.org/diseases-conditions/valley-fever/symptoms-causes/syc-20378761).

Valley of Love is a 2016 **film** that takes place in **Death Valley**, where a couple wanders around seeking some form of enlightenment.

Van Dyke, John C. See *The Desert*.

Ventifact (windkanter) is a rock that **wind**, through ice and sand particles, shapes into strange configurations.

Verde Canyon Railroad (southwest of **Sedona**). The only way to visit Arizona's Verde Canyon is via this four-hour round-trip train ride. One will see cliffs, **ruins**, **wildflowers**, and eagles (brochure).

Very Large Array (VLA) Radio Telescope in Soccoro, **New Mexico**, is an extremely important telescopic installation.

Villa, Pancho (1878–1923), was a general who led the Mexican Revolution. He is the hero of many **films** such as *Villa Rides.*

Virga (ghost rain) are streaks of **rain** (or **snow**) that fail to reach the ground because of evaporation.

Vleis is a clay **pan** in the **Namib**. (See **Sebkha**.)

Voices in the Desert: Writings and Photographs is a collection of 10 sometimes brief essays or excerpts by John Alcock, **Mary Austin**, John Steinbeck, Charles Bowden, and others. What makes this volume so esthetically attractive are the 61 stunning Jeff Garton **photographs**, most of which are quite large (Cheek). (See other **readers**: *The Nature of Desert Nature*, *the new desert reader*, and *The Sierra Club Desert Reader*.)

Voladores are men who climb a 100-foot pole and then simultaneously jump off, flying around as their ropes unpeel. In some instances, a much older man sits on high and beats a drum; it takes about 10 minutes to reach the ground. (See https://www.youtube.com/watch?v=0JyLCtOnaLU and **Pyramid**.)

Volcano is a **mountain** that when active spews out **lava**.

Voss is Patrick White's 1957 **novel** of spiritual redemption in the **Australian Desert**.

W

Wadi (oued), in the **Sahara** and Arabia, is a ravine that is dry except when it rains (a **river**). It is similar to an **arroyo**. (Also, in the **Sahel**, a kori; in **Namibia**, a reviere; in the **US**, a wash.) (Michael Martin, 312).

Wadi Arabah is a sandy depression that runs south from the **Dead Sea**.

Wadi Dra (Draa) is a **river** in **Morocco**.

Wadi Hajr (Hajar, Hijr) is the name of a **wadi** (**river**) in **Oman**, Syria, and **Yemen**.

Wadi Musa (Valley of Moses) is where **Petra** can be found.

Wadi Rum (Khawr Ramm, Valley of the Moon). Located in southwestern **Jordan**, not far from Aqaba, this **wadi** resembles a moonscape, and it is where **T. E. Lawrence** had his headquarters.

Wadi Rum traditional Bedouin camp in Jordan.

Wadi Rum dry desert, Jordan. VYACHESLAV ARGENBERG, 2007. CREATIVE COMMONS ATTRIBUTION-SHARE ALIKE 4.0

Wadi Shani (Nahal Shani, Red Canyon) is a dry **wadi** near **Eilat**, **Israel**.

Wahiba Sands (Sharqiya Sands). In northwestern **Oman**, this is a series of endless **dunes**.

Wahhabism is an Islamic sect prevalent in **Saudi Arabia**. Its adherents are extremely reactionary and fanatical.

Walkabout (derogatory outside of its context). For Australian **Aborigines**, it is a ritual for teenagers and an abrupt personal (spiritual) calling for adults who at a moment's notice will pick themselves up and wander into the **outback**. They might not return for six months. Also a 1971 **film**.

Walking. Traveling by foot across a **desert** can be as enjoyable as walking or **hiking** in cities, across countries, in the **mountains**, or over **glaciers**. But it does come with some peril. Most acute here is the **heat** (and lack of **water**). This author crossed **Death Valley** when the air **temperature** was 100 degrees Fahrenheit and the ground temperature hovered around 160. Our Italian hitchhiker removed virtually all of his **clothing**, but he did not seem to suffer any ill effects, whereas, at another time, my climbing partner failed to put sunblock on, to ward off reflected glacial light, and had to rush to a hospital with third-degree facial burns. A later Death Valley foray into a canyon saw the thermometer hit 115. A more pressing horror occurred as this author crossed the **Negev**. He was hitchhiking but cars passed at a rate of perhaps one per hour, so he walked from near **Eilat** north toward **Beersheba**. At last, a jeep pulled up; instead of greeting him, the driver yelled, "What are you doing out here? Don't you know that Jordanian horsemen come down out of those hills and shoot people? I just turned in my guns!"

Wall to Wall: From Beijing to Berlin by Rail. Mary Morris traveled, in comparative luxury, on the Trans-Siberian and Trans-Mongolian railroads across **Asia** and so she encountered the **Gobi Desert** as its **sand** infiltrated the cars; the windows remained closed so that it was stiflingly hot (79). Half a century ago, this encyclopedia's author had a similar experience crossing Turkey on a crawling train whose locomotive doubled back on its cars so that its burning coal's smoke came directly through the windows and nearly asphyxiated the travelers.

Walnut Canyon National Monument (southeast of Flagstaff). With interesting rocks and cliff **dwellings** of the Sinagua people, this is a typical desert site.

Warm-blooded. See **Cold-blooded**.

Wars (conflicts, coups) have been fought in **deserts** for thousands of years stretching back to the Israelites' incursion into Canaan, followed by Persians, **T. E. Lawrence**, **Rommel**, Iran, **Iraq**, the Gulf War, **Sudan**, Eritrea, **Western Sahara**, and so on. War creates destruction, death, **pollution**, and **refugees**. A substantial literature exists on desert warfare including the following. Russell Hill describes his experiences during the early days of World War I in the **Libyan Desert**, and Isla Forsyth discusses the activities of the British Long Range Desert Group (LRDG) during World War II, also in the Libyan Desert: Using modified vehicles, the soldiers waged "covert desert warfare" by appearing and disappearing behind enemy lines, giving the impression that they were ubiquitously present (Forsyth abstract). Bruce Allen Watson offers a comparative perspective on seven campaigns (including Rommel, Bonaparte, the Arab-Israeli War, and the Gulf War) in *Desert Battle*, although Lawrence does not make it into the index. He indicates that the desert is the enemy (despite the humans lobbing bombs and bullets at each other). Desert warfare is different, as he observes: As an example, tanks can hide behind **dunes** and still fire at the enemy (149). He notes six trends in the way war is waged in deserts: fighting occurs near **water**, mobility in penetrating the desert, larger battlefields, combined arms, lethality, and the same areas are used frequently (179ff.).

Novels of desert warfare include Michael Ondaatje's ***English Patient***, Cormac McCarthy's *Blood Meridian*, and David Zimmerman's *Sandbox*. Daniel P. Bolger has written a detailed volume on the army's National Training Center (NTC) at Fort Irvine in California's **Mojave Desert**, where battles are simulated in a very realistic manner. Nearby are China Lake Naval Weapons Center, Edwards Air Force Base, and Twentynine Palms Marine Corps Base (4–5). The military takes advantage of the isolated desert. (Note that there exist thousands of books on desert warfare.) (See **Desert Shield/ Desert Storm**; Cloudsley-Thompson, Keegan, Moorhead, and Morgan in the bibliography; https:/ /www.theguardian.com/books/2014/apr/16/top-10 -novels-desert-war-rupert-allison; and the "Desert War Novels" sidebar.)

DESERT WAR NOVELS
(partial listing)

- *Ice Cold in Alex*
- *Blood Meridian*
- *A Good Clean Fight*
- *Dune*
- *The Yellow Birds*
- *The Sands of Valour*
- *The English Patient*
- *Take These Men*
- *The Four Feathers*
- *Sandbox*

(In part from https://www.theguardian.com/books/2014/apr/16/top -10-novels-desert-war-rupert-allison.)

Water is life; its absence death. Water is the source and sustainer of all botanical and zoological life. It is sometimes hard to locate in **deserts**, often because there is little or none there. Nevertheless, wells, **rivers**, lakes, **flash floods**, and even the **ocean** influence desert **geography**, topography, and lives. Flowing water in a desert is called a **wadi** (Arabic), nahal (Hebrew), **arroyo** (Spanish), and wash (English); these gullies are usually dry but when it rains somewhere, they flow with powerful streams of water. (See **Aquifer**, **Desertification**, **Drought**, **Iceberg**, **Irrigation**, **Ladakh**, **Oasis**, **Rivers**, **Spain**, and **Wind**.)

Weather. See **Desert weather**.

Scalding water near Donner Lake, California. ROBERT HAUPTMAN

Weathering is **erosion** caused by weather phenomena such as **wind**, **rain**, and ice but also flowing **water**. Mechanical weathering simply breaks down rock, whereas chemical weathering alters its mineral content.

Weddings. See **Ceremonies**, **Clothing**, and **Jewelry**.

Well of Barhout (Yemen). This "Well of Hell" or "Prison for Demons" is a ca. 100-foot-wide, 360-foot-deep sinkhole that contains little more than **snakes** and falling **water** (https://www.you tube.com/watch?v=vx9-a7SFHcw).

Welwitsch, Friedrich Martin Josef (1806–1872) (**Namib**), collected many species of **plants**, **insects**, and **animals**. *Welwitschia mirabilis* is named for him (Goudie, *Great*, 243–45). (See **Explorers** and the "Desert Explorers" sidebar.)

Welwitschia mirabilis (tree tumbo). Nathaniel Harris describes this Namibian **plant** as a truly bizarre life form. Much of it is located underground, but its two large leaves last for 500 to as long as 3,000 years. (See https://www.youtube.com/watch ?v=CcsMY2schVI.)

Western Australian Desert is a broad area in western **Australia** that encompasses three of the country's primary deserts: the **Great Sandy**, the **Gibson**, and the **Great Victoria** (Nathaniel Harris, 158–59). (See **Central Australian Desert**.)

Western Desert. The **Sahara** in **Egypt** is divided into an eastern (86,000 square miles) and a western (169,000 square miles) branch. The **Nile** is the dividing line.

Western National Parks Association is a most unusual Tucson-based organization. Founded in 1938 and authorized by Congress, it promotes the educational and scientific activities of the National Park Service. It also publishes an extraordinary array of wonderful **books** and videos on Native Americans (Susanne and Jake Page's *Indian Arts of the Southwest*), **wildflowers** (Janice Emily Bower's *100 Desert Wildflowers of the Southwest*), **animals**

Plain near Paraburdoo, western Australia. S. DOWLING, 2003. CREATIVE COMMONS ATTRIBUTION 3.0

(Jonathan and Roseanne Hanson's *50 Common Reptiles and Amphibians of the Southwest*), sites (Scott Thybony's *Canyon de Chelly National Monument* and *Aztec Ruins National Monument*), and **deserts** (Rose Houk's *Sonoran Desert: An American Desert Handbook*)—hundreds of germane titles. (See https://wnpa.org and **Publishers**.)

Western Sahara

> 103,000 square miles | Morocco | Hot desert | Minimal precipitation (a few inches per year) | Sahrawi people

Characterized by rocky plateaus stretching from the **Atlas Mountains** to the Atlantic, this section of the **Sahara** contains phosphates, iron, and zinc. Flowing underground **water** rises in **oases** (and allows for gardens), and a bit of **rain** courses through **wadis**. There is some vegetation, and **antelope**, **gazelle**, and desert **fox** can be found here (Stoppato, 102–5). (See the "Sahara Sub-Deserts" sidebar.)

Wet desert occurs when bad drainage from faulty irrigation techniques waterlogs the **soil** (Jake Page, 153).

Whales. It is hard to believe, but whale skeletons have been found in the **Sahara** and **Atacama Deserts**.

When the Rains Come: A Naturalist's Year in the Sonoran Desert is John Alcock's account of a year spent in the **Sonoran Desert**. Very much like Joseph Wood Krutch's *Desert Year*, Alcock observes the plant and animal life, but his exquisite book is arranged by the month and many of the 335 large-format pages contain colorful **photographs**. Each month brings something different: January, two **ravens** and a dead **saguaro**; February, littering, which includes ancient arrowheads but also beer cans, bullet casings, Styrofoam, and lots of other junk that inconsiderate people leave behind and that he carts out (note that in some venues fines as high as $2,000 for littering are levied); June, **lizards**; July, a scorcher (118 degrees Fahrenheit) and "The Season of the Ants"; September, "The

End of the Monsoon"; December, "Winter 'Rain'" and "The End of a Year." The comparative photos of the same location at different times of the year or in differing years (as much as 25 years apart) (22) can look dramatically different (flowering and barren) because of lack of **rain** (6–7, 56–57) or **fire** (64–65). Alcock concludes his panegyric to the **desert** thus: "When the next rain does come, I will take this as an excuse to come here again to find out how the **plants** and **animals** are responding to their good fortune, each visit to this mountain increasing my understanding and appreciation of an always changing but most excellent place" (321). (See **Trash**.)

Where Mountains Roar: A Personal Report from the Sinai and Negev Desert. In this account, Lesley Hazleton discusses fear, **sand**, and **camels** as well as the people who conquer and live in ephemeral peace in this difficult land. (See **Negev Desert** and **Sinai**.)

Whirling Dervishes. Sufi Muslims whirl (**dance**) in a counterclockwise fashion that can go on for a long time. Whirling is a religious experience that may take place in a **desert** or in a city such as Istanbul. In some cases it is performed to entertain tourists. See https://www.youtube.com/watch?v=F-hIbq9KkkE8 and https://www.youtube.com/watch?v=cwLs7v0oejc. The latter video presents a more authentic example.

White Desert National Park (Sahara el Beyda) protects Egypt's White Desert with its extremely strange rock formations. (Also see **Ventifact**.)

White House Ruins are found within **Canyon de Chelly**. This famous **dwelling** is emplaced under a cliff and can be visited, but the **trail** leading to it is currently closed.

White-sand desert. Some **deserts** are covered with scrub, **plants**, **cactuses**, tumbleweed, rock, and other materials. A white-sand desert, such as parts of the **Sahara**, consists entirely of tan or reddish sand and plants only grow under unusual circumstances. (See **White Sands National Park**.)

White Sands National Park (275 square miles) is a small **white-sand desert** southwest of **Alamogordo, New Mexico**, in the **Chihuahuan Desert**. Here the **dunes** of **sand** are pure white, unlike the tan color of the **Sahara**. The **plants**, such as sand verbena and **yucca**, that manage to grow here often must put down roots 18 feet deep in order to reach the **soil** below the crystalline gypsum sand dunes. The bleached earless **lizard**, kit fox, **rabbit**, **coyote**, and introduced **oryx** also manage to survive here. The White Sands Missile Range completely surrounds the park (brochure).

Purple sand verbena. TERRY LUCAS, 2015. CREATIVE COMMONS ATTRIBUTION 3.0

Wild Horse and Burro Corral Facility (Mojave). This government facility manages the wild horse and burro populations in southern **California**, **Nevada**, and **Arizona**. Each year ca. 1,500 **animals** are cared for and prepared for adoption (handout).

Wild onions in the **Gobi Desert** have bunched white flowers and provide sustenance for wild and domestic **animals** as well as **food** and medicine for **nomads** (https://www.discovermongolia.mn/blogs/plants-of-the-gobi-desert).

Wild sunflower is a tall (8-foot) **plant** whose yellow flowers' seeds supply cardinals with **food** (*Desert Bird Gardening*, 15).

Wildebeest (gnu) is a peculiar-looking **antelope** in the **Kalahari** that migrates in large numbers.

Wilderness is a wild, uninhabited area—a forest or jungle or even a **desert**.

Wildfire. Out-of-control fires usually occur in forestland because that is where the fuel is found. But fire, on Alaska's **tundra**, for example, consumes vast quantities of vegetation. Indeed, this author crossed Alaska as seven out-of-control fires raged; additionally, he traveled along a Nevada interstate with fire leaping up along the **scrub** on both sides of the highway. During the past half century, Arizona, New Mexico, and Colorado **deserts** have suffered from fire. In 2022 extreme fires ravaged parts of New Mexico, causing irrevocable harm. In the Algerian Sahara, a gas well burned in what is known as "The Devil's Cigarette Lighter." Brush and other fires occur in **Australia**, the **Namib**, and the **Kalahari**, and the **Negev** burns because of terrorist attacks. (See **Joshua Tree National Park**.)

Wildflowers. Spring, summer, and fall wildflowers in any environment, including alpine meadows, open ski slopes, forest floors, or even desolate urban lots, are among nature's most overpowering manifestations. When yellow goldenrod and buttercup, red fireweed, white daisy and turtlehead, or blue chicory carpet acre after acre on the slopes of Vermont's Mount Snow or Stratton, there is nothing esthetically comparable in nature's botanical

Sonoran Desert cactus in bloom. BURKFAM6, 2018. CREATIVE COMMONS ATTRIBUTION-SHARE ALIKE 4.0

Desert wildflower super bloom, Diamond Valley Lake, California. ROB BERTHOLF, 2017. CREATIVE COMMONS ATTRIBUTION 2.0

DESERT WILDFLOWERS (US)
(partial listing)

- Antelope horns
- Brittlebush
- Common sowthistle
- Common sunflower
- Desert bedstraw
- Desert broomrape
- Desert lily
- Desert mariposa lily
- Desert rose-mallow
- Devil's claw
- New Mexico thistle
- Perennial snakeweed
- Rayless goldenrod
- Sahara mustard
- Silverleaf nightshade
- Sleepy catchfly
- Snapdragon penstemon
- Southwestern wild carrot
- Spreading fleabane
- Sweetbush
- Turpentine bush
- Winding mariposa lily
- Yellow columbine

THE MOST ABUNDANT WILDFLOWER LOCATIONS

- Antelope Valley California Poppy Reserve (California)
- Anza-Borrego Desert State Park (California)
- Badgingarra National Park (western Australia)
- Capital Reef National Park (Utah)
- Death Valley National Park (California)
- Iranian Desert (northwest) (Iran)
- Joshua Tree National Park (California)
- Namib Desert (southern) (Namibia)
- Red Rock Canyon National Conservation Area (Nevada)
- Saguaro National Park (Arizona)

armamentarium—except carpeted desert floors after powerful spring **rains**. This infrequent occurrence is so alluring (like an Arctic tern in North Dakota that draws hordes of fanatical bird-watchers) that thousands of people flock to a flowering desert when things are just right.

Stan Sesser points out that replete carpeting occurs in only a handful of places: western **Australia**, Iran, southern **Namibia**, and the **US**, including the **Mohave**, **Anza-Borrega**, and **Death Valley**. In the spring of 2005, abundant **rain** turned Death Valley into what might have been "the bloom of the century." In an otherwise barren landscape, the flowers' seeds provide sustenance for desert wildlife. Guides, from small brochures to large volumes, abound; especially useful are *Wildflowers of Death Valley National Park and the Mohave Desert*, a brochure with excellent illustrations of 60 flowers, and Gerald A. Rosenthal's *Sonoran Desert Life*. (See individual species and the "Desert Wildflowers (US)" and "The Most Abundant Wildflower Locations" sidebars as well as https://www.desertusa.com/wildflo/FieldGuide/fieldguide.html and https://www.davidsenesac.com.)

Wildlife refuges. In the western US **deserts**, there are many wildlife refuges in which **birds** and **animals** are protected, but only to a limited extent, since **hunting** and fishing are often allowed (this should be rethought). Throughout the world, there are almost 600 wildlife refuges, some of which are designated as **national parks** or sanctuaries. **India** alone claims to have 564. (See **Books**, individual refuges via the "National Wildlife Refuges" and "International Wildlife Refuges" sidebars, and https://ccfoodtravel.com/2017/05/6-great-national -parks-visit-sub-saharan-africa. When noting bird, mammal, and reptile species, this encyclopedia uses generic terms, e.g., eagle, rather than bald eagle.)

NATIONAL WILDLIFE REFUGES (NWR): USA
(partial listing)

- Bear River (Utah)
- Buenos Aires (Arizona)
- Cabeza Prieta (Arizona)
- Desert (Nevada)
- Fish Springs (Utah)
- Kern (California)
- Kofa (Arizona)
- Stillwater (Nevada)

INTERNATIONAL WILDLIFE REFUGES (OR PARKS)
(partial listing)

- Central Kalahari Game Reserve (Botswana)
- Desert National Park Jaisalmer (India)
- Etosha National Park (Namibia)
- Gemsbok (South Africa)
- Gobi Gurvansaikhan National Park (Mongolia)
- Great Gobi A Strictly Protected Area (Mongolia)
- Great Gobi B Strictly Protected Area (Mongolia)
- Indian Wild Ass Sanctuary (India)
- Karakorum (Nubra Shyok) Wildlife Sanctuary (India)
- Kutch Desert Wildlife Sanctuary (India)
- Kutse Game Reserve (Botswana)
- Moremi Game Reserve (Botswana)
- Ngorongoro National Park (Tanzania)
- Revillagigedo Archipelago (Mexico)
- Serengeti National Park (Tanzania)
- Tassili n'Ajjer National Park (Algeria)
- Yelyn Valley National Park (Mongolia)

Wildlife trade. Wildlife is caught and collected in the **desert** and sold illegally to those who wish to own an exotic pet. The skeptical will balk when told that someone wishes to have a Sonoran Desert tortoise living in his basement, but all one must do is to recall the man who owned a Siberian-Bengal tiger; he kept it in his apartment, his New York City apartment—a fully gown, 425-pound tiger named Ming! (See https://www.youtube.com/watch?v=m Wv3wmnxidk.) Capturing or killing illegally is called poaching, and sometimes it takes place under cover of darkness. Even when legal, it is, for some, unethical to capture desert **animals** and sell them. (See also **Cactus trade**, **Gebel Kamil Crater**, and **Ocucaje**.)

Wilfred Thesiger is Alexander Maitland's enormous 2011 biography of this great desert **explorer** and wanderer in **Africa**, **Arabia**, and Afghanistan. (See **Thesiger, Wilfred**.)

Willow. See **Desert willow**.

Wind is the movement of the air, which varies from a slight, barely perceptible flow through a moderate breeze to a gale, violent storm, and hurricane. The record is 253 miles per hour. Named desert winds include the ghibli, haboob, **harmattan**, **karaburan**, **khamsin**, loo, **nashi**, samiel, **Santa Ana**, simoon, **sirocco**, **shamal**, and zoboa. Persistent desert wind causes **erosion**, and therefore shapes the **desert**, and **sandstorms**. (See **Water** and de Villiers in the bibliography.)

Wind farms can be found in **deserts**, including the San Gorgonio Pass Wind Park in **Palm Springs**, where 2,700 wind turbines are located, and the Foum El Oued wind farm with 22 turbines in the **Western Sahara**.

The Wisdom of the Desert is a book of sayings of the **Desert Fathers** by Thomas Merton, the esteemed Catholic monk.

Wolf spider lives in various environments including the **desert**.

Woman in the Dunes is a 1964 Japanese existential **film** based on a **novel** by Kobo Abe. A man is trapped with a woman who cannot escape from her home among the desert **dunes**. (See the trailer at https://www.youtube.com/watch?v=S9xlqmyqftU and also *House of Sand*.)

Wright, Frank Lloyd (1867–1959). Perhaps the 20th century's most influential architect, Wright built Taliesin West, his masterpiece, in and into the **Sonoran Desert**.

Xerophyte is a **plant** that does not require a great deal of **water**.

Yak is a hardy, ox-like bovid that can be found in **Tibet**. It is used as a beast of burden. (See https://www.youtube.com/watch?v=9Qg4jnSD4VA.)

Yardang is a wavelike rock ridge created by wind **deflation**; **zeugen** is caused by **abrasion**. In Iran, when yardangs accumulate, they are called *shar lut*, desert towns, according to Michael Martin (308).

Yavapai Geology Museum, in **Grand Canyon National Park**, presents an account of the canyon's geology.

Yellow columbine. With a large yellow flower, this **shrub** favors moist areas, where it proliferates. It is found in the **Sonoran Desert** (Gerald A. Rosenthal, 169).

Yelyn Valley National Park (**Mongolia**) is in the southern **Gobi**, where cliffs contain condor (lammergeier) nests. (See the "International Wildlife Refuges (or Parks)" sidebar.)

Yemen is a small country to the south of **Saudi Arabia** and the west of **Oman**. It borders on both the **Red Sea** and the Indian Ocean. In 2009 it found that it was running out of **water** because farmers grow qat, a narcotic that is chewed by most Yemeni men. Its **irrigation** requires a great deal of water; additionally, almost a third of this land's water is wasted (Worth). To add to Yemen's burdens, it has at least 400 separate tribes and has been involved in a civil war for years. (See **Well of Barhout**.)

The Young Black Stallion. Based on *The Black Stallion*, an earlier **film**, this Imax production highlights the desert landscape in which a young girl trains and rides a racehorse.

Younghusband, **Sir Francis Edward** (1863–1942), was a military man and explorer of the **Gobi** and Turfan **deserts**, which he crossed and then continued over the Karakorum range all the way to Srinigar. He is responsible for *The Heart of a Continent* (Goudie, *Great*, 141–44), *India and Tibet*, and many other books. (See **Explorers** and the "Desert Explorers" sidebar.)

Yucca is a **succulent** that has sharp barbs that hurt if touched; its flowers open at night. The root is used as a sudsing agent and against arthritic pain and inflammation (Moore, 134–35). Various species include Mojave yucca (Spanish dagger), banana

Joshua tree blossoms. JOSHUA TREE NATIONAL PARK. ROBB HANNAWACKERM, 2008. PUBLIC DOMAIN

Joshua trees (Yucca brevifolia). KRZYSZTOF ZIARNEK, 2016. CREATIVE
COMMONS ATTRIBUTION-SHARE ALIKE 4.0

yucca, and the famous **Joshua tree**, which has short, twisted branches and white flowers and can reach 30 feet in height. Uses include **food**, basketry, and sandal manufacture (Rhode, 101ff.). (See **Joshua Tree National Park**.)

Yucca Mountain is a nuclear dump about 100 miles from **Las Vegas**. (See **Pollution**.)

Yuma. Located in the southwestern corner of **Arizona**, it is the hottest city in the **US**. The average July high **temperature** is 107 degrees Fahrenheit and the single-day record is 124. It is the "sunniest city on earth" and the site of cowboy movies. (See **Temperature** and the "Major Cities Associated with Deserts" sidebar.)

Yuma Desert is part of the **Sonoran Desert**.

Yurt. See **Ger**.

Zabriskie Point is a well-known location in **Death Valley**. It is also the title of a 1970 Michelangelo Antonioni **film** in which the beautiful scenery is featured and two young hippies make love. (See https://www.youtube.com/watch?v=Dl5L_qoVykM and the film's trailer at https://www.youtube.com/watch?v=iBaMHIBySpk.)

Zebu is a bovid used for work, **milk**, and meat.

Zeriba is a thorn fence used to protect **livestock**.

Zeugen. See **Yardang**.

Ziggurat is an ancient stepped building in the Near East.

Zoologists are **scholars** who deal with **animals**; ethologists study their behavior, often in the wild.

Zud (dzud), in the **Gobi**, is simultaneous **snow**, cold, and **wind** following a hot summer of **drought**: Cattle die.

Zuni Pueblo. Located west of **Albuquerque**, the Zuni are different: Their language is unrelated to the languages of the other Pueblo peoples, and they continue to practice their traditional shamanistic **religion** with its regular **ceremonies**, **dances**, and mythology. They produce silver and turquoise **jewelry**, baskets, beadwork, animal fetishes, and **pottery**.

Zuni Pueblo from a 1910 photo. PUBLIC DOMAIN

Bibliography

Items marked with an asterisk (*), most of which I have examined, did not contribute directly to this work, but they are included because readers may expect an encyclopedia's bibliography to offer a fairly comprehensive overview of *important* material; the websites of book purveyors such as Amazon or Goodreads may present only a limited number of currently available books, while large collections such as the New York Public Library or the Library of Congress (LC) will overwhelm with far too many possibilities. LC, for example, *only* lists the first 10,000 records for the term "desert"; the University of Vermont's general catalog lists more than 200,000 books and articles, Harvard's, astonishingly, almost 3,600,000! I only include a few foreign language titles because there are far too many of them. For example, LC lists 937 for French, 191 for German, 189 for Arabic, 137 for Chinese, 135 for Spanish, 73 for Italian, 16 for Afrikaans, and 12 for Mongolian. (These numbers alter as materials are added or weeded.) Including even a small percentage of these ca. 2,000 volumes would unnecessarily clutter the following pages.

Additionally, books merely cited within the entries as works authored by explorers, for example, are also elided here but are included if given a separate entry. Brochures and handouts, published by the US government and private entities, are noted as sources in the text but are usually not included in the following listing. Note that accessible online books are frequently quite dated (and so elided here), and some hard-copy volumes are extremely esoteric, e.g., Cathy Moser Marlett's *Shells on a Desert Shore: Mollusks in the Seri World* (and so Marlett does not make it into this bibliography). As for scientific papers and articles, there exist hundreds of thousands of sometimes *extremely abstruse* discussions of, for example, bacteria (a typical paper is titled "Asenjonamides A–C, antibacterial metabolites isolated from *Streptomyces asenjonii* strain KNN 42.f from an extreme-hyper arid Atacama Desert soil" [excluded]), insect activity, plant ecology, genomic particulars, hydrological manifestations, and so on. Another excellent example is M. I. Haverty's "The role of toilet paper in studies of desert subterranean termites (Isoptera) in Arizona, USA . . ." (which also is excluded here). Many of these, though examined, have been ignored, especially since the results noted in scholarly papers are often included in monographs and collections.

Abbasi, Sajjad, et al. "Microplastics in the Lut and Kavir Deserts, Iran." *Environmental Science and Technology* 55, no. 9 (2021): 5993–6000. https://primo.uvm.edu/primo-explore/fulldisplay?docid=TN_cdi_proquest_journals_2524954941&context=PC&vid=UVM.

Abbate, Francesco, ed. *Egyptian Art.* Translated by H. A. Fields. London: Octopus Books, 1972 [1966].

Abbey, Edward. *Desert Solitaire.* Tucson: University of Arizona Press, 1988.

*Abrahams, Athol D., et al. *Geomorphology of Desert Environments.* London and New York: Chapman & Hall, 1994.

"Abu Simbel—Unparalleled Relocation Project." Atlas Copco. https://www.atlascopco.com/en-eg/history-in-egypt/landmark-projects-abu-simbel.

*Adolph, E. F., and Associates. *Physiology of Man in the Desert.* New York: Wiley Interscience, 1947.

Aïtel, Fazia. *We Are Imazighen: The Development of Algerian Berber Identity in Twentieth-Century Literature and Culture.* Gainesville: University Press of Florida, 2014.

Albanov, Valerian Ivanovich. *In the Land of White Death: An Epic Story of Survival in the Siberian Arctic.* Translated by Alison Anderson. New York: Modern Library, 2000 [1917].

Alcock, John. *When the Rains Come: A Naturalist's Year in the Sonoran Desert.* Tucson: University of Arizona Press, 2009.

Alexander, Caroline. *The Endurance: Shackleton's Legendary Antarctic Expedition.* New York: Knopf, 1998.

Allaby, Michael. *Deserts.* New York: Facts on File, 2001.

*———. *Droughts.* New York: Facts on File, 1998.

Allan, Tony, and Andrew Warren, eds. *Deserts: The Encroaching Wilderness* (A World Conservation Atlas). New York: Oxford University Press, 1993.

Alloway, David. *Desert Survival Skills.* Austin: University of Texas Press, 2000.

Amrousi, Mohamed El, et al. "Are Garden Cities in the Desert Sustainable? The Oasis City of Al Ain in the Emirate of Abu Dhabi." *International Review for Spatial Planning and Sustainable Development* 6A, no. 1 (2018): 79–94. https://primo.uvm.edu/primo-explore/fulldisplay?docid=TN_cdi_crossref_primary_10_14246_irspsd_6A_1_79&context=PC&vid=UVM.

Amundsen, Roald. *The South Pole: An Account of the Norwegian Antarctic Expedition in the "Fram," 1910–1912.* 2 vols. London and New York: John Murray and Lee Keedick, 1913.

An Atlas of the Sahara-Sahel: Geography, Economics and Security. Paris: OECD Publishing, 2014.

Annerino, John. *Dead in Their Tracks: Crossing America's Desert Borderlands in the New Era.* Tucson: University of Arizona Press, 2009 [1999].

Antarctic Bibliography (1951–1961; 1965–1998) was a periodical published by the Library of Congress.

Archibold, Randal C. "Border Fence Must Skirt Objections from Arizona Tribe." *New York Times*, September 20, 2006, A24.

———. "28-Mile Virtual Fence Is Rising Along the Border." *New York Times*, June 26, 2007, A12.

Arid Land Research and Management (1987–). https://www.tandfonline.com/journals/uasr20.

Atkins, William. *The Immeasurable World: Journeys in Desert Places.* New York: Doubleday, 2018.

Austin, Mary Hunter. *Land of Little Rain.* New York: Modern Library, 2003 [1903].

Bachelet, D., et al. "Climate Change Effects on Southern California Deserts." *Journal of Arid Environments* 127 (2016): 17–29. https://primo.uvm.edu/primo-explore/fulldisplay?docid=TN_cdi_proquest_miscellaneous_1785253394&context=PC&vid=UVM.

Bagnold, Ralph A. *Libyan Sands: Travel in a Dead World.* London: Michael Haag, 1987 [1935].

Bainbridge, David A. *A Guide for Desert and Dryland Restoration: New Hope for Arid Lands.* Washington, DC: Island Press, 2007.

Bakalar, Nicholas. "A Lizard That Builds with the Family in Mind." *New York Times*, May 17, 2011, D3.

Baldizzone, Tiziana, and Gianni Baldizzone. *Wedding Ceremonies: Ethnic Symbols, Costumes and Rituals.* Translated by David Radzinowicz Howell. Paris: Flammarion, 2001.

Balme, Jane, et al. "Long-term Occupation on the Edge of the Desert: Riwi Cave in the Southern Kimberley, Western Australia." *Archaeology in Oceania* 54, no. 1 (2019): 35–52. https://primo.uvm.edu/primo-explore/fulldisplay?docid=TN_cdi_proquest_journals_2334616185&context=PC&vid=UVM.

Bantu Studies; a Journal Devoted to the Scientific Study of Bantu, Hottentot and Bushman (1921–1941).

Barker, Graeme, and David Gilbertson, eds. *The Archaeology of Drylands: Living at the Margin.* New York: Routledge, 2000.

Barker, Graeme, et al. "Living at the Margin: Themes in the Archaeology of Drylands." In *The Archaeology of Drylands: Living at the Margin*, edited by Graeme Barker and David Gilbertson, 3–18. New York: Routledge, 2000.

Barnard, Alan. *Anthropology and the Bushman.* Oxford: Berg, 2007.

*Baron, Stanley Richard. *Desert Locust.* New York: Charles Scribner's Sons, 1972.

Bar-Oz, Guy, et al. "Caravanserai Middens on Desert Roads: A New Perspective on the Nabataean–Roman Trade Network across the Negev." *Antiquity* 96, no. 387 (2022): 592–610. https://primo.uvm.edu/primo-explore/fulldisplay?docid=TN_cdi_proquest_journals_2673835994&context=PC&vid=UVM.

*Batanouny, Kamal H. *Plants in the Deserts of the Middle East.* New York: Springer, 2001

Bearak, Barry. "For Some Bushmen, a Homeland Worth the Fight." *New York Times*, November 5, 2010, A1, A13.

Bearak, Max. "The High-Tech Gamble on Green Hydrogen." *New York Times*, March 12, 2023, Y1, Y12–Y13.

Beattie, Owen, and John Geiger. *Frozen in Time: The Fate of the Franklin Expedition.* London: Bloomsbury, 1987.

Beckwith, Carol, and Angela Fisher. *African Ceremonies.* 2 vols. New York: Harry N. Abrams, 1999.

Behnke, Roy, et al., eds. *The End of Desertification? Disputing Environmental Change in the Drylands.* Berlin and Heidelberg: Springer Verlag, 2016.

Bellamy, David. *David Bellamy's Arabian Light: An Artist's Journey through Deserts, Mountains and Souks.* Tunbridge Wells, Kent, UK: Search Press, 2022.

Benanav, Michael. "A Drive with the Devil." *New York Times*, May 5, 2019, TR1, TR4–5.

———. *Men of Salt: Crossing the Sahara on the Caravan of White Gold.* Guilford, CT: Lyons Press, 2006.

Bender, Gordon L., ed. *Reference Handbook on the Deserts of North America.* Westport, CT: Greenwood Press, 1982.

Benoit, Peter. *Deserts.* New York: Children's Press, 2011.

*Berger, Bruce. *A Desert Harvest.* New York: Farrar, Straus & Giroux, 2019.

Berkofsky, Louis, and Morton G. Wurtele, eds. *Progress in Desert Research.* Totowa, NJ: Roman & Littlefield, 1987.

Bickel, Lennard. *Mawson's Will: The Greatest Polar Survival Story Ever Written.* New York: Penguin Random House, 2000.

Billington, Craig C., et al. "Implications of Illegal Border Crossing and Drug Trafficking on the Management of Public Lands." In *Southwestern Desert Resources*, edited by William Halvorson, 109–22. Tucson: University of Arizona Press, 2010.

*Blackmore, Charles. *Crossing the Desert of Death: Through the Fiercesome Taklamakan.* New ed. London: John Murray, 2000.

Bolger, Daniel P. *Dragons at War: 2-34 Infantry in the Mojave.* Novato, CA: Presidio Press, 1986.

*Bone, Michael, et al. *Steppes: The Plants and Ecology of the World's Semi-arid Regions.* Portland, OR: Timber Press, 2015.

Botha, Laurette Isabella. *The Namib Desert: A Bibliography.* Cape Town: University of Cape Town Libraries, 1970.

*Bourseiller, Philippe, et al. *Sahara: The Call of the Desert.* Translated by Simon Jones. New York: Abrams, 2004.

Bowers, Nora, et al. *Cactus of Arizona Field Guide.* Cambridge, MN: Adventure Publications, 2008.

*Bowman, Isaiah. *Desert Trails of Atacama.* New York: American Geographical Society, 1924. (UVM online.)

Bradbury, Ellen. "The Southwest." In *Native American Arts and Crafts*, edited by Colin F. Taylor, 22–41. Edison, NJ: Chartwell Books, 2000.

Branch, Oliver, and Volker Wulfmeyer. "Deliberate Enhancement of Rainfall Using Desert Plantations." *Proceedings of the National Academy of Sciences—PNAS* 116, no. 38 (2019): 18841–47.

Breckle, Siegmar-W., et al., eds. *Arid Dune Ecosystems: The Nizzana Sands in the Negev Desert.* Berlin and Heidelberg: Springer, 2008.

Brett, Michael, and Elizabeth Fentress. *The Berbers.* Cambridge, MA: Blackwell, 1996.

Briggs, Lloyd Cabot. *Tribes of the Sahara.* Cambridge, MA: Harvard University Press, 1960.

Broad, William J. "Black-Market Trinkets from Space." *New York Times*, April 5, 2011, D1, D4.

*Brown, G. W., Jr., ed. *Desert Biology: Special Topics on the Physical and Biological Aspects of Arid Regions.* 2 vols. New York: Academic Press, 1968–1974.

*Brown, John. *Journey into the Desert.* New York: Oxford University Press, 2002.

Brown, Patricia Leigh. "In Sonora Desert, Helping Save Prickly Victims of Development." *New York Times*, April 14, 2006, A1, A18.

Broyles, Bill, and Michael P. Berman. *Sunshot: Peril and Wonder in the Gran Desierto.* Tucson: University of Arizona Press, 2006.

Bryan, Rorke. *Ordeal by Ice: Ships of the Antarctic.* Dobbs Ferry, NY: Sheridan House, 2011.

Buma, Brian. *The Atlas of a Changing Climate.* Portland, OR: Timber Press, 2021.

Burt, Christopher C. *Extreme Weather: A Guide and Record Book.* New York: W. W. Norton, 2004.

Burton, Sir Richard Francis. *Personal Narrative of a Pilgrimage to Al-Medinah and Meccah.* New York: G. P. Putnam & Co., 1856.

Butelski, Kazimierz. "Public Buildings of North Chile's Desert Architecture." *Materials Science and Engineering* 960, no. 2 (2020): 22059. https://primo.uvm.edu/primo-explore/fulldisplay?docid=TN_cdi_proquest_journals_2562889257&context=PC&vid=UVM.

Butler, Henrietta, ed. *The Tuareg or Kel Tamasheq: The People Who Speak Tamasheq and a History of the Sahara.* Norwich, UK: Unicorn Press, 2015.

Byrd, Richard E. *Alone: The Classic Polar Adventure.* New York: G. P. Putnam's Sons, 1938.

*Cable, Mildred, et al. *The Gobi Desert.* London: Hodder and Stoughton, [1942].

*Campbell, David. *The Crystal Desert.* New York: Houghton Mifflin, 1992.

Campbell, Iain, et al. *Birds of Australia: A Photographic Guide.* Princeton, NJ: Princeton University Press, 2015.

*Campbell, John Martin. *Few and Far Between: Moments in the North American Desert.* Santa Fe: Museum of New Mexico Press, 1997.

Campbell, Joseph. *The Way of the Animal Powers.* Vol. 1, *Historical Atlas of World Mythology.* New York: Alfred Von der Marck Editions; San Francisco: Harper & Row, 1983.

Carranza, Rafael. "Tohono O'odham Historic Sites at Risk as Border Wall Construction Advances in Arizona." Halting and Healing the Border Disorder, January 22, 2020. https://www.healingtheborderdisorder.org/2020/01/22/tohono-oodham-historic-sites-at-risk-as-border-wall-construction-advances-in-arizona.

Casey, Harry, and Anne Morgan. *Geoglyphs of the Desert Southwest: Earthen Art as Viewed from Above.* El Cajon, CA: Sunbelt Publications, 2019.

Casey, Maura J. "The Little Desert That Grew in Maine." *New York Times*, September 22, 2006, D7.

*Cassuto, David N. *Dripping Dry: Literature, Politics and Water in the Desert Southwest.* Ann Arbor: University of Michigan Press, 2001.

Castellanos, Alejandro E., et al. "Termite Activity on Green Tissues of Saguaro (*Carnegiea gigantea*) in the Sonoran Desert." In *Southwestern Desert Resources*, edited by William Halvorson, 171–80. Tucson: University of Arizona Press, 2010.

Cather, Willa. *Death Comes for the Archbishop.* Boston: Houghton Mifflin, 1938.

Chang, Kenneth. "NASA Images Give New View of a Saturn Moon: No Oceans, but a Sea of Sand." *New York Times*, May 9, 2006, D3.

Chaplin, Julia. "Art Blooms in the Desert." *New York Times*, April 21, 2006, D1, D6.

Chatterjee, Manas. "Desert Dermatoses (Thar Desert, India)." *Indian Journal of Dermatology* 62, no. 1 (2017): 52–58. https://primo.uvm.edu/primo-explore/fulldisplay?docid=TN_cdi_doaj_primary_oai_doaj_org_article_b7a006efdb8a480db1558e2c88a37cd8&context=PC&vid=UVM.

Chatwin, Bruce. *Songlines.* New York: Viking, 1987.

Cheek, Lawrence W., ed. *Voices in the Desert: Writings and Photographs.* San Diego: Harcourt Brace, 1995.

Chlachula, Jiri. "Between Sand Dunes and Hamadas: Environmental Sustainability of the Thar Desert, West India." *Sustainability* 13, no. 7 (2021): 3602.

Clark, Bill. *Death Valley: The Story Behind the Scenery.* Las Vegas, NV: K. C. Publications, 1995.

Clavero, Miguel, et al. "Fish Invading Deserts: Non-native Species in Arid Moroccan Rivers." *Aquatic Conservation* 25, no. 1 (2015): 49–60. https://primo.uvm.edu/primo-explore/fulldisplay?docid=TN_cdi_proquest_miscellaneous_1660423630&context=PC&vid=UVM.

*Cloudsley-Thompson, John. *The Desert.* New York: Putnam's Sons, 1977.

*Cooke, Ronald U., et al. *Geomorphology in Deserts.* Berkeley: University of California Press, 1973.

Cotter, Holland. "Wonders Blossomed in the Desert." *New York Times*, January 31, 2020, C15, C20.

*Crace, Jim. *Quarantine.* New York: Farrar, Straus & Giroux, 1997.

Crosswhite, Frank S., et al. "The Sonoran Desert." In *Reference Handbook on the Deserts of North America*, edited by Gordon L. Bender, 163–295. Westport, CT: Greenwood Press, 1982.

Culver, David C., and William B. White, eds. *Encyclopedia of Caves.* Amsterdam: Elsevier, 2005.

Dargis, Manohla. "A Teacher Caught between the Desert and a Deep Quandary." Review of *Far From Men. New York Times*, May 1, 2015, C6.

Davenport, Coral, et al. "Biden Vows Protection, but Not Monument Status, for Sacred Land." *New York Times*, December 2, 2022, A10.

Davidson, Robyn. *Desert Places.* New York: Viking, 1996.

———. *Tracks.* New York: Pantheon, 1980.

Davis, Diana K. *The Arid Lands: History, Power, Knowledge.* Cambridge, MA: MIT, 2016.

*Dean, W. Richard J. *Nomadic Desert Birds.* New York: Springer, 2004.

De Roy, Tui, et al. *Penguins: The Ultimate Guide.* 2nd ed. Princeton, NJ: Princeton University Press, 2022.

*de Saint-Exupéry, Antoine. *Wind, Sand, and Stars.* New York: Reynal & Hitchcock, [1939].

the desert. New York: Thames & Hudson, 2000.

Desert Bird Gardening. Tucson: Arizona Native Plant Society and Tucson Audubon Society, 1997.

The Desert Realm: Lands of Majesty and Mystery. Washington, DC: National Geographic Society, 1982.

Desert Wildflowers. Tucson: Arizona Native Plant Society, 1991.

*de Villiers, Marq. *Windswept: The Story of Wind and Weather.* New York: Walker and Co., 2006.

Dorfman, Ariel. *Desert Memories: Journeys through the Chilean North.* Washington, DC: National Geographic Society, 2004.

Doughty, Charles M. *Travels in Arabia Deserta.* 2 vols. Cambridge: Cambridge University Press, 1888.

Durou, Jean-Marc. *Sahara: The Forbidden Sands [Sahara: La Passion du Desert].* New York: Abrams, 2000.

Dutch, Steven I., ed. *Encyclopedia of Global Warming.* 3 vols. Pasadena, CA: Salem Press, 2010.

Eboch, M. M. *Desert Biomes around the World.* Mankato, MN: Capstone Press, 2020.

Ecological Management & Restoration (1999–).

Edwards, E. I. *Desert Harvest.* Los Angeles: Westernlore Press, 1962. (UVM online.)

———. *Desert Voices: A Descriptive Bibliography.* Los Angeles: Westernlore Press, 1958. (UVM online.)

*Ehrenreich, Ben. *Desert Notebooks: A Road Map for the End of Time.* Berkeley, CA: Counterpoint Press, 2020.

Einhorn, Catrin. "From Tiny Geckos to King Cobras, 20% of Reptiles Face Extinction Risk." *New York Times*, April 28, 2022, A9.

———. "Nations Approve U.N. Pact Aiming to Protect Nature." *New York Times*, December 20, 2022, A10.

Ekernas, L. Stefan, et al. "Desert patoralists' negative and positive effects on rare wildlife in the Gobi." *Conservation Biology* 31, no. 2 (2017): 269–77. https://primo.uvm.edu/primo-explore/fulldisplay?docid=TN_cdi_proquest_miscellaneous_1881749332&context=PC&vid=UVM.

*El Daly, Okasha, et al., eds. *Desert Travellers: From Herodotus to T. E. Lawrence.* Durham, NC: Astene, 2000.

Elbein, Asher. "Invasive Donkeys Have Become Prey in Death Valley." *New York Times*, August 16, 2022, D4.

"11 Incredible Deserts in South America." Wildlife Explained, July 15, 2022. https://www.wildlifeexplained.com/deserts-in-south-america.

Elkin, A. P. *The Australian Aborigines.* Garden City, NY: Doubleday, 1964 [1938].

Englebert, Victor. *Wind, Sand & Silence: Travels with Africa's Last Nomads.* San Francisco: Chronicle Books, 1992.

Etherton, P. T. *Across the Great Deserts.* New York: McGraw Hill, 1948. (UVM online.)

European Space Agency Globcover Portal, http://due .esrin.esa.int/page_globcover.php.

Findley, Rowe. *Great American Deserts.* Washington, DC: National Geographic Society, 1972.

Finkelman, Paul, ed. *Macmillan Encyclopedia of World Slavery.* 2 vols. New York: Simon and Schuster Macmillan, 1998.

Fisher, Angela. *Africa Adorned.* New York: Abrams, 1984.

Flegg, Jim. *Deserts: Miracle of Life.* New York: Facts on File, 1993.

Fluehr-Lobban, Carolyn. *Ethics and Anthropology: Ideas and Practice.* Lanham, MD: Altamira Press, 2013.

Folch, Raymon, et al., eds. *Encyclopedia of the Biosphere: Humans in the World's Ecosystem.* Vol. 4, *Deserts.* Translated by Trevor Foskett. Detroit: Gale Group, 2000.

Forsyth, Isla. "Piracy on the High Sands: Covert Military Mobilities in the Libyan Desert, 1940–1943." *Journal of Historical Geography* 58 (2017): 61–70.

Fountain, Henry. "Western Drought Is the Worst in 1,200 Years." *New York Times*, February 15, 2022, A12.

Fowler, Brenda. "Scientists Explore Ancient Lakefront Life, in the Sahara." *New York Times*, January 27, 2004, D3.

*Fretwell, Peter. *Antarctic Atlas: New Maps and Graphics That Tell the Story of A Continent.* London: Particular Books, 2020.

Gautam, Ritesh, et al. "Satellite Observations of Desert Dust-induced Himalayan Snow Darkening." *Geophysical Research Letters* 40, no. 5 (2013): 988–93. https://primo.uvm.edu/primo-explore/ fulldisplay?docid=TN_cdi_proquest_journals _1760152497&context=PC&vid=UVM.

*Gearon, Eamonn. *The Sahara: A Cultural History.* Oxford: Oxford University Press, 2011.

*George, Alexander S., et al., eds. *Flora of Australia.* 58 vols. Canberra, AU: AGPS Press, 1981–1992.

*George, Uwe. *In the Deserts of this Earth.* Translated by Richard and Clara Winston. New York: Harcourt Brace Jovanovich, 1977.

Gilley, Bruce. "The Case for Colonialism." *Third World Quarterly* 8 (September 2017). https://www.tandfon line.com/doi/abs/10.1080/01436597.2017.1369037: retracted. Nevertheless, the article may be read at https://www.nas.org/academic-questions/31/2/the _case_for_colonialism.

Glantz, Michael H., ed. *Desertification: Environmental Degradation in and around Arid Lands.* Boulder, CO: Westview Press, 1977.

*Glen, Simon, et al. *Sahara Handbook.* London: Roger Lascelles, 1987.

Glueck, Nelson. *Rivers in the Desert: A History of the Negev.* New York: Grove Press, 1960.

Goudie, Andrew S. "Desert Dust and Human Health Disorders." *Environment International* 63 (2014): 101–13. https://primo.uvm.edu/primo-explore/full display?docid=TN_cdi_proquest_miscellaneous _1685770707&context=PC&vid=UVM.

———, ed. *Encyclopedia of Geomorphology.* 2 vols. New York: Routledge, 2004. https://courses.ess .washington.edu/ess-306/links/Goudie_Encyclopedia _of_Geomorphology.pdf.

———. *Great Desert Explorers.* London: Society for Libyan Studies, 2016.

*The Green Planet, "Desert World" (TV episode), 2022. Directed by Paul Williams.

Grimal, Pierre, ed. *Larousse World Mythology.* Translated by Patricia Beardsworth. Secaucus, NJ: Chartwell Books, 1976 [1963].

Guide to Three Rivers Petroglyph Site and Picnic Area (brochure). Las Cruces, NM: Bureau of Land Management, n.d.

Halvorson, William, et al., eds. *Southwestern Desert Resources.* Tucson: University of Arizona Press, 2010.

Ham, Anthony. "'Ghost Bird' Haunts Those Searching for It." *New York Times*, January 4, 2022, D1, D4.

Hammer, Joshua. *The Bad-Ass Librarians of Timbuktu: And Their Race to Save the World's Most Precious Manuscripts.* New York: Simon & Schuster, 2016.

———. " Hiking Deep into Dogon Country in Mali." *New York Times*, March 7, 2010, TR8.

Hanbury-Tenison, Robin, ed. *The Great Explorers.* New York: Thames and Hudson, 2010.

Harding, Adella. "Nevada Leads US in Toxic Releases." *Elko Daily Free Press*, March 9, 2022. https://elko daily.com/mining/nevada-leads-us-in-toxic-releases/ article_5df955ca-8970-5fbd-9242-3d0c0ead7aa1.html.

Harris, Nathaniel. *Atlas of the World's Deserts.* New York: Fitzroy Dearborn, 2003.

Harris, Rick. *Easy Field Guide to Rock Art Symbols of the Southwest.* Phoenix, AZ: American Traveler Press, 2003 [1995].

*Harrison, Peter. *Fortress Monasteries of the Himalayas: Tibet, Ladakh, Nepal, and Bhutan.* Oxford and New York: Osprey, 2011.

Hayden, Julian D. *Field Man: Life as a Desert Archeologist.* Edited by Bill Broyles and Diane E. Boyer. Tucson: University of Arizona Press, 2011.

*Hayes, Derek. *Historical Atlas of the Arctic.* Seattle: University of Washington Press, 2003.

Haynes, Roslynn D. *Desert: Nature and Culture.* London: Reaktion Books, 2013.

*———. *Seeking the Centre: The Australian Desert in Literature, Art and Film.* Cambridge: Cambridge University Press, 1999.

Hazleton, Lesley. *Where Mountains Roar: A Personal Report from the Sinai and Negev Desert.* New York: Holt, Rinehart and Winston, 1980.

Heacox, Kim. *Antarctica: The Last Continent.* Washington, DC: National Geographic Society, 1998.

*Hedin, Sven. *Across the Gobi Desert.* London: G. Routledge and Sons, [1931?].

Henni, Samia, ed. *Deserts Are Not Empty.* New York: Columbia University Press, 2023.

*Herbert, Wally. *Polar Deserts.* London: Collins; New York: F. Watts, [1971].

Hill, Russell. *Desert War.* New York: Knopf, 1942. (UVM online.)

Hillel, Daniel. *Negev: Land, Water, and Life in a Desert Environment.* New York: Praeger, 1982.

Hoffman, Katherine E., and Susan Gilson Miller. Introduction to *Berbers and Others: Beyond Tribe and Nation in the Maghrib*, edited by Katherine E. Hoffman and Susan Gilson Miller, 1–2. Bloomington: Indiana University Press, 2010.

Hogue, Lawrence. *All the Wild and Lonely Places: Journeys in a Desert Landscape.* Washington, DC: Island Press/Shearwater Books, 2000.

Holland, Glenn S. *Gods in the Desert: Religions of the Ancient Near East.* Lanham, MD: Rowman & Littlefield, 2009.

*Howard, Richard A. *Sun, Sand and Survival: An Analysis of Survival Experiences in Desert Areas.* Maxwell Air Force Base, AL: Arctic, Desert, Tropic Information Center, Air University, [1953].

Hubbard, Ben, and Asmaa Al-Omar. "Screeching Across the Desert: A Relic of a Railway Dream." *New York Times*, June 12, 2022, A4.

Hunka, George. "In 'Desert Sunrise,' a West Bank 'Godot.'" Review of *Desert Sunrise. New York Times*, April 18, 2006. https://www.nytimes.com/2006/04/18/theater/reviews/in-desert-sunrise-a-west-bank-godot.html.

Huntford, Roland. *The Last Place on Earth: Scott and Amundsen's Race to the South Pole.* New York: Modern Library, 1999.

Hussein, Ethar A., et al. "Do Anthropogenic Activities Affect Floristic Diversity and Vegetation Structure More Than Natural Soil Properties in Hyper-Arid Desert Environments?" *Diversity* (Basel) 13, no. 4 (2021): 157. https://primo.uvm.edu/primo-explore/fulldisplay?docid=TN_cdi_doaj_primary_oai_doaj_org_article_533545ec9e46472593368e8db3887693&context=PC&vid=UVM.

Hutchinson, Alex. "A Walkabout at the Center of Down Under." *New York Times*, November 29, 2009, TR9.

Hyde, Philip. *Drylands: The Deserts of North America.* San Diego: Harcourt Brace Jovanovich, 1987.

Isaacson, Rupert. *The Healing Land: The Bushmen and the Kalahari Desert.* New York: Grove Press, 2001.

*Iverson, Peter. *Diné: A History of the Navajos.* Albuquerque: University of New Mexico Press, 2002.

Jabbur, Jibrail S. *The Bedouins and the Desert: Aspects of Nomadic Life in the Arab East.* Translated by Lawrence I. Conrad. Albany: State University of New York Press, 1995.

Jacobs, Jessica. *Sex, Tourism and the Postcolonial Encounter: Landscapes of Longing in Egypt.* Farnham, Surrey, UK: Ashgate, 2010.

Jaeger, Edmund C. *Desert Wildlife.* Stanford, CA: Stanford University Press, 1968 [1950].

*James, George Wharton. *The Indians of the Painted Desert Region: Hopis, Navahoes, Wallapais, Havasupais.* Boston: Little, Brown, and Company, 1903.

Janz, Lisa, et al. "Zaraa Uul: An Archaeological Record of Pleistocene-Holocene Palaeoecology in the Gobi Desert." *PloS one* 16, no. 4 (2021): e0249848. https://primo.uvm.edu/primo-explore/fulldisplay?docid=TN_cdi_plos_journals_2510234039&context=PC&vid=UVM.

Jasper, David. *The Sacred Desert: Religion, Literature, Art, and Culture.* Malden, MA: Blackwell, 2004.

Johnson, Kirk. "A Drier and Tainted Nevada May be Legacy of a Gold Rush." *New York Times*, December 30, 2005, A1, A20.

Journal of Arid Environments (1978–). https://www.sciencedirect.com/journal/journal-of-arid-environments.

Journal of Arid Land (2009–). http://jal.xjegi.com/EN/1674-6767/current.shtml.

*Judd, Tony. *Rock Art of the Eastern Desert of Egypt: Content, Comparisons, Dating and Significance.* Oxford: Archaeopress, 2009.

Juskalian, Russ. "The Fort and the Desert." *New York Times*, March 30, 2014, TR11.

Kappel-Smith, Diana. *Desert Time: A Journey through the American Southwest.* Tucson: University of Arizona Press, 1996.

Karabell, Zachary. *Parting the Desert: The Creation of the Suez Canal.* New York: Alfred A Knopf, 2003.

Katanich, Deloresz. "How Climate Change Is Turning Once Green Madagascar into a Desert." Euronews, March, 20, 2022. https://www.euronews.com/green/2022/03/20/how-climate-change-is-turning-once-green-madagascar-into-a-desert.

Keating, Jennifer. *On Arid Ground: Political Ecologies of Empire in Russian Central Asia.* New York: Oxford University Press, 2022.

*Keegan, John. *A History of Warfare.* New York: Knopf, 1994.

*Kennedy, David, et al. *Rome's Desert Frontier: From the Air.* Austin: University of Texas Press, 1990.

Kershner, Isabel. "Is It a Burning Bush? Mt. Sinai? Solstice Bolsters a Claim." *New York Times*, December 31, 2021, A6.

Khalifa, Mohamed Mounir, and Mohamed Ibrahim Abdelall. "Ecological Desert Settlement Egypt

Western Desert" [*sic*]. *Alexandria Engineering Journal* 58, no. 1 (2019): 291–301. https://primo.uvm.edu/primo-explore/fulldisplay?docid=TN_cdi_doaj_primary_oai_doaj_org_article_61745ebcdd624219b14afeaf2d38d466&context=PC&vid=UVM.

Kimmelman, Michael. "One Visionary's Desert Dream." *New York Times*, August 21, 2022, AR1, AR10–16.

Kingsley, Patrick. "'It's the Eye of Sauron.' A Solar Tower Dazzles and Annoys Its Neighbors." *New York Times*, October 10, 2022, A6.

*Kirk, Ruth. *Deserts: The American Southwest.* Boston: Houghton Mifflin, 1973.

Klein, Joanna. "On a Mars-Like Landscape, Some Not-So-Alien Life Forms." *New York Times*, July 30, 2019, D2.

Kraft, Dina. "From Far Beneath the Israeli Desert, Water Sustains a Fertile Enterprise." *New York Times*, January 2, 2007, D3.

Krebbs, Karen. *Desert Life: A Guide to the Southwest's Iconic Animals and Plants and How They Survive.* Cambridge, MN: Adventure Publications, 2017

*Kroeber, A. L. *Handbook of the Indians of California.* Smithsonian Institution, Bureau of American Ethnology, Bulletin No. 78. Washington, DC: Government Printing Office, 1925.

Krutch, Joseph Wood. *The Desert Year.* New York: Viking Press, 1971 [1951].

———. *The Voice of the Desert: A Naturalist's Interpretation.* New York: William Sloane Associates, 1967 [1954].

Kuletz, Valerie, L. *The Tainted Desert: Environmental Ruin in the American West.* New York: Routledge, 1998.

*Kumar, Arun, et al. *Geo Spatial Atlas for the Wetland Birds of Thar Desert Rajasthan.* Kolkata, India: ZSI, 2006.

*Lamb, Anthony. *Sand: A Photographic Depiction of the Arabian Desert.* Dubai: Motivate Media Group, 2021.

Langloh Parker, K. *Wise Women of the Dreamtime: Aboriginal Tales of the Ancestral Powers.* Edited by Johanna Lambert. Rochester, VT: Inner Traditions International, 1993.

Lansing, Alfred. *Endurance: Shackleton's Incredible Voyage.* New York: McGraw-Hill, 1959.

Larson, Peggy. *Deserts of America.* Englewood Cliffs, NJ: Prentice-Hall, 1970.

*Larson, Peggy Pickering, with Lane Larson. *The Deserts of the Southwest.* San Francisco: Sierra Club Books, [2000].

*Lattimore, Owen. *The Desert Road to Turkestan.* Boston: Little, Brown, and Company, 1929.

*Lawrence, T. E. *Seven Pillars of Wisdom: A Triumph.* Garden City, NY: Doubleday, 1966 [1935].

Layne, Ken. *Desert Oracle: Strange True Tales from the American Southwest.* Vol. 1. New York: MCD, Picador, 2021.

Lazagabaster, Ignacio A., et al. "Cave Paleozoology in the Judean Desert: Assembling Records of Holocene Wild Mammal Communities." *Journal of Quaternary Science* 37, no. 4 (2022): 651–63.

Leary, Warren E. "New Map of Antarctica Brings Frozen Landscape into Focus." *New York Times*, December 4, 2007, D2.

Lee, W. Storrs. *The Great California Deserts.* New York: G. P. Putnam's Sons, 1963.

*Le Clézio, J. M. G. *Desert.* Translated by C. Dixon. Boston: David R. Godine, 2009.

Lehmann, Philipp. *Desert Edens: Colonial Climate Engineering in the Age of Anxiety.* Princeton, NJ: Princeton University Press, 2022.

*Leopold, A. Starker. *The Desert.* Alexandria, VA: Time-Life Books, 1978 [1961].

Lewis, Paul. "Wilfred Thesiger, 93, Dies; Explored Arabia." *New York Times*, August 27, 2003, C14.

Lim, Sophak, et al. "50,000 Years of Vegetation and Climate Change in the Southern Namib Desert, Pella, South Africa." *Palaeogeography, Palaeoclimatology, Palaeoecology* 451 (2016): 197–209. https://primo.uvm.edu/primo-explore/fulldisplay?docid=TN_cdi_hal_primary_oai_HAL_hal_01817628v1&context=PC&vid=UVM.

Lingenfelter, Richard E. *Death Valley and the Amargosa: A Land of Illusion.* Berkeley: University of California Press, 1986.

Lopez, Barry. *Arctic Dreams: Imagination and Desire in a Northern Landscape.* New York: Vintage, 1986.

López Torres, Núria. "An Intimate Look at Mexico's Indigenous Seri People." *New York Times*, June 6, 2022, A6.

Loti, Pierre. *Le Désert.* Paris: Calmann Lévy, 1895.

Lovegrove, Barry G. "Fog Basking by Namib Desert Weevils." *Frontiers in Ecology and the Environment* 18, no. 9 (2020): 495. https://primo.uvm.edu/primo-explore/fulldisplay?docid=TN_cdi_proquest_journals_2468385723&context=PC&vid=UVM.

*Luard, Nicholas. *The Last Wilderness: A Journey across the Great Kalahari Desert.* New York: Simon and Schuster, 1981.

Luo, Lihui, et al. "The Hidden Costs of Desert Development." *Ambio* 49, no. 8 (2020): 1412–22. https://primo.uvm.edu/primo-explore/fulldisplay?docid=TN_cdi_pubmedcentral_primary_oai_pubmedcentral_nih_gov_7239957&context=PC&vid=UVM.

Lyons, Gary. *Desert Plants: A Curator's Introduction to the Huntington Desert Garden.* San Marino, CA: Huntington Library Press, 2007.

Mabbutt, J. A. *Desert Landforms.* Cambridge, MA: MIT Press, 1977.

Mabry, T. J., et al. *Creosote Bush: Biology and Chemistry of* Larrea *in New World Deserts.* Stroudsburg, PA: Dowden, Hutchinson & Ross, 1977.

Mack, John E. *A Prince of Our Disorder: The Life of T. E. Lawrence.* Boston: Little, Brown, 1976.

*Mack, Susanne, and Anthony Ham. *Deserts of the World* [*Wüsten der Welt*].Cologne: Koenemann, 2019.

MacLeod, Calum. "Chinese Railroad Link to Tibet Launches." *USA Today*, July 3, 2006, 4A.

MacMahon, James A., ed. *Deserts.* New York: Knopf, 1985.

MacQuitty, Miranda. *Desert.* New York: Alfred A. Knopf, 1994.

Madadin, Mohammed, et al. "Desert Related Death." *International Journal of Environmental Research and Public Health* 18, no. 21 (2021): 11272. https://primo.uvm.edu/primo-explore/fulldisplay?docid=TN_cdi_doaj_primary_oai_doaj_org_article_91aa215af6a64e-6c9e7ab5ad324135f5&context=PC&vid=UVM.

Mallos, Tess. *The Complete Middle East Cookbook.* Sydney: Landsdowne Press, 1982.

Man, John. *Gobi: Tracking the Desert.* New Haven, CT: Yale University Press, 1997.

Mandaville, James P. *Bedouin Ethnobotany: Plant Concepts and Uses in a Desert Pastoral World.* Tucson: University of Arizona Press, 2011.

Manley, Deborah, et al., eds. *Traveling through Egypt: From 450 B.C. to the Twentieth Century.* Cairo: American University in Cairo Press, 2004.

Mares, Michael A. *A Desert Calling: Life in a Forbidding Landscape.* Cambridge, MA: Harvard University Press, 2002.

Mares, Michael A., ed. *Encyclopedia of Deserts.* Norman: University of Oklahoma Press, 1999.

Martin, Genna. "Shipwrecks and a Treasure, Solitude, in Namibia." *New York Times*, January 24, 2022, A10.

Martin, Michael. *Deserts of the Earth: Extraordinary Images of Extreme Environments.* Translated by David H. Wilson. New York: Thames and Hudson, 2004.

Massey, Lyle, et al., eds. *The Invention of the American Desert: Art, Land, and the Politics of Environment.* Oakland: University of California Press, 2021.

McCarthy, Christopher, et al. "Preserving the Gobi: Identifying Potential UNESCO World Heritage in Mongolia's Gobi Desert." *Journal of Asia-Pacific Biodiversity* 15, no. 4: 500–517. https://primo.uvm.edu/primo-explore/fulldisplay?docid=TN_cdi_nrf_kci_oai_kci_go_kr_ARTI_10108689&context=PC&vid=UVM.

*McGinnies, William G. *Discovering the Desert: Legacy of the Carnegie Desert Botanical Laboratory.* Tucson: University of Arizona Press, 1987.

McNamee, Gregory, ed. *The Sierra Club Desert Reader: A Literary Companion.* San Francisco: Sierra Club Books, 1995.

Medina, Jennifer. "In Nevada, a Millennium of Messages in Jeopardy." *New York Times*, January 6, 2011, A14.

Meher, Ramona. *Shifting Sands: The Story of Sand Dunes.* New York: John Day, 1968.

Mei, Ying, et al. "On the Mountain-River-Desert Relation." *Thermal Science* 25, no. 6, Part B (2021): 4817–22. https://primo.uvm.edu/primo-explore/fulldisplay?docid=TN_cdi_doaj_primary_oai_doaj_org_article_9e84e36ddb7c4416ad7ec8ec05510f15&context=PC&vid=UVM.

*Mendelsohn, John, et al. *Atlas of Namibia: A Portrait of the Land and its People.* Cape Town, South Africa: David Philip Publishers, 2002.

Merlin, Pinau. *A Guide to Southern Arizona Bird Nests & Eggs.* Tucson: Arizona Sonora Desert Museum, 2001.

*Merton, Thomas. *The Wisdom of the Desert: Sayings of the Desert Fathers of the Fourth Century.* Translated by Thomas Merton. New York: New Directions, [1961].

Meskell, Lynn, ed. *Archeology Under Fire: Nationalism, Politics and Heritage in the Eastern Mediterranean and Middle East.* New York: Routledge, 1998.

Meteor Crater: Experience the Impact (brochure). Flagstaff, AZ: Meteor Crater Enterprises, n.d.

*Middleton, Nick, and David Thomas, eds. *World Atlas of Desertification.* 2nd ed. New York: Edward Arnold, 1997.

Miller, David. *Deserts: A Panoramic Vision.* Edison, NJ: Chartwell Books, 2006.

*Mills, Barbara, et al., eds. *The Oxford Handbook of Southwest Archaeology.* New York: Oxford University Press, 2017.

*Millward, James A. *The Silk Road: A Very Short Introduction.* Oxford: Oxford University Press, 2013.

*Moerman, Daniel E. *Native American Ethnobotany.* Portland, OR: Timber Press, 1998.

Moffitt, Ian, and the Editors of Time-Life Books. *The Australian Outback.* Amsterdam: Time Life International, 1976.

Moore, Michael. *Medicinal Plants of the Desert and Canyon West.* Santa Fe: Museum of New Mexico Press, 1989.

*Moorhead, Alan. *The Desert War: The North African Campaign, 1940–1943.* New York: Penguin Books, 2001 [1945].

*Morgan, Mike. *Sting of the Scorpion: The Inside Story of the Long Range Desert Group.* Stroud, Gloucestershire, UK: Sutton, 2000.

*Morhardt, Sia, and Emil Morhardt. *California Desert Flowers: An Introduction to Families, Genres, and Species.* Berkeley: University of California Press, 2004.

Morris, Mary. *Wall to Wall: From Beijing to Berlin by Rail.* New York: Penguin Books, 1992.

*Morton, Steve. *Australian Deserts: Ecology and Landscapes.* Clayton South, AU: CSIRO Publishing, [2022].

Mulvihill, Keith. "A Road Trip on the Edge of America." *New York Times*, January 11, 2008, D1, D6.

Murphy, Heather. "A Short-Term Rental Rush Fuels Debate." *New York Times*, April 11, 2022, B8.

Murray, Peter. *Deserts.* Plymouth, MN: Child's World, 1997.

Nabhan, Gary Paul. *Gathering the Desert.* Tucson: University of Arizona Press, 1997 [1985].

———, ed. *The Nature of Desert Nature.* Tucson: University of Arizona Press, 2020.

Nachtigal, Gustav. *Sahara and Sudan.* 3 vols. Translated by Allan G. B. Fisher, et al. Atlantic Highlands, NJ: Humanities Press, 1974–1987 [1879–1881].

Naiman, Robert J., and David L Soltz. *Fishes in North American Deserts.* New York: John Wiley & Sons, 1981.

Nansen, Fridtjof. *Farthest North: The Incredible Three-Year Voyage to the Frozen Latitudes of the North.* New York: Penguin Random House, 1999 [1897].

Nevada: Official State Map. Carson City: Department of Transportation, 2002.

New Mexico Road and Recreation Atlas. Santa Barbara, CA: Benchmark Maps, 2020 [1995].

Nicas, Jack. "Transforming Brazil's Fertile Northeast to Desert, in Slow Motion." *New York Times*, December 3, 2021, A1, A10–11.

Nicolaisen, Johannes, and Ida Nicolaisen. *The Pastoral Tuareg: Ecology, Culture, and Society.* 2 vols. New York: Thames & Hudson, 1997.

Nomadic Peoples (1997–).

Norment, Christopher J. "In a Biologist's Heart, Facts and Feelings." *Chronicle of Higher Education* 3 (October 2014): B20.

*———. *Relicts of a Beautiful Sea: Survival, Extinction, and Conservation in a Desert World.* Chapel Hill: University of North Carolina Press, 2014.

Norte, Federico. "Climate." *Encyclopedia of Deserts.* Edited by Michael A. Mares. Norman: University of Oklahoma Press, 1999.

Núñez, Lautaro, et al. "The Temple of Tulán-54: Early Formative Ceremonial Architecture in the Atacama Desert." *Antiquity* 91, no. 358 (2017): 901–15. https://primo.uvm.edu/primo-explore/fulldisplay?docid=TN_cdi_proquest_journals_1977738543&context=PC&vid=UVM.

Nuwer, Rachel. "In 'Circle' Mystery, a Theory That Points to Termites Crumbles." *New York Times*, November 22, 2022, D2.

*Oldfield, Sara. *Deserts: The Living Drylands.* Cambridge, MA: MIT Press, 2004.

Page, David. "Rock Art Redefines 'Ancient.'" *New York Times*, December 18, 2009, C35.

Page, Jake, and the Editors of Time-Life Books. *Arid Lands.* Alexandria, VA: Time-Life Books, 1984.

Pailes, Richard A. "Desert Peoples." In *Encyclopedia of Deserts*, edited by Michael A. Mares, 156–67. Norman: University of Oklahoma Press, 1999.

*Palgrave, W. G. *Personal Narrative of a Year's Journey through Central and Eastern Arabia.* London: Macmillan and Co., 1868.

Parrott-Sheffer, Chelsey. "Mogollon Culture." *Encyclopedia Britannica*, March 4, 2020. https://www.britannica.com/topic/Mogollon-culture.

*Passarge, Siegfried. *The Kalahari Ethnographies (1896–1898) of Siegfried Passarge: Nineteenth Century Khoisan- and Bantu-Speaking Peoples.* Cologne: Rüdiger Köppe, 1997.

Pavlik, Bruce. M. *The California Deserts: An Ecological Rediscovery.* Berkeley: University of California Press, 2008.

Paylore, Patricia, ed. *Sonoran Desert: A Retrospective Bibliography.* Tucson: University of Arizona, Office of Arid Lands Studies, [1976].

Paylore, Patricia, and J. A. Mabbutt, eds. *Desertification: World Bibliography Update 1976–1980.* Tucson: University of Arizona, Office of Arid Land Studies, 1980.

Paylore, Patricia, and W. G. McGinnies, comps. *Desert Research: Selected References 1965–1968.* Natick, MA: Earth Sciences Laboratory, US Army, 1969.

*Pedró-Alió, Carlos. *Las plantas de Atacama: El desierto cálido más árido del mundo.* Madrid: CSIC, Catarata, 2021.

Petroglyph (brochure). Albuquerque, NM: National Park Service, n.d.

Petrov, M. P. *Deserts of the World.* Translated by IPST staff. New York: John Wiley & Sons, 1976 [1973].

Piantadosi, Claude A. *The Biology of Human Survival: Life and Death in Extreme Environments.* New York: Oxford University Press, 2003.

Plog, Stephen. *Ancient Peoples of the American Southwest.* 2nd ed. London and New York: Thames & Hudson, 2008 [1997].

*Polis, Gary A., ed. *The Ecology of Desert Communities.* Tucson: University of Arizona Press, 1991.

Polo, Marco. *The Travels.* Translated by Nigel Cliff. London: Penguin, 2015.

Pond, Alonzo W. *The Desert World.* Westport, CT: Greenwood, 1975 [1962].

Pourkhorsandi, Hamed, et al. "Meteorites from the Lut Desert (Iran)." *Meteoritics & Planetary Science* 54, no. 8 (2019): 1737–63. https://primo.uvm.edu/primo-explore/fulldisplay?docid=TN_cdi_hal_primary_oai_HAL_hal_02144596v1&context=PC&vid=UVM.

Powell, Lawrence Clark. *The Creative Literature of the Arid Lands: Essays on the Books and Their Writers* (Southwest Classics). Pasadena: Ward Ritchie Press, 1975.

Prelorenzo, Claude. "The Desert as a (Movie) Myth." In *Desert Tourism: Tracing the Fragile Edges of Development*, edited by Virginie Picon-Lefebvre, et al., 20–33. Cambridge, MA: Harvard University Graduate School of Design, 2011.

Price, Tom. "From the Depths, a Cathedral Emerges." *New York Times*, April 8, 2005, D1, D2.

Przywolnik, Kathryn. "Long-Term Transitions in Hunter-Gatherers of Coastal Northwestern Australia." In *Desert Peoples: Archeological Perspectives,* edited by Peter Veth, et al., 177–205. Malden, MA: Blackwell Publishing, 2005.

Quézel, Pierre. *La Végétation du Sahara: du Tchad à la Mauritanie.* Stuttgart: G. Fischer; Paris: Masson et Cie, 1965.

Ramsey, Nancy. "In the Gobi, Camel Family Values." *New York Times*, May 30, 2004, AR9.

Rao, Nina. *Himalayan Desert: Tibet, Ladakh, Lahul, Spiti, Mustang.* New Delhi: Lustre Press, Roli Books, 1999.

Regenold, Stephen. "Lonesome Highway to Another World?" *New York Times*, April 13, 2007, D1, D7.

Reichard, Gladys A. *Navaho Medicine Man Sandpaintings.* New York: Dover Publications, 1977 [1939].

Rex, Markus. *The Greatest Polar Expedition of All Time: The Arctic Mission to the Epicenter of Climate Change.* Translated by Sarah Pybus. Vancouver, BC: Greystone Books, 2022.

Rhode, David. *Native Plants of Southern Nevada: An Ethnobotany.* Salt Lake City: University of Utah Press, 2002.

Roberts, David. *Alone on the Ice: The Greatest Survival Story in the History of Exploration.* New York: W. W. Norton, 2013.

Rocha, Joana L., et al. "Life in Deserts: The Genetic Basis of Mammalian Desert Adaption." *Trends in Ecology and Evolution* 36, no. 7 (2021): 637–50. https://www.cell.com/trends/ecology-evolution/fulltext/S0169-5347(21)00074-4.

Rock Art Research (1984–).

Rodrigues, Daniel. "Taking a Desert Road Trip along the King's Highway to See Jordan's Treasures." *New York Times*, April 3, 2022, Y18.

Rodriguez, Junius P. *The Historical Encyclopedia of World Slavery.* 2 vols. Santa Barbara, CA: ABC-CLIO, 1997.

Romero, Aldemaro. *Cave Biology: Life in Darkness.* New York: Cambridge University Press, 2009.

Romero, Simon, Roni Caryn Rabin, and Mark Walker. "Life Expectancy Plunge Puts Number to Misery in Native Communities." *New York Times*, September 1, 2022, A14.

Romero, Simon, and Andrea Zarate. "Buried in Peru's Desert, Marine Fossils Draw Smugglers." *New York Times*, December 12, 2010, Y6.

*Rooyen, Noel Van. *Flowering Plants of the Kalahari Dunes.* Lynnwood, South Africa: Ekotrust, 2001.

Rosen, Steven A. "The Decline of Desert Agriculture: A View from the Classical Period Negev." In *The Archaeology of Drylands: Living at the Margin,* edited by Graeme Barker, et al., 45–62. New York: Routledge, 2000.

Rosenthal, Elisabeth. "Water Is a New Battleground in a Drying Spain." *New York Times*, June 3, 2008, A1, A12.

Rosenthal, Gerald A. *Sonoran Desert Life: Understanding, Insights, and Enjoyment.* Scottsdale, AZ: Academic Insights, 2007.

Rothberg, Daniel. "The Coming Crisis on the Colorado River." *New York Times*, August 7, 2022, SR8.

*Rothman, Hal K., and Sara Dant Ewert, eds. *Encyclopedia of American National Parks.* Armonk, NY: M. E. Sharpe, 2004.

Rubin, Alissa J. "Mesopotamia, Once Verdant, Is Running Dry." *New York Times*, July 30, 2023, Y1, Y10–13.

Rubin, Alissa J., et al. "Extreme Heat Will Change Us." *New York Times*, November 25, 2022, A8–11.

Sahner, Christian C. "Temple of 'The Bride of the Desert.'" *Wall Street Journal*, August 27–28, 2011, C13.

Salopek, Paul. "Lost in the Sahel." *National Geographic* 213, no. 4 (April 2008): 34–67.

Salzman, Philip Carl. *Pastoralists: Equality, Hierarchy, and the State.* Boulder, CO: Westview Press, 2004.

Santana-Sagredo, Francisca, et al. "'White Gold' Guano Fertilizer Drove Agricultural Intensification in the Atacama Desert from AD 1000." *Nature Plants* 7, no. 2 (2021): 152–58.

Schaafsma, Polly. *Images and Power: Rock Art and Ethics.* New York: Springer, 2013.

*———. *Indian Rock Art of the Southwest.* Santa Fe: School of American Research, 1980.

Scheele, Judith, and James McDougall. Introduction to *Saharan Frontiers: Space and Mobility in Northwest Africa*, edited by James McDougall and Judith Scheele, 1–21. Bloomington: Indiana University Press, 2012.

Schmidt-Nielsen, Knut. *The Camel's Nose: Memoirs of a Curious Scientist.* Washington, DC: Island Press, Shearwater Books, 1998.

*———. *Desert Animals: Physiological Problems of Heat and Water.* New York: Oxford University Press, 1964.

Schneider, Keith. "No Water? Developers Take On the Challenge." *New York Times*, December 28, 2022, B5.

Schreiber, Katharina, and Josué Lancho Rojas. *Irrigation and Society in the Peruvian Desert: The Puquios of Nasca.* Lanham, MD: Lexington Books, 2003.

Scott, Chris. *Sahara Overland: A Route and Planning Guide.* Hindhead, Surrey, UK: Trailblazer Publications, 2000.

Scott, Patrick. "The Austere Beauty of Egypt's Hiking Trails." *New York Times*, December 31, 2022, C7.

*Sears, Paul B. *Deserts on the March.* Washington, DC: Island Press, 1988 [1935].

*Serventy, Vincent. *Desert Walkabout.* Sydney: Collins, 1973.

Sesser, Stan. "Deserts in Bloom." *Wall Street Journal*, March 1–2, 2008, W1, W4–5.

Seth, Vikram. *From Heaven Lake.* New York: Vintage Books, 1987 [1983].

*Shackleton, Sir Ernest Henry. *South: The Story of Shackleton's Last Expedition, 1914–1917.* New York: Macmillan, 1920.

Shannon, Terry. *Desert Dwellers.* Chicago: Albert Whitman, 1958.

*Sharma, Dilip. *Ethnomedicinal Plants of Thar Desert.* Chennai, India: Notion Press, 2022.

Sherman, Anna. "The Endless Desert." *T: The New York Times Style Magazine*, May 17, 2020, 35–36, 38, 40, 50–56.

Shirihai, Hadoram. *The Complete Guide to Antarctic Wildlife: Birds and Marine Mammals of the Arctic Continent and the Southern Ocean.* 2nd ed. Princeton, NJ: Princeton University Press, 2008.

Shulevitz, Judith. "An Israeli Oasis as a Passage to Ancient Times." *New York Times*, September 12, 2010, TR5.

Sides, Hampton. *In the Kingdom of Ice: The Grand and Terrible Polar Voyage of the USS* Jeannette. New York: Knopf Doubleday, 2015.

Siliotti, Alberto. *Sinai* (Egypt Pocket Guide). Translated by Richard Pierce. Cairo: American University in Cairo Press, et al., 2000.

The Silk Road (2003–).

Sinclair, Ian, et al. *Birds of Southern Africa.* 4th ed. Princeton, NJ: Princeton University Press, 2011 [1993].

Sink, Mindy. "Thirsty Shrub Is a Target as the West Fights Drought." *New York Times*, May 15, 2004, A10.

Smith, Craig S. "Rain on Sahara's Fringe Is Lovely Weather for Locusts." *New York Times*, July 21, 2004, A3.

Smith, Mike. *The Archaeology of Australia's Deserts.* New York: Cambridge University Press, 2013.

Smith, Roberta. "Desert Visionary." *New York Times*, March 13, 2020, C11, C16.

"Smooth Desert Boulders May Be Quakes' Work." *New York Times*, October 25, 2011, D3.

Snell, Marilyn Berlin. "From Dudes to Detox in the Arizona Desert." *New York Times*, May 28, 2004, D6.

Sonoran Quarterly (1947–).

Sowell, John. *Desert Ecology: An Introduction to Life in the Arid Southwest.* Salt Lake City: University of Utah Press, 2001.

*Spooner, Brian, and H. S. Mann, eds. *Desertification and Development: Dryland Ecology in Social Perspective.* London: Academic Press, 1982.

Squire, Vicki. "Desert 'Trash': Posthumanism, Border Struggles, and Humanitarian Politics." *Political Geography* 39 (2014): 11–21. https://primo.uvm.edu/primo-explore/fulldisplay?docid=TN_cdi_proquest_miscellaneous_1542001476&context=PC&vid=UVM.

Stark, Freya. *East Is West.* London: John Murray, 1945.

Stein, Marc Aurel. *Ruins of Desert Cathay: Personal Narrative of Explorations in Central Asia and Westernmost China.* 2 vols. London: Macmillan, 1912.

*Steinbeck, John. *Travels with Charley: In Search of America.* New York: Penguin Books, 1997 [1962].

*Steinmetz, George. *Desert Air.* New York: Abrams, 2012.

Stoppato, Marco C., and Alfredo Bini. *Deserts.* Translated by Linda M. Eklund. Buffalo, NY: Firefly Books, 2003.

*Streever, Bill. *Heat: Adventures in the World's Fiery Places.* New York: Little, Brown, 2013.

Streitberger, Kiki. "Where Beauty Queens Have Curves, Long Eyelashes and 4 Legs." *New York Times*, December 21, 2021, A10.

Strochlic, Nina. "The Lost Libraries of the Sahara." *Daily Beast*, September 11, 2014 (updated April 14, 2017). https://www.thedailybeast.com/the-lost-libraries-of-the-sahara.

Sutter, John D. "Slavery's Last Stronghold." CNN Freedom Project, 2012. https://www.cnn.com/interactive/2012/03/world/mauritania.slaverys.last.stronghold/index.html.

Swift, Jeremy, et al. *The Sahara.* Amsterdam: Time-Life Books, 1976.

Táíwò, Olúfémi. *Against Decolinsation: Taking African Agency Seriously.* London: Hurst & Co., 2022.

Tamisiea, Jack. "In Australian Fossils, a Rainforest Beckons." *New York Times*, January 18, 2022, D2.

Tanaka, Jiro. *The San, Hunter-Gatherers of the Kalahari: A Study in Ecological Anthropology.* Translated by David W. Hughes. Tokyo: University of Tokyo Press, 1980.

*Teague, David W. *The Southwest in American Literature and Art: The Rise of a Desert Aesthetic.* Tucson: University of Arizona Press, 1997.

Teiwes, Helga. *Kachina Dolls: The Art of Hopi Carvers.* Tucson: University of Arizona Press, 1991.

Tejero - Cicuéndez, Héctor, et al. "Desert Lizard Diversity Worldwide: Effects of Environment, Time, and Evolutionary Rate." *Global Ecology and Biogeography* 31, no. 4 (2022): 776–90. https://primo.uvm.edu/primo-explore/fulldisplay?docid=TN_cdi_proquest_journals_2639112401&context=PC&vid=UVM.

*Thesiger, Wilfred. *Across the Empty Quarter.* London: Penguin UK, 2007.

———. *Arabian Sands.* New York: E. P. Dutton, 1959.

*———. *The Last Nomad.* New York: E. P. Dutton, 1980.

*Thomas, Elizabeth Marshall. *The Old Way: A Story of the First People.* New York: Farrar, Straus & Giroux, 2006.

Thomas, Natalie, and Sumant Nigam. "Twentieth-Century Climate Change over Africa: Seasonal Hydroclimate Trends and Sahara Desert Expansion." *Journal of Climate* 31, no. 9 (2018): 3349–70. https://primo.uvm.edu/primo-explore/fulldisplay?docid=TN_cdi_proquest_journals_2117981103&context=PC&vid=UVM.

Tonkinson, Robert. *The Mardu Aborigines: Living the Dream in Australia's Desert.* 2nd ed. Fort Worth, TX: Holt, Rinehart and Winston, 1991 [1978].

Torr, Geordie. *The Silk Roads: A History of the Great Trading Routes Between East and West.* London: Arcturus, 2021.

Tyler, Patrick E. " Libya's Vast Pipe Dream Taps into Desert's Ice Age Water." *New York Times*, March 2, 2004, D3.

Ure, John. *In Search of Nomads: An Anglo-American Obsession from Hester Stanhope to Bruce Chatwin.* New York: Carroll & Graf, 2003.

van der Post, Laurens. *The Lost World of the Kalahari.* New York: William Morrow, 1958.

*van der Walt, Pieter, et al. *The Kalahari and Its Plants.* Pretoria: Info Naturae, 1999.

Van Dyke, John C. *The Desert: Further Studies in Natural Appearances.* Introduction by Peter Wild. Baltimore: Johns Hopkins University Press, 1999 [1901].

*van Wyk, Ben-Erik, and Michael Wink. *Medicinal Plants of the World.* Portland, OR: Timber Press, 2004.

Vanhoenacker, Mark. "Left at the Dune, Right at the Palace." *New York Times*, February 17, 2013, TR9.

Veth, Peter, et al., eds. *Desert Peoples: Archeological Perspectives.* Malden, MA: Blackwell Publishing, 2005.

Viljoen, Russel. *Khoikhoi, Microhistory, and Colonial Characters at the Cape of Good Hope.* Lanham, MD: Lexington Books, 2023.

*Vollmann, William T. *Imperial.* New York: Viking, 2009.

Wade, Nicholas. "A Host of Mummies, a Forest of Secrets." *New York Times*, March 16, 2010, D1, D4.

Wald, Matthew L. "In the Desert, Harnessing the Power of the Sun by Capturing Heat Instead of Light." *New York Times*, July 17, 2007, C3.

Walker, Dale. *Fool's Paradise.* New York: Vintage Books, 1988 [1980].

*Walker, Jenny. *Arabian Desert in English Travel Writing Since 1950: A Barren Legacy?* New York: Routledge, 2022.

Wall, Dennis. *Western National Wildlife Refuges: Thirty-Six Ecological Havens from California to Texas.* Santa Fe: Museum of New Mexico Press, 1996.

*Wallace, David Rains. *Chuckwalla Land: The Riddle of California's Desert.* Berkeley: University of California Press, 2011.

Walls, James. *Land, Man, and Sand: Desertification and Its Solution.* New York: Macmillan, 1980.

Walsh, Declan, and Vivian Yee. "A New Capital Rises in Egypt, but at What Price?" *New York Times*, October 9, 2022, Y4.

*Ward, David. *The Biology of Deserts.* 2nd ed. Oxford: Oxford University Press, 2016.

Waters, Frank. *Masked Gods: Navaho and Pueblo Ceremonialism.* Albuquerque: University of New Mexico Press, 1950.

Watson, Bruce Allen. *Desert Battle: Comparative Perspectives.* Westport, CT: Praeger, 1995.

*Watson, Jane Werner. *Deserts of the World: Future Threat or Promise?* New York: Philomel Books, 1981.

Webb, Robert H., et al., eds. *The Mojave Desert: Ecosystem Processes and Sustainability.* Reno: University of Nevada, 2009.

Welland, Michael. *The Desert: Lands of Lost Borders.* London: Reaktion Books, 2015.

———. *Sand: The Never-Ending Story.* Berkeley: University of California Press, 2009.

Wengrow, David. *The Archaeology of Early Egypt: Social Transformations in North-East Africa, 10,000 to 2,650 BC.* New York: Cambridge University Press, 2006.

*West, Neil E., ed. *Temperate Deserts and Semi-deserts.* Amsterdam and New York: Elsevier Scientific, 1983.

Wheeler, Sara. *Terra Incognita: Travels in Antarctica.* New York: Modern Library, 1999.

*White, Patrick. *Voss.* New York: Alfred A. Knopf, [2012].

Whitford, Walter. *Ecology of Desert Systems.* San Diego: Academic Press, 2002.

Widlok, Thomas. "Theoretical Shifts in the Anthropology of Desert Hunter-Gatherers." In *Desert Peoples: Archaeological Perspectives*, edited by Peter Veth, et al., 17–33. Malden, MA: Blackwell Publishing, 2005.

Wild, Peter, ed. *the new desert reader: Descriptions of America's Arid Regions.* Salt Lake City: University of Utah Press, 2006.

———. *The Opal Desert: Explorations of Fantasy and Reality in the American Southwest.* Austin: University of Texas Press, 1999.

Wilford, John Noble. "Desert Roads Lead to Discovery in Egypt." *New York Times*, September 7, 2010, D1, D3.

Wilford, John Noble. "For Fossil Hunters, Gobi Is No Desert." *New York Times*, September 13, 2005, D1, D4–5.

———. "Mysterious Circles Dotting an African Desert May be the Work of Termites." *New York Times*, March 29, 2013, A7.

Williams, Paul (director). "The Green Planet," *Desert Worlds* episode. PBS, 2022.

Wilson, Michael. "In Phoenix, Even Cactuses Wilt in Clutches of Record Drought." *New York Times*, March 10, 2006, A1, A15.

Winter, Tim. *The Silk Road: Connecting Histories and Futures.* New York: Oxford University Press, 2022.

Wolfe, Alexandra. "The Two-Thousand-Year-Old Plant." *The Wall Street Journal*, April 26–27, 2014, C12.

Wood, Nancy, ed. *The Serpent's Tongue: Prose, Poetry, and Art of the New Mexico Pueblos.* New York: Dutton, 1997.

Woodin, Ann. *Home Is the Desert.* New York: Macmillan, 1965.

Woody, Todd. "Concerns as Solar Installations Join a Desert Ecosystem." *New York Times*, November 17, 2010, F7.

The World's Biomes: The Deserts Biome. UC Museum of Paleontology, University of California, Berkeley. https://ucmp.berkeley.edu/exhibits/biomes/deserts.php.

Worth, Robert F. "Thirsty Plant Steals Water in Dry Yemen." *New York Times*, November 1, 2009, Y9.

*Wyss, Johann David. *The Swiss Family Robinson.* Lewisville, TX: School of Tomorrow, 1992 [1812].

Yardley, Jim. "A Crescent of Water Is Slowly Sinking into the Desert." *New York Times*, May 27, 2005, A1, A4.

Yee, Vivian. "A Cave Village Emptied by Drought and Modernity." *New York Times*, January 19, 2023, A4.

———. "Climate Change and Human Activity Erode Egypt's Antiquities." *New York Times*, November 13, 2022, Y4.

Yetman, David. *The Great Cacti: Ethnobotany and Biogeography.* Tucson: University of Arizona Press, 2007.

*Yousef, Mohamed K., et al., eds. *Physiological Adaptions: Desert and Mountain.* New York: Academic Press, 1972.

Zahid, Nisha. "4,500-Year-Old Sumerian Palace Discovered in Iraq Desert." *Greek Reporter*, February 21, 2023. https://greekreporter.com/2023/02/21/4500-year-old-sumerian-palace-discovered-iraq-desert.

Zahran, M. A., and A. J. Willis. *The Vegetation of Egypt.* London: Chapman & Hall, 1992.

*Zell, Len, and Ian Glover. *Australian Deserts and Savannah: Atlas and Guide.* Ashgrove, AU: Wild Discovery, 2014.

Zerubavel, Yael. *Desert in the Promised Land.* Stanford, CA: Stanford University Press, 2019.

Zhong, Raymond. "One Thing Those Spines Won't Protect Them Against: Climate Change." *New York Times*, April 15, 2022, A17.

Zhuang, Yan. "Alcohol Ban for Aboriginal Australians Returns. So Do Questions." *New York Times*, March 13, 2023, A4.

Zraick, Karen. " In Jordan, a Grim Glimpse of a Parched Future." *New York Times*, November 10, 2022, A4.

Zwinger, Ann Haymond. *The Mysterious Lands: An Award-Winning Naturalist Explores the Four Great Deserts of the Southwest.* New York: Truman Talley Books, Plume, 1990.

Index

Italicized page numbers refer to images.

abal, 3
Abe, Kobo, 222
ablation, 3
Aborigine, 3; arts and languages,
    4; Australian culture of, *3*
Abstract Expressionism, 4
Abu Simbel, 4
abydos, 5
acacia plants, 5, *5*
Acoma (Sky City), 5, *5*
adaption, 6
addax, 6
adder snake, 6
adiabatic cooling, 6
adobe, 6–7; contemporary
    structure, *7*; wall, *6*
The Adventures of Priscilla,
    Queen of the Desert (film), 7
aerial observation, 7
aerial tramway, 7
afar depression, 7
afforestation, 7
Africa, 7
The African Origin of Civilization
    (art exhibit), 22
African peyote cactus, 7
African sycamore tree, 8
Afton Canyon, 8
agave, 8, *8*
Aïr Mountains, 8
Ait Atta, 8
Aïtel, Fazia, 31
ajo lily, 8
Albanov, Valerian Ivanovich, 108
Alcock, John, 219
Al Dhafra festival, 8
Alexander, Caroline, 80
Algeria, 8
Alice Springs, 8
Allaby, Michael, 71
All- American canal, 8, *9*
Allen, Benedict (1960–), *9*
Allenby Bridge (King Hussein
    Bridge), 9
alluaudia montagnacii, *129*
alluvial fan, 9
Almásy, Count László
    (1865–1951), 9
Alone (Byrd), 9
Alone on the Ice (Roberts), 10
Altai mountain range, 10
Altiplano (Puna), 10

Amen-Ra, in Karnak, *128*
amphibians, 10
Amrousi, Mohamed El, 19
Amundsen, Roald, 10, *10*, 193
Anasazi, 10
ancient desert civilizations, 10–11
ancient desert peoples, 11
Andes mountain range, 11
Andrews, Roy Chapman
    (1884–1960), 11
animal rights, 11–12
animals, 12
Anna Creek station, 12
Annerino, John, 108
ant (insect), 12–13
Antarctic, 13–14;
    convergence map, *13*
antelope, 15
antelope horns plant, 15
anthropogenic influence, 15
anthropology, 15
antiquities, 15
antivenin, 15
Anza- Borrego Desert, 15, *15*
Anza- Borrego Desert
    State Park, 16
Apache, 16
aquifer, 16
Arabah, 16
Arabian Desert, *16*, 16–17; Empty
    Quarter in, *79*
Arabian Sands (Thesiger), 17
arachnids, 17
Aral Sea, 17
ARAMCO oil company, 17
Arboretum at Flagstaff, 17
Archaeology of Australia's
    Deserts (Smith), 17–18
Archeology, 17–18
Archeology Under Fire
    (Meskell), 18
Arches National Park, *18*, 19
architecture, 19
Arctic, 19; North Pole, *20*
Arctic National Wildlife
    Refuge, 21, *21*
area 51, *21*
aridity, 21
Arid Land Research and
    Management (1987–), 21
The Arid Lands (Davis), 71
Arizona, 21

Arizona Highways magazine, 21
Arizona-Sonora Desert
    Museum, *21*, 21–22
Arizona State Museum
    (Tucson), 22
arroyo, 22
arthropod, 23
arts: of aboriginals, 4; desert, 22;
    indigenous, 22, *22*
Asia, 23
Aswan High Dam, 23
Atacama Desert: flowers in, *23*;
    map, *23–24*; snow in, *23*;
    Vicuñas, in, *12*
Atkins, William, 108
atlases, 24
Atlas Mountains, 24
The Atlas of a Changing
    Climate, 24
Austin, Mary Hunter, 24
Australia, 24; deserts in, 25, *25*
Australian Arid Lands Botanic
    Garden, 25
automobiles/trucks, 25
Aymara, 25
Azalai, 25
Aztec Ruins National
    Monument, 25

Bab'Aziz (film), 27
baboon, 27
bacteria, 27
Badain Jaran Desert, 27
Badiya (Badu), 27
badlands, 27
Badwater Basin, 27, *27*
Bagnold, Ralph Alger (1896–
    1990), 28, 125
Bainbridge, David A., 71
Baja California, 28
bajada, 28
Bandelier National Monument, 28
baobab, 28
barrel cactus, 28, *28*
Barth, Heinrich (1821–1865), 28
basin and range, 29
bats, 29
bauxite, 29
beach grass, 29
Bear River National Wildlife
    Refuge, 29
bears, 29

Beattie, Owen, 90
Beau Geste (film), 29
beavertail cactus, 29
Bedouins (Bedu), 29–30, *30*;
    Wadi Rum camp of, *215*
Beersheba, 31
beetle, 31
Behnke, Roy, 71
Bell, Gertrude Margaret Lowthian
    (1868–1926), 31, *31*
Bender, Gordon L., 171
Berber (indigenous peoples), 31
Berman, Michael P., 196
Besh-Ba-Gowah, 31
Bible, 31–32
bibliography, desert, 32
Bi! bulb plant, 31
Bickel, Lennard, 132
Big Bend National Park, 32
Billy button daisy, 32
bindweed, 32
Bingham Canyon Mine
    (in Utah), 32
biodiversity, 32
biology, 32
biome, 32
biospeliology, 33
biosphere reserves, 33
birds, *147*, 161; desert, 33;
    Kangaroo Flat, *25*; in Sonoran
    Desert, 192
bizarre manifestations, 33
black blizzards, 33
Black Gobi, 33
Black Rock Desert, High Rock
    Canyon, 33
The Black Stallion (film), 33
Boab tree, 33
Bonneville Salt Flats, 34
Boojum tree (cirio), 34
books, 34
borax, 34, *210*
borders, 34
Bradbury, Ellen, 54
Brett, Michael, 31
Broyles, Bill, 196
A Brush with Georgia O'Keeffe
    (play), 36
Buddhism, 36
Buenos Aires National Wildlife
    Refuge, 36
Bukhara, 36

Burckhardt, Johann Ludwig (1784–1817), 4, 36
Bureau of Land Management (BLM), 36
Burning Man Festival, 36
Burton, Sir Richard F. (1821–1890), 36, *36*, 158
Butelski, Kazimierz, 19
Buzzati, Dino, 68
Byrd, Richard E., Jr. (1888–1957), 9, 37, *37*

Cabeza Prieta National Wildlife Refuge (Arizona), 39
cactus, 39–40; Cardón, 43; in Sonoran Desert, *220*
Cactus and Succulent Society of America (CSSA), 40
cactus longhorn beetle, 40
cactus rescue crew, 40
Caillié, René-Auguste (1799–1838), 40
cairn, 40
Cairo, 40
caldera, 40
Calderan, Max (1967–), 40
caliche, 40
California, 40, 72; deserts, 41, *41*; poppies, 41, *41*
California Desert Protection Act (of 1994), 41
The California Deserts (Pavlik), 41
California Desert Studies Consortium (CDSC), 41
camels, 42, *181*; Desert Masileh Qom, *42*
camel thorn tree, 42
Camus, Albert, 196
Canadian desert, 42
Canyon de Chelly National Monument, 42
Canyonlands National Park, 43
Capulin Volcano National Monument, 43
caravan, 43
Cardón cactus, 43
Carlsbad Caverns National Park, 43
Casablanca, 43–44
Casa Grande Ruins National Monument, 43, *43*
Casas Grandes (Paquimé) Ruins, 44
casbah, 44
"The Case for Colonialism" (Gilley), 53
Cathedral in the Desert (Utah), 44
cats, 44
caves, *44*, 44–45
cenote, 45
centipede, 45
Central Australian Desert, 45
Central Kalahari Game Reserve, 45
ceremonies, 45–46

Chaco Culture National Historical Park, 46–47
Chad, 47
Chandler, Raymond, 171
Chang Tang, 47
chaparral plant, 47
charco, 47
Charles, Jack (1943–2022), 47
cheetah, 47
Chihuahuan Desert Nature Center and Botanical Gardens, *47*, 48
children's literature, 48
Chile, 48
Chimayo, New Mexico, 48
China, 49
China-Tibet railway, 49
Chlachula, Jiri, 70–71
cholla, 49, *49*
Chott lake, 49
Christianity, 49
chuckwalla lizard, 49
chuparosa, 49
cicada, 49
Cima Dome, 50
circles, mysterious, 50
cities, 50
Clifford, Sir Bede Edmund Hugh (1890–1969), 50
climate change, 51
clothing, 51
clouds, 51
Coachella Valley Music and Arts Festival, 51
Cochise, 52
cold-blooded, 52
Colorado Desert, 52
Colorado River, 52
Colossal Cave Mountain Park, 52
Colum, Padraic, 64
Committee on Desert and Arid Zone Research, 52
common sowthistle, 52
common sunflower, 52
compass barrel cactus, 52
Complete Middle East Cookbook (Mallos), 88
condor, 52
conservation, 52
controversies, 52–53
Coober Pedy, 53
copper, 53
Copper Mountain College, 53
coriolos effect, 53
Coronado State Monument, 53
corroboree, 53
cosmos plant, 53
cottonwood, 53
countries, 53
COVID-19, 53
coyote, 53
crabeater seals, *14*
Crace, Jim, 167
crafts, 54; pottery, *54*
Craters of the Moon, 55
creosote bush, 55
Crescent Lake, 55

crows, 55
cryptogamic crust, 55
cuesta, 55
Cure Salee Festival, 55
current, 55
cypress, 55
Cyrenaica, 55

Dadd, Richard, 101
Dades Gorge, 57
daisy, 57
Dakar Rally, 57
dam, 57
damaras, 57
Dana Biosphere Reserve, 57
Danakil Desert, 57
dance, indigenous, 57
Dandan-Uiliq, 57
dangers, 57–58
Darwin, Charles (1809–1882), 58
dasht, 58
Dasht-Margo, 58
data palm, 58, *58*
David Bellamy's Arabian Light, 58
Davidson, Robyn, 207, *207*
Davis, Diana K., 71
Dead in Their Tracks (Annerino), 108
Dead Sea, 58
Dead Sea Scrolls, 58
death adder, 58
Death Valley, *181*; lake, 58; Scotty's Castle, 185, *185*
Death Valley (film), 61
Death Valley National Park, 58–59; flowers and tracks in, *60*; landscape and rolling dunes of, *59*; wildflowers in, 61
deer, 61
deflation (erosion), 61
dehydration, 61
demographics, 61
desalination plants, 61
Desert (Haynes), 22
Desert (journal), 63
Desert (Lé Clezio), 64
Desert (MacQuitty), 64
Le Désert (The Desert) (Loti), 64
The Desert (Colum), 64
The Desert (Van Dyke), 64
desert animal species, endangered, 64
desert anomaly, 64
desert bedstraw, 64
desert bighorn sheep, 64–65
The Desert Biome website, 65
Desert Botanical Garden, 65
desert broomrape, 65
A Desert Calling (Mares), 65
desert changes, 65
desert chicory, 65
desert climate, 65
desert color, 65
desert death, 65

desert dermatoses, 66
desert development, negative aspects, 66
Desert Discovery Center, 66
desert documentaries, 66
Desert Dream (Hyazgar) (film), 66
desert ecology, 66
The Deserted Station (film), 70
desert ethic, 66
desert formations, 66
desert futures, 66
desert glass, 66
desert grasses, 67
Desert Heat (film), 67
desertification, 70–71
desert implications, 67
Desert Institute at Joshua Tree National Park, 67
desert iris, 67
desert kangaroo rat, 67
desert kites, 67
Desert Land Act of 1877, 67
desert landscape, 67
Desert Life (Krebbs), 67
desert lily, 67
Desert Lily Sanctuary, 67
desert marigold, 68
desert mariposa lily, 68
Desert Masileh Qom camels, *42*
Desert Memories (Dorfman), 68
Desert National Park Jaisalmer (India), 68
Desert National Wildlife Range (Nevada), 68
Desert Notebooks (Ehrenreich), 68
Desert of Maine, 68
The Desert of the Tartars (Buzzati), 68
Desert Oracle (collection), 68
desert peoples, 68
Desert Plants (journal), 68–69
deserts, 61–63; in America, 72, *72*; in Australia, 25, *25*; demographics, *62*; distances in, 73; explorers, 82; extraterrestrial, 83; locust swarm in, *127*; mammals, *130*; novels, 125; of Peru, *158*; plant in white sands of, *63*; population, 161; pseudo, 163; resources, 172; snow in, 190, *190*; survival, 196–97; temperatures in, 201–2; wildflowers in, *170*; world's most important, *64*
Deserts (Allaby), 71
Deserts (MacMahon), 71
Déserts (musical composition), 71
Deserts Are Not Empty (Henni), 72
Deserts: Miracle of Life (Flegg), 72
Deserts of America (Larson), 72
Deserts of the World (Petrov), 72

desert spine lizards, *126*
Desert Tortoise Natural Area, 70
desert types, 70
DesertUSA website, 72
desert weather, 70
desert willow, 70
Desert Wind (documentary), 70
"The Desert World" (Pond), 70
Desert X, 70
desert zinnia, 70
desiccation theory, 72
Devil's claw, 72
Devil's Cornfield, 72
Devil's Garden (California), 72
Devil's Golf Course, 72
Devil's Hole, 72
Devil's Marbles
    (Australia), 72–73
Devil's Postpile National
    Monument, 73
Dhahran, 73
Diamond Sutra, 73
dingo, 73, *73*
dingo fence, 73
dinosaur eggs, *90*
Dinosaur National Monument, 73
dinosaurs, 73
distances, in desert, 73
diyafa, 72
Djenné (Mali), 73–74, *206*
donkey, 74
Donner Lake, *217*
Door to Hell, 74
Dorfman, Ariel, 68
dot paintings, 74
Double Mountain, *13*
Doughty, Charles Montagu
    (1843–1926), 74, *74*, *208*
Doum palm, 74
draa sand dune, 74
dreamtime, 74
drought, 74
drumming, Indigenous
    peoples, 74
druze people, 75
dryland archeology, 75
drylands, 75
Dubai Desert Conservation
    Reserve, 75
Dumont Sand Dunes, 75
Dune (Herbert), 75
dune grass, 75
dunes: petrified, 75; types, 75
dung, 75
Dunhuang, China, 75
dust, 76
dust bowl, 76
dust devil, 76
dust storm, *200*
dwellings, 76
Dzungaria, 76

Earthships community, 77
East is West (Stark), 77
Eberhardt, Isabelle
    (1877–1904), 77

echidna (spiny anteater), 77
echinacea (purple coneflower), 77
ecosystem, 77
Eggers, Dave, 104
Egypt, 77–78; Marsa Alam Desert
    in, *77*; nomad from, *78*
The Egyptian (Waltari), 78
Egyptian art, 78
Egyptian sculpture, 78
Ehrenreich, Ben, 68
eid, 78
Eilat, 78
Ein Gedi, 78–79
Ein Gedi Botanical Gardens, 79
eland, 79
El Camino del Diablo (the Devil's
    Highway), 79
electrical production, 79
elephants, 79
El Gran Desierto del Altar, 79
El Malpais National
    Monument, 79
El Moro National Monument, 79
Empty Quarter, 79, *79*
emu, 80
Encyclopedia of Deserts
    (Mares), 80
*Encyclopedia of Global
    Warming*, 80
Encyclopedia of the Biosphere, 80
Endurance (Lansing), 80
The Endurance (Alexander), 80
The English Patient
    (Ondaatje), 80
eremophobia, 80
erg, 80
Erg Murzuk, 80
Erg of Admer, 80
Erg of Bilma, 80
Erg of Chech Desert, 81
erosion, 81
Erta Ale, 81
estivation, 81
ethics, 81
ethnobotany, 81
ethnographic, 81
Etosha National Park
    (Namibia), 81
euphorbias, 81
European Southern
    Observatory, 81
evaporation ponds, 81
exploration, 81–82
explorers, 82
extraterrestrial deserts, 83
extraterrestrial highway, 83
Eye of the Sahara, 83

Far from Men (film), 85
Farthest North (Nansen), 85
Fauna & Flora International, 85
Fentress, Elizabeth, 31
Festival au Desert, 85
Fezzan (Fazzan) Desert, 85
*The Fighting Dervishes of the
    Desert* (film), 85

films, desert
    documentaries, 85–86
films, westerns, 86
firewheel, 87
first people, indigenous people, 87
FiSahara, 87
fish, 87
Fishes in North American Deserts
    (Naiman and Soltz), 87
Fish River Canyon (Namibia), 87
Fish Springs National Wildlife
    Refuge, 87
Five Civilized Tribes, 87
flamingo, 87, *87*
flash flood, 88
Flegg, Jim, 72
flowers, in Atacama Desert, *23*
fog, 88
fog basking, 88
foggara, 88
food, 88–89
food desert, 89
food gathering, 89
Fool's Paradise, 89
Fort Defiance (Arizona), 89
Fort Sumner State Monument
    (New Mexico), 89
Fossil Falls (Mojave), 89
fossils, 89
foxes, 90
freeze/thaw cycle, 90
Frémont, John Charles
    (1813–1890), 90
French Foreign Legion, 90
From Heaven Lake (Seth), 90
Frozen in Time (Beattie and
    Geiger), 90
Fulani, 90
Furnace Creek, 90

gahwa, 91
Gambel's oak tree, 91
Gambel's quail, 91
Gander's cholla, 91
Gaochang, 91
garua, 91
gazelle, 91
gebel, 91
Gebel Kamil Crater, 91
gecko, 91
Geiger, John, 90
gems, 91
gemsbok, 91
Gemsbok National Park, 91
geodiode, 91
geoglyphs, 91–92
geography, 92
geology, 92
geomorphology, 92
ger, 92–93
Germa, 93
Geronimo (1829–1909), 93, *93*
Ghaf tree, 93
Ghan Railway, 93
ghat, 93
ghazu, 93

Ghost gum tree, 93
ghost towns, 93
ghutra, 93
gibber, 93
Gibson Desert, 93
Gila Cliff Dwellings National
    Monument, 93
gila monster, 93
Gilf Kebir, 93
Gilley, Bruce, 53
Giza, 94
glacier, 94
Glantz, Michael H., 71
global warming, 51, 94
goat, 94
Gobekli Tepe, 94
Gobero, 94
gobi, 94
gobi bear, 94
Gobi Desert, 94–95, *95*; dinosaur
    eggs in, *90*; map, *94*
gold, 96
Goldeneye, 96
Goudie, Andrew, 92
Grand Canyon National Park, 96
Grand Egyptian Museum, 96
Grand Erg Occidental
    (Algeria), 96
Grand Erg Oriental (Algeria), 96
The Grapes of Wrath
    (Steinbeck), 96
grass, 96
grasshoppers, 96
Great Artesian Basin, 96
Great Basin Desert, 97, *97*
Great Basin National Park, 97
The Great Cacti (Yetman), 97
*The Great California
    Deserts* (Lee), 97
great desert skink, 97
Great Dividing Range, 98
Great Eastern Erg, 98
Great Gobi B Strictly
    Protected Area, 98
Great Green Wall, 98
Great Karoo, 98
Great Kavir, 98
The Great Kiva of Pueblo Bonito
    and diagram, *46*
Great Man-Made River
    Project, 98
Great Nafud, 98
Great Salt Lake (Utah), 98
Great Salt Lake Desert, 98
Great Sand Dunes National Park
    and Preserve, 98
Great Sand Sea, 98
Great Sandy Desert, 98
Great Victoria Desert, 99
Great Western Erg, 99
Greenland Ice Sheet, 99
Grey, Zane, (1872–1939), 99
Guadalupe Mountains
    National Park, 99
guano, 99
guayule, 99

Guban Desert, 99
guelta, 99
Guelta d'Archei, 99
guidebooks, 100
A Guide to Desert and Dryland Restoration (Bainbridge), 71
gully, 100
gullywasher, 100

Hadley cell, 101
hajj, 101
halophyte plant, 101
The Halt in the Desert (artwork), 101
hamada, 101
Hamada du Draa (Dra Hamada), 101
Hami oasis, 101
Han Desert Ruins, 101
Hanson, Erin (1981–), 101
harm, 101
harmattan, 101
hartani, 101
Hassanein, Ahmed (Bey) (1889–1946), 101
Hassani tribe, 101
Havasu Falls, 101
Hayden, Julian D. (1911–1998), 101–2
Hayduke Trail, 102
Hazleton, Lesley, 219
healing, 102
Heard Museum, 102
hedgehog cactus, 102
Hedin, Sven Anders (1865–1952), 102
Hejaz Railway, 102
hemotoxic, 102
Henni, Samia, 72
Herbert, Frank, 75
hermit, 102
Herzog, Werner, 167
Hexi Corridor (Gansu Corridor), 102–3
Hidalgo (film), 103
Hi-Desert Nature Museum, 103
High Desert Museum, 103
HIGW. *See* human-induced global warming
hijab, 103
Himalayan desert, 103
historical overview, 103
hogan, 103
Hoggar Desert, 104
Hohokam tribe, 104
A Hologram for the King (film), 104
Hoodia gordonii, 104
Hoover Dam, 104
Hopi snake dance, 105, *105*
Hopi tribe, 104; chief, *160*; pueblo and children of, *104*
horned desert viper, *147*
horned melon tree, 105
Horned melon tree (film), 105
Hotan oasis, 105

hot desert, 105
housing development, 105
Hovenweep National Monument, 105
Huacachina, Peru, *107*
Hueco Tanks State Historic Site, 105
human-induced global warming (HIGW), 106
humpy, 106
Huntford, Roland, 124
hunting, 106
Huntington Botanical Gardens, 106
hyena, 106
hyperthermia, 106

ibex, 107
Ibn Battuta, Muhammad (1304–1368?), 29
Ibn Khaldun, 31
Ica Desert, 107, *107*
iceberg, 108
Iceland deserts, 108
iguana, 108
The Immeasurable World (Atkins), 108
immigration, 108
Imperial Valley Desert Museum, 108
India, 108
Indian paintbrush, 108
Indian Petroglyph State Park, 108
Indian Pueblo Cultural Center, 108
Indian Wild Ass Sanctuary (Gujarat, India), 108
indigenous art, 22, *22*; music, 139–40
indigenous peoples, 109, *109*
Indio, 110
insects, 110
Institute of American Indian Arts (Santa Fe), 110
international ruins, 175
In the Kingdom of Ice (Sides), 108
In the Land of White Death (Albanov), 108
Inuit, 110; women, *20*
The Invention of the American Desert (Massey and Nisbet), 110
Iranian Desert, 110, *110*
Iraq, 110
ironwood tree, 110
irrigation, 110
Islam, 111
Israel, 111
ivory gulls, in Arctic, *20*

Jabbur, Jibrail S., 30
Jabès, Edmond (1912–1991), 113
jackal, 113
Jacob Blaustein Institutes for Desert Research, 113
Jaguar, 113

Jaisalmer, 113
Janz, Lisa, 94
Jasper, David, 4
javelina, 113
Jebel Musa, *189*
Jemez Mountain Trail, 113
jerboa, 113
Jericho, 113
jewelry, 113–14
jojoba shrub, 114
Jordan, 114; Wadi Rum in, *215–16*
Jornada del Muerto, 114
Joshua Tree blossoms, *223–24*
Joshua Tree National Park, 114, *114*
Judaism, 115
Judean Desert, 115
juniper tree, 115

kachina, 117
kaffir, 117
Kailash, Mount, 117
Kalahari Desert, 117–19; children, *118*; language in, *123*; map, *117*; San people, *182*; yellow-billed hornbill in, *118*
kangaroo, 119, *119*
Kangaroo Flat (bird), *25*
Kaokoveld, 119
karaburan dust storm, 119
Karakorum (Nubra Shyok) Wildlife Sanctuary, 119
Karakum Canal, 119
Karakum Desert, 120
karez, 120
Karkom, Mount, 120
Karnak, *128*
karst, 120
Kartchner Caverns State Park, 120
Kasha-Katuwe Tent Rocks National Monument, 120
Kashgar oasis, 120
Kelso Dunes, 120
Kern National Wildlife Refuge, 120
Kharga Oasis, 120
Khartoum, 120
Khoikhoi, 120–21
Kibbutz, 121
Kimberley Plateau, 121
Kitt Peak National Observatory, 121
Kiva, 121, *121*
Kofa National Wildlife Refuge, 122
Kokopelli, 122, *174*
Krebbs, Karen, 67
Krutch, Joseph Wood (1893–1970), 122
!Kung people, 122
Kunoth-Monks, Rosalie (1937–2022), 122
Kutch Desert Wildlife Sanctuary, 122
Kutse Game Reserve, 122
Kyzylkum (Kyzyl Kum), 122

Ladakh, 123
Laguna chief, Pueblo, *163*
Lake Chad, 123
Lake Mead, 123
Lake Nasser, 123
Lake Powell, 123
Land of Little Rain (Austin), 24
language, 123, 164; of aboriginals, 4; in Kalahari Basin area, *123*
Lansing, Alfred, 80
Lanzarote, 123
Larapinta Trail, 123
Larson, Peggy, 72
The Last Place on Earth (Huntford), 124
Las Vegas, 124
latitudinal belts, 124
lava, 124, *124*
Lawrence, T. E., 124, *124*, 186
Lawrence of Arabia (film), 124
Layne, Ken, 68
Leave No Trace, 124
Lé Clezio, J. M. G., 64
Lee, W. Storrs, 97
Lençóis Maranhenses National Park, 125
libraries, 125
Libya, 125
Libyan Desert, 125
Libyan Sands (Bagnold), 28, 125
lichen, 125
Lilies of the Field (film), 125
lions, 126, 137
literature, 125; for children, 48
lithop, 126
Little Sahara, 126
livestock, 126
The Living Desert (film), 126
Living Desert Zoo and Gardens, 126
lizards, 126; desert spine, *126*
llama, 127
locust, 127, *127*
loess, 127
Lop Desert, 127
Lop Nor, 127
Los Alamos, 127
The Lost World of the Kalahari (Van der Post), 128
Loti, Pierre, 64
lupine, 128; grape soda, *128*
Luxor, 128

Mabbutt, J. A., 92
MacMahon, James A., 71
MacQuitty, Miranda, 64
Madagascar, 129, *129*
Mada'in Saleh, 129
Maghreb (Maghrib), 129
Maitland, Alexander, 222
Malevich, Kazimer, 4
Mali, 129
Mallos, Tess, 88
Malpais-Valley of Fires, 129
mammals, 129–30, *130*

Mandaville, James P., 81
Manley, Deborah, 208
mano and metate, 130
manuscripts, rescued, 130
Manzanar National
    Historic Site, 130
manzanita, 130
maps, 130–31
Marco Polo, 131
Marden, Brice, 4
Mardu people, 131
Mares, Michael A., 65, 80
Al Marmoom Desert Conservation
    Reserve, 9
Marree Man, 131
Mars, 131
Marsa Alam Desert in Egypt, *77*
Marshall, John (1932-2005), 131
marsupials, 131
Martin, Agnes, 4
Martinez, Maria, 54
Masada, 131
Masdar City, 131
masonry, 132
Massey, Lyle, 110
Mawson. Douglas, *9*, 10
Mawson's Will (Bickel), 132
McGraths Flat, 132
Mecca, 132
medicinal plants, 132
Meherengarth Fort, 132
memoirs, 133
Merton, Thomas, 222
Mesa Verde National Park, 133
Meskell, Lynn, 18
mesquite tree, 133
Meteor Crater, 133
Mexico, 133
microclimate, 133
microorganisms, 133
midden, 133
migration, 133–34
milk, 134
mines, 134
mirage, 134
missions, 134
Mission San Xavier del Bac, 134
mistreatment, 134
Mogao Caves, 134
Mogollon people, 134
Mohave Museum of History
    and Arts, 134
Mojave Desert, 135; map, *135*
Mojave Mystery
    (documentary), 135
monastery, St. Catherine, *189*
Mongolia, 136; Nomads Day
    Festival, *136*; women, *136*
Mongolian Ping Pong
    (documentary), 136
monkeys, 136
Monod, Théodore André
    (1902–2000), 137
monsoons, 137
Monte Desert, 137

Montezuma Castle National
    Monument, 137
Montezuma's Castle, *11*
Montezuma Well, 137
Monument Valley, 137
Morenci Copper Mine, 137
Mormonism, 137
Morocco, 137
Morris, Mary, 216
moss, 137
Mound Springs, 137
mountain lions, 137
mountains, 137
mourning wheatear, *147*
mummies, 138
Murgab Oasis, 138
Museum of Indian Arts and
    Culture, 138
Museum of Indigenous
    People, 138
Museum of New
    Mexico, 138, *138*
Museum of Northern Arizona, 138
museums: international, 138–39;
    in United States, 139
music, 139–40
musk oxen, *21*
Mustangs, 140
The Mysterious Lands
    (Zwinger), 140
myth of the desert, 140
myths, 140

Naadam, 141
Nabataea.net, 141
Nabataeans, 141
Nachtigal, Gustav (1834–
    1885), 141, 177
Naiman, Robert J., 87
Namaqua sand grouse, 141
Namib Desert, 141, *160*; drought,
    142; dune, *142*; ostriches and
    oryx, *142*; quiver tree, *167*
Namibia, 143
Namib-Naukluft
    National Park, 143
Nansen, Fridtjof (1861–1930),
    85, 143, *143*
National parks and
    monuments, 143–44
National Park Service app, 144
National trail systems, 144
Native Americans, *138*
Native Plants of Southern Nevada
    (Rhode), 144
natural areas, 144
The Nature of Desert Nature
    (collection of authors), 144–45
Navaho National Monument, 145
Navaho tribe, 145, *145*
Nazca Lines (Peru), 145
Nebraska Sand Hills, 145
Negev Desert, 145–47; birds and
    snakes, *147*; Exodus route,
    146; north, *146*; Palm trees,
    155; panoramic view, *147*

Negev Desert Botanical
    Garden, 148
Negev rock art, 148
Nevada, 148
Nevada State Museum, 148
The new desert reader (essay
    collection), 148
New Mexico, 148; landscape,
    148; perpetual ice cave, *158*
Ngorongoro National Park, 149
Nicolaisen, Johannes and Ida, 157
Nisa (Shostak), 150
Nisbet, James, 110
nomad, from Egypt, *78*
Nomadic Peoples (journal), 150
nomads, 150
Nomads Day Festival, *136*
North America, 150
northern desert nightsnake, 150
North Pole, in Arctic, *20*
novels, 150; on wars, 217
Nubian Desert, 150
Nullarbor Plain, 150
nunatak, 150

Oak Creek Canyon, 151
oasification, 151
Oasis de Huacachina (Peru), *151*
oasis sanctuary, 151
obelisks, 152
oceans, 152
ocotillo shrub, 152
Ocucaje Desert, 152
off-road adventures, 152
Okavango Delta, 152
O'Keeffe, Georgia
    (1887–1986), 152
Oman, 152
Ondaatje, Michael, 80
The Opal Desert (Wild), 152
operas, 152
Ordos Desert, 152
ore, 152
owl, in Arizona-Sonora Desert
    Museum, *21*

Pacific Crest Trail, 155
Pailes, Richard A., 68
Painted Desert, 155
Paiute tribe, 155
Pakistan, 155
paleoclimatology, desert
    climate, 155
paleomagnetism, 155
Palladio, Andrea, 19
Palm Desert, 155
Palm Springs, 155
Palm trees, 155, *155*
Palmyra, 156; Temple of Bel, *156*
palo verde tree, 156
Pan de Azúcar National Park
    (Chile), 156
Papago, 156
Paraburdoo plains, *218*
Parakeelya, 156
Parker, K. Langloh, 4

Pasolini, Pier Paolo
    (1922–1975), 156
passion flowers, 156
Passion in the Desert (film), 157
pastoralist, 157
The Pastoral Tuareg
    (Nicolaisen), 157
Patagonian Desert, 157
Pavlik, Bruce M., 41
Peary, Robert Edwin
    (1856–1920), 157
Pecos National
    Historical Park, 157
Pelton, Agnes (1881–1961), 157
penguins, *14*, 157
penitentes, 157
*Peoples of the American
    Southwest* (Plog), 11
perennial snakeweed, 157
perpetual ice cave, 157, *158*
*Personal Narrative of a
    Pilgrimage to Al-Medinah and
    Meccah* (Burton), 158
Peru, *107*, 145, 158; deserts of,
    158; Oasis de Huacachina, *151*
Peruvian Coastal Desert, 158
Petra, 158
Petrified Forest
    National Park, 158
Petroglyph National
    Monument, 159
petroglyphs, 158–59, *159*, *205*
petroleum, 159
Petrov, M. P., 72
Philip L. Boyd Deep Canyon
    Desert Research Center, 159
photographs, 159
phreatophyte plant, 159
pictographs, 160
Pima Air and Space Museum
    (Tucson), 160
pincushion cactus, 160
Pinnacles Desert, 160
plants, 160; salt desert, *181*
playa, 161
Plog, Stephen, 11
poems, 161
polar depression, 161
pollution, 161
Polo, Marco (1254–1324), 161
Pond, Alonzo W., 70
poor-will bird, 161
poppy, California, *41*
population, desert, 161
pottery, 161–62; crafts, *54*
Powell, John Wesley
    (1834–1902), 162
Powell, Lawrence Clark, 193
The Power of the Dog (film), 162
precipitation, 162
preservation, 162
prickly pear, 162–63, *163*
primrose plant, 163
prodigiosa, 163
pronghorn antelope, 163
Przewalski's horse, 163

Pueblo, 163; Laguna chief, *163*; languages, 164
Pueblo Bonito, 164, *164*
Pueblo Grande Museum and Archeological Park, 164
Pueblo Indian Cultural Center, 164
purple sand verbena, *220*

Qaidam desert, 167
Qanat, 167
Qatar, 167
Quarantine (Crace), 167
Queen of the Desert (Herzog), 167
quicksand, 167
Quillagua (Chile), 167
quiver tree, 167, *167*

races, 169
The Racetrack, 169
railway, 169
Rainbow Basin Natural Area, 169
Rainbow Bridge National Monument, 169
rain shadow, 169
Rajasthan (Indira Gandhi) Canal, 169
Ramadan, 169
Ramsey Canyon Preserve, 169
Ras Mohammed National Park, 169–70
rattlesnakes, 170
Rauscheberg, Robert, 4
ravens, 170
rayless goldenrod, 170
readers, 170
Red Desert, 170
The Red Desert (Il Deserto Rosso) (film), 170–71
Red Rock Canyon National Conservation Area (Nevada), 171
Red Sea, 171
"Red Wind" (Chandler), 171
reference books, 171
Reference Handbook on the Deserts of North America (Bender), 171
religion, 171
Reno, 171–72
reptiles, 172
research, 172
resources, desert, 172
resurrection plants, 172
Revillagigedo Archipelago (Mexico), 172–73
Rhode, David, 144
"The Rime of the Ancient Mariner" (Coleridge), 173
rimth saltbush, 173
rivers, 173
roadrunner, 173
road train, 173
Roberts, David, 10
rock art symbols, 173–74
rodents, 174

Rommel, Erwin (1891–1944), 174
Rosenthal, Gerald A., 192
Ross Ice Shelf, *14*
Roswell (New Mexico), 174
Rub' al Khali, *204*
rugs, 174
ruins, 174–75
Ruins of Desert Cathay (Stein), 175
Russian thistle, 175

sacred datura, 177
The Sacred Desert (Jasper), 4
safari, 177
sagebrush, 177
saguaro cactus, 177, *192*
Saguaro-Juniper Covenant, 177
Saguaro National Park, 177
Sahara (film), 177
Sahara and Sudan (Nachtigal), 177
Sahara Desert, 178–79; landscape of, *178*
Sahara mustard, 179
Saharan art, 179
Saharan tribes, 179–80
Sahara Overland (guide), 179
Sahel, 180
Salada, 180
Salinas Pueblo Missions National Monument, 180
Salmon Ruins and Heritage Park, 180
salt, 180–81
saltation, 182
salt bush, 181
salt desert, 181, *181*
Salton Sea, 182
saltwort plant, 182
San Andreas Fault, 182
sand, 182–83; paintings, 183
sandboarding, 183
Sands of the Kalahari (film), 183
sandstorms, 183
Sangre de drago plants, 183–84
San people, 182, *182*
Santa Ana, 184
Santa Fe, 184
Saqqara, 184
Saudi Arabia, 184
saxaul shrub, 184
Schumacher, Carol, 53
scorpion., 184
Scott, Robert Falcon (1868–1912), 184, *184*
Scotty's Castle (Death Valley Ranch), 185, *185*
sculpture, 185–86
seals, crabeater, *14*
Searles Lake, 186
sebkha, 186
Sechura Desert, 186, *186*
Sedona, 186
Seedskadee National Wildlife Refuge (Wyoming), *170*
Selima Sand Sheet, 186

Seri (indigenous group), 186
serir, 186
Seth, Vikram, 90
Seven Cities of Gold, 186
Seven Pillars of Wisdom (Lawrence), 186
Shackleton, Sir Ernest (1874–1922), 80, 186, *187*
shamal, 187
Shawka Wadi, 212
The Sheltering Sky (Bowles), 187
Sherlock Holmes, 187
Shostak, Marjorie, 150
showy milkweed, 187
shrub, 187
side-blotched lizard, 187
Sides, Hampton, 108
sidewinder, 187
The Sierra Club Desert Reader (collection), 187
Silk Road, 187; exhibitions and organizations, 188
silver, 188
silver cholla cactus, 188
silverleaf nightshade, 188
Simpson Desert, 188
Sinagua people, 188
Sinai (Peninsula), 188–89
Sinai, Mount, 189
Sinkiang desert, 189
sipapu, 189
el-Sisi, Abdel Fattah, 78
Siwa oasis, 189
Skeleton Coast, 190
Skeleton Coast National Park, 190
Skeletons on the Sahara (documentary), 190
Sky Islands, 190
slavery, 190
sleepy catchfly, 190
Smith, Mike, 17–18
snakes, 190
snapdragon penstemon, 190
snow, 190, *190*; in Atacama Desert, *23*
snow leopard, 191
soil, 191
solar installation, 191
solar salt pond, 191
Soleri, Paolo, 19
Soltz, David L., 87
songlines, 191
Sonoran Desert (Yuma), *135*, 191; birds, 192; cactus, *220*; map, *191–92*; saguaro cactus, *192*
Sonoran Desert Life (Rosenthal), 192
Sonoran Desert National Monument, 192
Sonoran Quarterly (journal), 192
sophora flavescens plant, 193
Sossusvlei, 193
South Africa, 193
South Africa San Institute, 193
South America, 193
The South Pole (Amundsen), 193

Southwest Classics (Powell), 193
southwestern wild carrot, 193
southwest US Indian ruins, 175
sovereign nations, 193
Spain, 193
Sphinx, 194, *194*
spider, 194
spiny mimosa, 194
Spirit Mountain, 194
sports, 194
spreading fleabane, 194
squirrel, 194
Stark, Freya (1893–1993), 77, 195
state parks, 195
St. Catherine monastery, *189*
steatopygous, 195
Stein, Marc Aurel (1862–1943), 175, 195, *195*
Steinbeck, John, 96
Stillwater National Wildlife Refuge, 195
St. Jerome in the Desert (painting), 194
stolen generations, 195
storms, 196
The Story of the Weeping Camel (documentary), 196
Stovepipe Wells, 196
The Stranger (Camus), 196
Strzelecki Desert is a Central Australian Desert, 196
Sturt, Charles (1795–1869), 196
Sturt Stony Desert, 196
Sudan, 196
Suez Canal, 196
Sumer, 196
Sunset Crater Volcano National Monument, 196
Sunshot (Broyles), 196
survival, 196–97
sweetbush, 197
The Swiss Family Robinson (Wyss), 198
Syrian Desert, *188*, 198, *198*

taboo, 199
tagelmoust, 199
Taklamakan Desert, 199–200; dust storm in, *200*; map, *199*
Tamanrasset, 200
Tamarisk, 200
Tanami Desert, 200
Tanezrouft Desert, 200
Taos Pueblo, 200
tarantula, 200
Tassili n'Ajjer Desert, 200
Tassili n'Ajjer National Park (Algeria), 201
Tataouine, 201
Teda of the Tibesti, 201
teddy bear cholla, 201, *201*
Tehama, 201
Tehuelche people, 201
Telescope Peak, 201
Tellem Burial Caves, 201
temperatures, 201–2

Temple of Bel, Palmyra, *156*
Temple of Luxor, *185*
Temple of Tulán-54, 202
Ténéré Desert, 202
Ten Tall Men (film), 202
termites, 202, *202*
terrain, 202
Terra Incognita (Wheeler), 202
Thaj ruins, 202
Thar Desert, 203; map, *203*
Thebian Desert Road Survey, 203
theft, 203–4
thermoregulation, 204
Thesiger, Wilfred P. (1910–2003), 17, 204, *204*
thistle, 204
thobe, 204
Thomas, Bertram Sidney (1892–1950), 204
thorn tree, 204
threatened species, 204
Three Rivers Petroglyph Site, 204–5, *205*
thyme, 205
Tibesti, 205
Tibet, 205
Tien Shan, 205
Timbuktu, 206
Tinaja, 206
tobacco tree, 206
Tohono O'odham, 34, 206
Tottori Sand Dunes, 206
tourism, 206
toxins, 206–7
Tracks (Davidson), 207, *207*
trails, 207
travel, 207–8
travel guiding companies, 208
Traveling through Egypt (collection), 208
Travels in Arabia Deserta (Doughty), 208
Trinity Site, 208, *208*
truffles, 209
tsamma melon, 209
Tsodilo Hills (Kalahari), 209
Tuareg people, 209, *209*

Tucson Botanical Gardens, 209
Tularosa Valley (Basin), *209*; lava in, *124*
Tumacácori National Historic Park, 209
tundra, 210
Tunisia, 210
Turpan Desert, 210
turpentine bush, 210
Tuzigoot National Monument, 210
twenty-mule team, 210, *210*

uluru (Ayers Rock), 211, *211*
Uluru-Kata Tjuta National Park, 211
Umm al Samim, 211
United Arab Emirates (UAE), 212
United States, museums in, 139
United States Geological Survey (USGS), 212
Ur of the Chaldees, 212
Utah, 212
Utah juniper, 212

Valencia, Pablo, 213
Valley fever, desert rheumatism, 213
Valley of Love (film), 213
Van der Post, Laurens, 128
Van Dyke, John C., 64
Varèse, Edgard, 71
ventifact rocks, 213
verbena, purple sand, *220*
Verde Canyon Railroad, 213
Very Large Array (VLA) Radio Telescope, 213
Vicuñas, in Atacama Desert, *12*
Villa, Pancho (1878–1923), 213
virga, 213
VLA. *See* Very Large Array Radio Telescope
vleis, 213
Voices in the Desert (collection), 213
voladores, 213
volcano, 213
Voss (White), 213

wadi, 215
Wadi Arabah, 215
Wadi Dra, 215
Wadi Hajr, 215
Wadi Rum, 215, *215–16*
Wadi Shani, 216
Wahhabism, 216
Wahiba Sands, 216
walkabout, 216
walking, 216; water requirements, *197*
Wall to Wall (Morris), 216
Walnut Canyon National Monument, 216
Waltari, Mika, 78
wars, 217
water, 217; scalding hot, *217*
water requirements, walking, *197*
We Are Imazighen (Aïtel), 31
weathering, 218
well of Barhout, 218
Welwitsch, Friedrich Martin Josef (1806–1872), 218
Welwitschia mirabilis plant, 218
Western Australian Desert, 218
Western Desert, 218
Western National Parks Association, 218–19
Western Sahara, 219
wet desert, 219
whales, 219
Wheeler, Sara, 202
When the Rains Come (Alcock), 219
Where Mountains Roar (Hazleton), 219
Whirling Dervishes, 219
White, Patrick, 213
White Desert National Park, 219
White House Ruins, 219
white-sand desert, 219
White Sands National Park, 220
Wild, Peter, 152
wildebeest, 220
wildfire, 220
wildflowers, *170*, 220–21, *220–21*

Wild Horse and Burro Corral Facility (Mojave), 220
wildlife refuges, 221–22
wild onions, 220
wild sunflowers, 220
Wilfred Thesiger (Maitland), 222
wind, 222
wind farms, 222
The Wisdom of the Desert (Merton), 222
Wise Women of the Dreamtime (Parker), 4
Woman in the Dunes (film), 222
Wright, Frank Lloyd (1867–1959), 222
Wyss, Johann David, 198

xerophyte plant, 223

yak, 223
yardang, 223
Yavapai Geology Museum, 223
yellow columbine, 223
Yelyn Valley National Park (Mongolia), 223
Yemen, 223
Yetman, David, 97
The Young Black Stallion (film), 223
Younghusband, Sir Francis Edward (1863–1942), 223
yucca, 223–24
Yucca Mountain, 224
Yuma, 224

Zabriskie Point, 224
zebu, 224
zeriba, 224
ziggurat, 224
zoologists, 224
zud, 224
Zuni Pueblo, 224, *224*
Zwinger, Ann Haymond, 140

Acknowledgments and Thanks

I offer my thanks to Fred Hartemann, with whom I have climbed mountains in many countries and this led to *The Mountain Encyclopedia* (2005), which, in turn, has led here. In addition, Fred and I have also descended into our southwestern deserts.

Thanks to the University of Vermont for its collection and to Trina Magi for help; to Jessica Joyal and Logan Selkirk of the South Burlington, Vermont, Library Interlibrary Loan Department for acquiring so many wonderful desert books for me; to Greg Glade for his suggestions as always; and finally to my wife, Terry, and daughter, Kira, for allowing me to spend thousands of hours secreted in my rooms writing, ceaselessly writing.

About the Author

Robert Hauptman holds a doctorate in comparative literature, is a doctoral candidate (ABD) in library science, and is an emeritus full professor (St. Cloud State University). His ca. 700 scholarly and popular publications (including 19 books) in five disciplines deal with a variety of topics. He is diversely knowledgeable in most areas, especially the humanities, including 10 languages, art history, philosophy, and musicology, but also the social and hard sciences (the latter to a very limited extent), and has traveled widely and frequently in all 50 states and ca. 40 countries. But he also has practical skills that he used to good advantage when he cleared the land and built his large house basically by himself and with hand tools. He is married to the artist and poet Dr. Terry Hauptman, is 82, has a 22-year-old daughter, and climbs mountains. His most recent books are *Deadly Peaks* (2016) (with Frederic Hartemann), *The Scope of Information Ethics* (2019), *A Popular Handbook of the Emotions* (2021), *Debunking Scholarly Nonsense* (2022), and *Travel Ruminations* (2023). He is the founding and current editor of the *Journal of Information Ethics.*